Chemoinformatics Approaches to Virtual Screening

Charles Seale-Hayne Library
University of Plymouth
(01752) 588 588
LibraryandITenquiries@plymouth.ac.uk

Chemoinformatics Approaches to Virtual Screening

Edited by

Alexandre Varnek
Louis Pasteur University of Strasbourg, Strasbourg, France

Alex Tropsha
School of Pharmacy, University of North Carolina at Chapel Hill, Chapel Hill, NC, USA

RSCPublishing

ISBN: 978-0-85404-144-2

A catalogue record for this book is available from the British Library

Published by The Royal Society of Chemistry,
Thomas Graham House, Science Park, Milton Road,
Cambridge CB4 0WF, UK

Registered Charity Number 207890

For further information see our web site at www.rsc.org

Preface

Early stages in modern drug discovery often involve screening small molecules for their effects on a selected protein target or a model of a biological pathway. In the past 15 years, innovative technologies that enable rapid synthesis and high-throughput screening of large libraries of compounds have been adopted in almost all major pharmaceutical and biotech companies. As a result, there has been a huge increase in the number of compounds available on a routine basis to quickly screen for novel drug candidates against new targets/pathways. In contrast, such technologies have rarely become available to the academic research community, thus limiting its ability to conduct large-scale chemical genetics or chemical genomics research. However, the landscape of publicly available experimental data collection methods for chemoinformatics has changed dramatically in very recent years. In 2005, the National Health Institute (NIH) launched a Molecular Libraries Initiative (MLI) that included the formation of the national Molecular Library Screening Centers Network (MLSCN). MLSCN aims to offer to the research community the results of testing about a million compounds against hundreds of biological targets.

Due to the broad application of high-throughput synthetic and analytical chemical technologies, scientists who generate large volumes of data are no longer equipped with adequate tools and approaches to manage and analyze this data. At the same time, the revolutionary development of information and communication technologies during the last few decades has changed dramatically our capabilities of collecting and accessing all sorts of molecular data and has afforded the creation of huge heterogeneous data depositories. For instance, the PubChem database has been developed by the NIH as the central repository for chemical structure–activity data. PubChem currently contains over 18 million chemical compound records, more than 700 bioassay results, and links from chemicals to bioassay description, literature, references and assay data for each entry. This data requires the development and application of sophisticated mathematical and statistical tools for the discovery of new patterns and structures in large chemical datasets and to achieve a deeper

Chemoinformatics Approaches to Virtual Screening
Edited by Alexandre Varnek and Alex Tropsha
© Royal Society of Chemistry, 2008
Published by the Royal Society of Chemistry, www.rsc.org

understanding of the relationship between chemical structure and physical and biological properties. From the computational knowledge mining prospective, the availability of large collections of chemical structures affords the opportunity for *virtual screening* of these collections *in silico* to identify and prioritize promising candidate compounds for experimental validation.

Virtual screening has been typically considered an area of computer-aided drug discovery where three-dimensional protein structures are used to discover small molecules that fit into the active site (docking) and have high predicted binding affinity (scoring). Traditional docking protocols and scoring functions rely on explicitly defined three-dimensional coordinates and standard definitions of atom types of both receptors and ligands. Albeit reasonably accurate in many cases, conventional structure-based virtual screening approaches are relatively computationally inefficient, which has precluded them from screening really large compound collections. Significant progress has been achieved over many years of research in developing many structure-based virtual screening approaches. However, several recent publications comparing many available scoring and docking approaches suggest that their accuracy still needs to be improved considerably to afford their automated and successful application to solve practical problems in drug design.[3,4] Yet the availability of millions of compounds in chemical databases and billions of compounds in synthetically feasible virtual chemical libraries for virtual screening calls for the development of approaches that are both fast and accurate in their ability to identify a small number of viable computational hits that deserve subsequent experimental investigation.

Here we discuss the use of chemoinformatics as a powerful virtual screening methodology that presents both an *alternative* as well as *complement* to traditional structure-based docking and scoring approaches. The first published definition of chem[o]informatics[i] defined it as:

the use of information technology and management has become a critical part of the drug discovery process. Chemoinformatics is the mixing of those information resources to transform data into information and information into knowledge for the intended purpose of making better decisions faster in the area of drug lead identification and organization.

This definition introduced by an industrial pharmaceutical scientist was obviously biased towards pharmaceutical applications. However, many years of research in multiple areas of chemistry, computational chemistry, chemometrics, molecular modeling, computer science and statistics, both before and after that publication, provide clear evidence that modern chemoinformatics appeals to almost any area of chemical research, including organic, physical, analytical chemistry and, more recently, systems biology.[5] In this sense, following an early definition by G. Paris[6] we describe chemoinformatics broadly

---

[i] Both cheminformatics and chemoinformatics are used in the literature interchangeably and both spellings will be found in this book, depending on the personal preferences of the authors.

as a scientific discipline encompassing the design, creation, organization, management, retrieval, analysis, dissemination, visualization and use of chemical information.

We note that chemoinformatics is distinct from other computational molecular modeling approaches in that it uses unique representations of chemical structures in the form of multiple chemical descriptors, has its own metrics for defining similarity and diversity of chemical compound libraries, and applies a wide array of statistical, data mining and machine learning techniques to very large collections of chemical compounds in order to establish robust relationships between a chemical structure and its physical or biological properties. Chemoinformatics addresses a broad range of problems in chemistry and biology; however, the most commonly known applications of chemoinformatics approaches have arguably been in the area of drug discovery, where chemoinformatics tools have played a central role in the analysis and interpretation of structure–property data collected by the means of modern high-throughput screening.

Owing to the broad nature of chemoinformatics, several monographs have appeared recently that discuss various aspects of chemoinformatics research.[7–14] The present book presents a unique focus on chemoinformatics approaches that are used for virtual screening of available collections of chemical compounds to identify novel biologically active molecules. The approaches discussed by the contributors rely on chemoinformatics concepts such as representation of molecules using multiple descriptors of chemical structures, advanced chemical similarity calculations in multidimensional descriptor spaces, the use of advanced machine learning and data mining approaches for building quantitative and predictive structure–activity models, the use of chemoinformatics methodologies for the analysis of drug-likeness and property prediction, and the emerging trend of combining chemoinformatics and bioinformatics concepts in structure-based drug discovery.

The chapters are organized in a logical flow that a typical chemoinformatics project would follow, *i.e.*, from structure representation and comparison to data analysis and model building to applications of structure–property relationship models for hit identification and chemical library design. Chapter 1, by I. Baskin and A. Varnek, discusses the fundamental chemoinformatics concept of chemical structure representation by the means of molecular descriptors, focusing on fragment descriptors and their use in Quantitative Structure–Activity Relationship (QSAR) studies and database mining. This introductory chapter is followed by chapters by D. Horvath (Chapter 2) and by T. Langer and colleagues (Chapter 3) that discuss recent advances in pharmacophore identification and their use in virtual screening. Naturally, the pharmacophore is the major concept in medicinal chemistry and computational drug discovery, and many research papers and monographs have been published on this subject over the years. Still, these two chapters that have different focuses on pharmacophores derived from (two-dimensional) chemical graphs (Chapter 2) and, the more common, three-dimensional pharmacophores (Chapter 3) offer unique perspective on pharmacophore identification as a tool for knowledge discovery and mining in molecular databases.

Whereas pharmacophore identification can be viewed as an example of chemical data mining approaches focusing on specific descriptors of chemical structures, much information about structure–activity relationships can be obtained by using another major concept of chemoinformatics, *i.e.*, that of chemical (or molecular) similarity. Chapter 4 by L. Peltason and J. Bajorath summarizes recent advanced studies into this fundamental chemoinformatics problem and discusses the use of molecular similarity calculations in virtual screening. The next two chapters focus on recent methodologies that establish and explore quantitative structure–activity relationships (QSAR). E. Radchenko, V. Palyulin and N. Zefirov (Chapter 5) cover the use of topological molecular fields in drug design and virtual screening whereas D. Filimonov and V. Poroikov (Chapter 6) present an analysis of promising probabilistic approaches in QSAR modeling.

Chemoinformatics approaches are finding growing and important application in developing a better understanding of the chemical features that distinguish drugs and drug-like molecules from other organic molecules. In fact, this area so far has almost exclusively relied on ligand based approaches. Chapter 7 by G. Schneider and colleagues addresses the issue of drug-likeness and discusses ligand-based methodologies that can be used in designing viable drug candidates. Chapter 8, by I. Tetko and T. Oprea, presents an overview of chemoinformatics methods that are used in early stages of drug discovery to identify and prioritize compounds with optimal ADMET (Adsorption, Distribution, Metabolism, Excretion and Toxicity) properties.

Chemical library design has always been an important component of chemoinformatics studies and it could be viewed in fact a special case of virtual screening. Chapter 9 by W. Zheng and S.R. Johnson provides an expert overview of computational approaches that are employed in the design of targeted and diverse chemical libraries, including the use of property (*i.e.*, ADMET) filters.

The final chapter by A. Tropsha looks into chemoinformatics methodologies that rely on compound representation in multidimensional chemical descriptor space and chemical similarity searches that could be employed in structure based drug discovery. These approaches could enrich traditional structure-based virtual screening and docking methodologies. The chapter may serve to illustrate the importance of building natural bridges between structural bioinformatics and structural chemoinformatics approaches in addressing the common problem of virtual screening that is the major theme of this book.

In conclusion, we believe that the focus on extending the experiences accumulated in chemoinformatics research towards virtual screening makes the theme of this monograph highly attractive for all computational and experimental researchers in the area of drug discovery or, more broadly, chemical biology. We stated at the beginning that virtual screening is one important area of modern chemoinformatics research that deserves special attention, which motivated us to develop this monograph. We believe, however, that due to its generic data-analytical focus we will see growing application of chemoinformatics approaches in multiple areas of chemical and biological research such as synthesis planning, nanotechnology, proteomics, physical and analytical chemistry and, of course, chemical genomics.

Alexandre Varnek and Alexander Tropsha

Acknowledgements

The editors are grateful to Dr Dennis Fourches for designing the front cover illustration.

References

1. C. P. Austin, L. S. Brady, T. R. Insel and F. S. Collins, *Science*, 2004, **306**, 1138–1139
2. PubChem, http://pubchem.ncbi.nlm.nih.gov/, 2007.
3. G. L. Warren, C. W. Andrews, A. M. Capelli, B. Clarke, J. LaLonde, M. H. Lambert, M. Lindvall, N. Nevins, S. F. Semus, S. Senger, G. Tedesco, I. D. Wall, J. M. Woolven, C. E. Peishoff and M. S. Head, *J. Med. Chem.*, 2006, **49**, 5912–5931.
4. H. Chen, P. D. Lyne, F. Giordanetto, T. Lovell and J. Li, *J. Chem. Inf. Model.*, 2006, **46**, 401–415.
5. T. I. Oprea, A. Tropsha, J. L. Faulon and M. D. Rintoul, *Nat. Chem. Biol.*, 2007, **3**, 447–450.
6. G. Paris, in *Meeting of the American Chemical Society, August 1999*, quoted by W. Warr at http://www.warr.com/warrzone.htm.
7. B. A. Bunin, B. Siesel, G. Morales and J. Bajorath, *Chemoinformatics: Theory, Practice, & Products*, Springer, New York, 2006.
8. *Chemoinformatics: A Textbook*, Willey-VCH, Weinheim, 2003.
9. A. R. Leach and V. J. Gillet, *An Introduction to Chemoinformatics*, Kluwer Academic Publishers, Dordrecht, 2003.
10. *Handbook of Chemoinformatics: From Data to Knowledge*, Willey-VCH, Weinheim, 2003.
11. *Chemoinformatics: Concepts, Methods, and Tools for Drug Discovery*, (Methods in Molecular Biology, Vol. 275), J. Bajorath (ed.), Humana Press, Totowa, New York, 2004.
12. *Chemoinformatics In Drug Discovery*, Willey-VCH, Weinheim, 2005.
13. *Chemometrics and Chemoinformatics*, B. Levine (ed.), ACS Symposium Series Vol. 894, American Chemical Society, Washington DC, 2005.
14. *Cheminformatics Developments*, IOS Press, Amsterdam, 2004.

Contents

Chemoinformatics Approaches to Virtual Screening
Edited by Alexandre Varnek and Alex Tropsha
© Royal Society of Chemistry, 2008
Published by the Royal Society of Chemistry, www.rsc.org

Fragment Descriptors in SAR/QSAR/QSPR Studies, Molecular Similarity Analysis and in Virtual Screening

IGOR BASKIN[a] AND ALEXANDRE VARNEK[b]

[a] Department of Chemistry, Moscow State University, Moscow 119992, Russia; [b] Laboratoire d'Infochimie, UMR 7177 CNRS, Université Louis Pasteur, 4, rue B. Pascal, Strasbourg 67000, France

1.1 Introduction

Chemoinformatics[1-5] is an emerging science that concerns the mixing of chemical information resources to transform data into information, and information into knowledge. It is a branch of theoretical chemistry based on its molecular model, and which uses its own basic concepts, learning approaches and areas of application. Unlike quantum chemistry, which considers molecules as ensemble of electrons and nuclei, or force field molecular mechanics or dynamics simulations based on a classical molecular model ("atoms" and "bonds"), chemoinformatics represents molecules as objects in a chemical space defined by molecular descriptors. Among thousands of descriptors, fragment descriptors occupy a special place. Fragment descriptors represent selected subgraphs of a 2D molecular graph; structure–property approaches use their occurrences in molecules or binary values (0, 1) to indicate their presence or absence in the given graph.

The unique properties of fragment descriptors are related to the fact that (i) any molecular graph invariant (*i.e.*, any molecular descriptor or property)

Chemoinformatics Approaches to Virtual Screening
Edited by Alexandre Varnek and Alex Tropsha
© Royal Society of Chemistry, 2008
Published by the Royal Society of Chemistry, www.rsc.org

can be uniquely represented as a linear combination of fragment descriptors;[7–9] (ii) any symmetric similarity measure can be uniquely expressed in terms of fragment descriptors;[10,11] and (iii) any regression or classification structure–property model can be represented as a linear equation involving fragment descriptors.[12,13]

An important advantage of fragment descriptors is related to the simplicity of their calculation, storage and interpretation (see review articles[14–18]). They belong to information-based descriptors,[19] which tend to code the information stored in molecular structures. This contrasts with knowledge-based (or semi-empirical) descriptors derived from consideration of the mechanism of action. Owing to their versatility, fragment descriptors can efficiently be used to build structure–property models, perform similarity search, virtual screening and *in silico* design of chemical compounds with desired properties.

This chapter reviews fragment descriptors with respect to their use in structure–property studies, similarity search and virtual screening. After a short historical survey, different types of fragment descriptors are considered thoroughly. This is followed by a brief review of the application of fragment descriptors in virtual screening, focusing mostly on filtering, similarity search and direct activity/property assessment using quantitative structure–property models.

1.2   Historical Survey

Among a multitude of descriptors currently used in Structure–Activity Relationships/Quantitative Structure–Activity Relationships/Quantitative Structure–Property Relationships (SAR/QSAR/QSPR) studies,[20] fragment descriptors occupy a special place. Their application as atoms and bonds increments in the framework of *additive schemes* can be traced back to the 1930–1950s; Vogel,[21] Zahn,[22] Souders,[23,24] Franklin,[25,26] Tatevskii,[27,28] Bernstein,[29] Laidler,[30] Benson and Buss[31] and Allen[32] pioneered this field. Smolenskii was one of the first, in 1964, to apply graph theory to tackle the problem of predictions of the physico-chemical properties of organic compounds.[33] Later on, these first additive schemes approaches have gradually evolved into *group contribution methods*. The latter are closely linked with thermodynamic approaches and, therefore, they are applicable only to a limited number of properties.

The epoch of QSAR (Quantitative Structure–Activity Relationships) studies began in 1963–1964 with two seminal approaches: the σ-ρ-π analysis of Hansch and Fujita[34,35] and the Free–Wilson method.[36] The former approach involves three types of descriptors related to electronic, steric and hydrophobic characteristics of substituents, whereas the latter considers the substituents themselves as descriptors. Both approaches are confined to strictly congeneric series of compounds. The Free–Wilson method additionally requires all types of substituents to be sufficiently present in the training set. A combination of these two approaches has led to QSAR models involving *indicator variables*, which indicate the presence of some structural fragments in molecules.

The non-quantitative SAR (Structure–Activity Relationships) models developed in the 1970s by Hiller,[37,38] Golender and Rosenblit,[39,40] Piruzyan, Avidon *et al.*,[41] Cramer,[42] Brugger, Stuper and Jurs,[43,44] and Hodes *et al.*[45] were inspired by the, at that time, popular artificial intelligence, expert systems, machine learning and pattern recognition paradigms. In those approaches, chemical structures were described by means of indicators of the presence of structural fragments interpreted as topological (or 2D) pharmacophores (biophores, toxophores, *etc.*) or topological pharmacophobes (biophobes, toxophobes, *etc.*). Chemical compounds were then classified as active or inactive with respect to certain types of biological activity.

Methodologies based on fragment descriptors in QSAR/QSPR studies are not strictly confined to particular types of properties or compounds. In the 1970s Adamson and coworkers[46,47] were the first to apply fragment descriptors in multiple linear regression analysis to find correlations with some biological activities,[48,49] physicochemical properties,[50] and reactivity.[51]

An important class of fragment descriptors, the so-called *screens* (or structural *keys*, *fingerprints*), were also developed in 1970s.[52–56] As a rule, they represent the bit strings that can effectively be stored and processed by computers. Although their primary role is to provide efficient substructure searching in large chemical structure databases, they can be efficiently used also for similarity searching,[57,58] clustering large chemical databases,[59,60] assessing their diversity,[61] as well as for SAR[62] and QSAR[63] modeling.

Another important contribution was made in 1980 by Cramer who invented BC(DEF) parameters obtained by means of factor analysis of the physical properties of 114 organic liquids. These parameters correlate strongly with various physical properties of diverse liquid organic compounds.[64] On the other hand, they could be estimated by linear additive-constitutive models involving fragment descriptors.[65] Thus, a set of QSPR models encompassing numerous physical properties of diverse organic compounds has been developed using only fragment descriptors.

One of the most important developments of the 1980s was the CASE (Computer-Automated Structure Evaluation) program by Klopman *et al.*[66–69] This "self-learning artificial intelligent system"[69] can recognize activating and deactivating fragments (biophores and biophobes) with respect to the given biological activity and to use this information to determine the probability that a test chemical is active. This methodology has been successfully applied to predict various types of biological activity: mutagenicity,[67,70,71] carcinogenicity,[66,69,71–73] hallucinogenic activity,[74] anticonvulsant activity,[75] inhibitory activity with respect to sparteine monooxygenase,[76] β-adrenergic activity,[77] μ-receptor binding (opiate) activity,[78] antibacterial activity,[79] antileukemic activity,[80] *etc.* Using the multivariate regression technique, CASE can also build quantitative models involving fragment descriptors.[72,77]

Starting in the early 1990s, various approaches and related software tools based on fragment descriptors have been developed and are listed in several conceptual and mini-review papers.[14–18] Because of the wide scope and large variety of different approaches and applications in this field, many important ideas were reinvented many times and continue to be reinvented. In this review we try to present a clear state-of-the-art picture in this area.

1.3   Main Characteristics of Fragment Descriptors

In this section different types of fragments are classified with respect to their topology and the level of abstraction of molecular graphs.

1.3.1   Types of Fragments

A tremendous number of various fragments are used in structure–property studies: atoms, bonds, "topological torsions", chains, cycles, atom- and bond-centered fragments, maximum common substructures, line notation (WLN and SMILES) fragments, atom pairs and topological multiplets, substituents and molecular frameworks, basic subgraphs, *etc.* Their detailed description is given below.

Depending on the application area, two types of values taken by fragment descriptors are considered: binary and integer. Binary values indicate the presence (*true, yes,* 1) or the absence (*false, no,* 0) of a given fragment in a structure. They are usually used as screens and elements of fingerprints for chemical database management and virtual screening using similarity-based approaches as well as in SAR studies. Integer values corresponding to the occurrences of fragments in structures are used in QSAR/QSPR modeling.

1.3.1.1   Simple Fixed Types

Disconnected atoms represent the simplest type of fragments. They are used to assess a chemical or biological property P in the framework of an additive scheme based on atomic contributions:

$$P \approx \sum_{i=1}^{N} n_i \cdot A_i \tag{1.1}$$

where n_i is the number of atoms of i-type, A_i is corresponding atomic contributions. Usually, the atom types account for not only the type of chemical element but also hybridization, the number of attached hydrogen atoms (for heavy elements), occurrence in some groups or aromatic systems, *etc.* Nowadays, atom-based methods are used to predict some physicochemical properties and biological activities. Thus, several works have been devoted to assess the octanol–water partition coefficient log P: the ALOGP method by Ghose-Crippen,[81–83] later modified by Ghose and co-workers,[84,85] and by Wildman and Crippen,[86] the CHEMICALC-2 method by Suzuki and Kudo,[87] the SMILOGP program by Convard and co-authors,[88] and the XLOGP method by Wang and co-authors.[89,90] Hou and co-authors[91] used Equation (1.1) to calculate aqueous solubility. The ability of this approach to assess biological activities was demonstrated by Winkler *et al.*[92]

Chemical bonds are another type of simple fragment. The first bond-based additive schemes, such as those of Zahn,[22] Bernstein[29,93] and Allen,[32,94] appeared almost simultaneously with the atom-based ones and dealt, presumably, with predictions of some thermodynamic properties.

"Topological torsions" invented Nilakantan *et al.*[95] are defined as a linear sequence of four consecutively bonded non-hydrogen atoms. Each atom there is described by the type of corresponding chemical element, the number of attached non-hydrogen atoms and the number of π-electron pairs. Molecular descriptors indicating the presence or absence of topological torsions in chemical structures have been used to perform qualitative predictions of biological activity in structure–activity (SAR) studies.[95] Later on, Kearsley *et al.*[96] recognized that characterizing atoms by element types can be too specific for similarity searching and, therefore, it does not provide sufficient flexibility for large-scaled virtual screening. To solve this problem, they suggested assigning atoms in the Carhart's atom pairs and Nilakantan's topological torsions to one of seven classes: cations, anions, neutral hydrogen bond donors, neutral hydrogen bond acceptors, polar atoms, hydrophobic atoms and other.

The above-mentioned structural fragments – atoms, bonds and topological torsions – can be regarded as *chains* of different lengths. Smolenskii[33] suggested using the occurrences of chains in an additive scheme to predict the formation enthalpy of alkanes. For the last four decades, chain fragments have proved to be one of the most popular and useful type of fragment descriptors in QSPR/QSAR/SAR studies. Fragment descriptors based on enumerating chains in molecular graphs are efficiently used in many popular structure–property and structure–activity programs: CASE[66–69] and MULTICASE (MultiCASE, MCASE) by Klopman[97,98] NASAWIN[99] by Baskin *et al.*, BIBIGON[100] by Kumskov, TRAIL[101,102] and ISIDA[18] by Solov'ev and Varnek. "Molecular pathways" by Gakh and co-authors,[103] and "molecular walks" by Rücker,[104] represent chains of atoms.

In contrast to chains, cyclic and polycyclic fragments are relatively rarely applied as descriptors in QSAR/QSPR studies. Nevertheless, *implicitly* cyclicity is accounted for by means of: (i) introducing special "cyclic" and "aromatic" types of atoms and bonds, (ii) "collapsing" the whole cycles and even polycyclic systems into "pharmacophoric" pseudo-atoms and (iii) generating cyclic fragments as a part of large fragments [Maximum Common Substructure (MCS), molecular framework, substituents]. Besides, the cyclic fragments are widely used as screens for chemical database processing.[105,106]

1.3.1.2 WLN and SMILES Fragments

WLN and SMILES fragments correspond respectively to substrings of the Wiswesser Line Notation[107] or Simplified Molecular Input Line Entry System[108,109] strings used for encoding the chemical structures. Since simple string operations are much faster than processing of information in connection tables, the use of WLN descriptors was justified in the 1970s when computers were still very slow. At that time Adamson and Bawden published some linear QSAR models based on WLN fragments.[48,50,51,110,111] They have also applied this kind of descriptor for hierarchical cluster analysis and automatic classification of chemical structures.[112] Qu *et al.*[113,114] have developed AES (Advanced Encoding System), a new WLN-based notation encoding chemical information for group contribution methods. Interest in line notation descriptors has not disappeared completely with the advent of powerful computers. Thus, SMILES fragment descriptors are used in the SMILOGP program to predict log P,[88] whereas the recently developed LINGO system for assessing some biophysical properties and intermolecular similarities uses holographic representations of canonical SMILES strings.[115]

1.3.1.3   Atom-centered Fragments

Atom-Centered Fragments (ACF) consist of a single central atom surrounded by one or several shells of atoms separated from the central one by the same topological distance. This type of structural fragments was introduced in the early 1950s by Tatevskii,[27,28,116–119] and then by Benson[31] to predict some physicochemical properties of organic compounds in the framework of additive schemes.

ACF fragments containing only one shell of atoms around the central one (*i.e.*, atom-centered neighborhoods of radius 1) were introduced into chemoinformatics practice in 1971 under the names "atom-centered fragments" and "augmented atoms" by Adamson,[120,121] who studied their distribution in large chemical databases with the intention of using them as screens in chemical database searching. Hodes used, in SAR studies, both "augmented atoms"[45] and "ganglia augmented atoms"[325] representing ACF fragments with radius 2 and generalized second-shell atoms. Subsequently, ACF fragments with radius 1 were implemented in NASAWIN,[122–124] TRAIL[101,102,125] and ISIDA[18] programs. ACF fragments with arbitrary radius were implemented by Filimonov, Poroikov and co-authors in the PASS[126] program under the name Multilevel Neighborhoods of Atoms (MNA),[127] by Xing and Glen as "tree structured fingerprints",[128] by Bender and Glen as "atom environments"[129,130] and "circular fingerprints"[131–133] (Figure 1.1), and by Faulon as "molecular signatures".[134–136]

Several types of ACF fragments were designed to store local spectral parameters (chemical shifts) in spectroscopy data bases. Thus, Bremser has developed Hierarchically Ordered Spherical Environment (HOSE), a system of substructure codes aimed at characterizing the spherical environment of single atoms and complete ring systems.[137] The codes are generated automatically from 2D graphs and describe structural entities corresponding to chemical shifts. A very similar idea has also been implemented by Dubois *et al.* in the DARC system based on FREL (Fragment Réduit à un Environment Limité) fragments.[138,139] Xiao *et al.* have applied Atom-Centered Multilayer Code

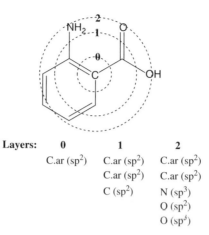

Layers:	0	1	2
	C.ar (sp^2)	C.ar (sp^2)	C.ar (sp^2)
		C.ar (sp^2)	C.ar (sp^2)
		C (sp^2)	N (sp^3)
			O (sp^2)
			O (sp^3)

Figure 1.1 Circular fingerprints with Sybyl mol2 atom typing. An individual finger-print is calculated for each atom in the molecule, considering those atoms up to two bonds from the central atom (level 2). The molecular fingerprint consists of the individual atom fingerprints of all the heavy atoms in the structure. (Adapted from ref. 132.)

(ACMC) fragments for structural and substructural searching in large data-bases of compounds and reactions.[140] An important recent application of ACF fragments concerns target prediction ("target fishing") in chemogenomic data analysis.[126,141,142]

1.3.1.4   *Bond-centered Fragments*

Bond-centered fragments (BCF) consist of two atoms linked by the bond and surrounded by one or several shells of atoms separated by the same topological distance from this bond. Although these fragments are rather rarely used in structure–property studies, they can be efficiently used as screens for chemical database processing.[143] BCF have been used as a part of MDL keys[144,145] for substructure search in chemical databases, database clustering[60] and for SAR studies of 17 different types of biological activity.[62] Bond-centered fragments have also been used in the DARC system.[138,139]

1.3.1.5   *Maximum Common Substructures*

For a set of molecular graphs, a Maximum Common Substructure (MCS) is defined as a largest substructure in all graphs belonging to the given set. In most practical applications, only MCS for graph pairs are considered, *i.e.*, for sets containing only two graphs. MCS can be found by *intersecting* molecular graphs using several different algorithms (for a review see ref. 146), the best known of which involve clique detection in so-called compatibility graphs.

Notably, a pair of graphs can have more than one MCS. The main advantage of MCS fragments is related to the fact that their complexity is not limited and therefore they can be used to detect property-relevant features that could not be detected by fragments (subgraphs) of limited complexity.

MCSs were first applied to SAR studies in the early 1980s by Rozenblit and Golender in the framework of their logical-combinatorial approach.[40,41,147] Since at that time computer power was limited, the authors suggested the use of reduced graphs (Section 1.3.5) built on pharmacophoric centers. The MCS fragments were subsequently applied to perform a similarity search,[148] to cluster chemical databases[149,150] as well to assess biological activities of organic compounds.[99,151,152]

1.3.1.6 Atom Pairs and Topological Multiplets

Characterizing atoms only by element types is too specific for similarity searching and, therefore, does not provide sufficient flexibility for large-scale virtual screening. For that reason, numerous studies have been devoted to increase the informational content of fragment descriptors by adding some useful empirical information and/or by representing a part of the molecular graph implicitly. The simplest representatives of such descriptors were "atom pairs and topological multiplets" based on the notion of a "descriptor center" representing an atom or a group of atoms that could serve as centers of intermolecular interactions. Usually, descriptor centers include heteroatoms, unsaturated bonds and aromatic cycles. An *atom pair* is defined as a pair of atoms (**AT**) or descriptor centers separated by a fixed topological distance: $AT_i\text{-}Dist\text{-}AT_j$, where $Dist_{ij}$ is the shortest path (the number of bonds) between AT_i and AT_j. Analogously, a topological multiplet is defined as a multiplet (usually triplet) of descriptor centers and topological distances between each pair of them. In most of cases, these descriptors are used in binary form to indicate the presence or absence of the corresponding features in studied chemical structures.

Atom pairs were first suggested for SAR studies by Avidon as Substructure Superposition Fragment Notation (SSFN).[41,153] They were then independently reinvented by Carhart and co-authors[154] for similarity and trend vector analysis. In contrast to SSFN, Carhart's atom pairs are not necessarily composed only of descriptor centers but account for the information about element type, the number of bonded non-hydrogen neighbors and the number of π electrons. Nowadays, Carhart's atom pairs are popular in virtual screening. Topological Fuzzy Bipolar Pharmacophore Autocorrelograms (TFBPA)[155] by Horvath are based on atom pairs, in which real atoms are replaced by pharmacophore sites (hydrophobic, aromatic, hydrogen bond acceptor, hydrogen bond donor, cation, anion), while $Dist_{ij}$ corresponds to different ranges of topological distances between pharmacophores. These descriptors were successfully applied in virtual screening against a panel of 42 biological targets using a similarity search based on several fuzzy and non-fuzzy metrics,[156] performing only slightly less

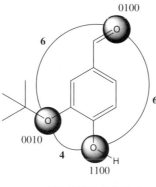

0010-**4**-1100-**6**-0100-**6**-

Figure 1.2 Example of a Similog key. (Adapted from ref. 158.)

well than their 3D counterparts.[155] Fuzzy Pharmacophore Triplets (FPT) by Horvath[157] is an extension of FBPF[156] for three-site pharmacophores. An important innovation in the FPT concerns accounting for proteolytic equilibrium as a function of pH.[157] Owing to this feature, even small structural modifications leading to a pK_a shift may have a profound effect on the fuzzy pharmocophore triples. As a result, these descriptors efficiently discriminate structurally similar compounds exhibiting significantly different activities.[157]

Some other topological triplets should be mentioned. Similog pharmacophoric keys by Schuffenhauer *et al.*[158] represent triplets of binary coded types of atoms (pharmacophoric centers) and topological distances between them (Figure 1.2). Atomic types are generalized by four features (represented as four bits per atom): potential hydrogen bond, donor or acceptor, bulkiness and electropositivity. The "topological pharmacophore-point triangles" implemented in the MOE software[159] represent triplets of MOE atom types separated by binned topological distances. Structure–property models obtained by a support vector machine method with these descriptors have been successfully used for virtual screening of COX-2 inhibitors[160] and D_3 dopamine receptor ligands.[161]

1.3.1.7 Substituents and Molecular Frameworks

In organic chemistry, decomposition of molecules into substituents and molecular frameworks is a natural way to characterize molecular structures. In QSAR, both the Hansch–Fujita[34,35] and the Free–Wilson[36] classical approaches are based on this decomposition, but only the second one explicitly accounts for the presence or the absence of substituent(s) attached to molecular framework at a certain position. While the multiple linear regression technique was associated with the Free–Wilson method, recent modifications of this approach involve more sophisticated statistical and machine-learning approaches, such as the principal component analysis[162] and neural networks.[163]

In contrast to substituents, molecular frameworks are rarely used in SAR/ QSAR/QSPR studies. In most cases, they are implicitly involved as indicator variables discriminating different types of molecular motifs (see, for example, ref. 164). The distributions of different molecular frameworks and substituents (side chains) in the databases of known drug molecules has been thoroughly studied by Bemis and Murcko.[165,166]

1.3.1.8   Basic Subgraphs

Regarding fragment descriptors, one could imagine a huge number of possibilities to split a molecular graph into constituent fragments. Making a parallel with the decomposition of vectors into a limited number of basis functions, Randič[326] suggested the existence of a small set of *basic subgraphs* representing any structure and which could be used to calculate any molecular property. In particular, for small alkanes a set of disconnected graphs representing paths (chains) of different length has been proposed (Figure 1.3).

However, later it has since been found that this set is not sufficient to differentiate any two structures. Skvortsova *et al.* have extended the set of Randič basic subgraphs by including cyclic fragments and more complex subgraphs consisting of single node attached to a cyclic fragment.[167] This set exhibits good coding uniqueness (*i.e.*, different vectors of descriptors correspond to different structures) and coding completeness (*i.e.*, they can approximate a numerous structure–property functions). Basic fragment descriptors of this kind were used in several QSPR studies.[168]

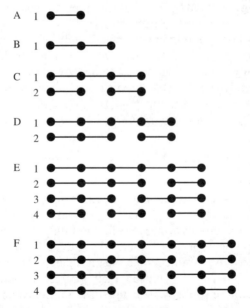

Figure 1.3   Randič basic graphs for a maximum number of nodes of 7.

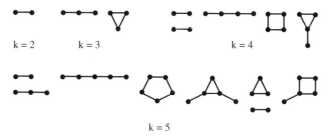

Figure 1.4 Skvortsova's basic graphs for a maximum number of nodes of 5.

In fact, a rigorous solution of the problem of finding a set of basic graph invariants was obtained by Mnukhin[169] for simple graphs and then extended to molecular graphs by Baskin, Skvortsova et al.[7–9] (Figure 1.4). It has been shown that the complete set of basic graph invariants could be built on all possible subgraphs, and hence one can not to confine this to any subset of limited size. Nonetheless, for many practical tasks the application of a limited number of basic subgraphs and the corresponding fragment descriptors could be useful.

Another application of basic subgraphs arises from the possibility[8,169] of relating the invariants of molecular graphs to the occurrence numbers of some basic subgraphs. Estrada has developed this methodology for *spectral moments* of the edge-adjacency matrix of molecular graphs – defined as the traces of the different powers of such matrix:[170–172]

$$\mu_k = \text{tr}(\boldsymbol{E}^k) \tag{1.2}$$

where μ_k is the k-th spectral moment of the edge-adjacency matrix $\boldsymbol{E}$ (which is a symmetric matrix whose elements e_{ij} are 1 only if edge i is adjacent to edge j) and tr is the trace, *i.e.* the sum of the diagonal elements of the matrix. On the other hand, spectral moments can be expressed as linear combinations of the occurrence numbers of certain structural fragments in the molecular graph. These linear combinations for simple molecular graphs not containing hetero-atoms have been reported for acyclic[170] and cyclic[172] chemical structures.

To illustrate these notions, consider a correlation between the boiling points of alkanes and their spectral moments reported in ref. 170:

$$\text{bp}(^\circ\text{C}) = -76.719 + 23.992\mu_0 + 2.506\mu_2 - 2.967\mu_3 + 0.149\mu_5 \tag{1.3}$$

$$R = 0.9949, \ s = 4.21, \ F = 1650$$

The first six spectral moments of the edge-adjacency matrix $\boldsymbol{E}$ are expressed as linear combinations of the occurrence numbers of fragments listed in Figure 1.5:

$$\mu_0 = |F_1| \tag{1.4}$$

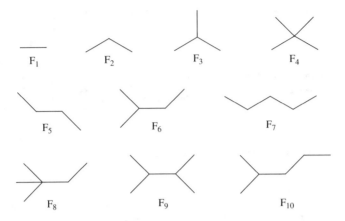

Figure 1.5 First ten structural fragments contained in molecular graphs of alkanes. (Adapted from ref. 170.)

$$\mu_2 = 2 \times |F_2| \tag{1.5}$$

$$\mu_3 = 6 \times |F_3| \tag{1.6}$$

$$\mu_4 = 2 \times |F_2| + 12 \times |F_3| + 24 \times |F_4| + 4 \times |F_5| \tag{1.7}$$

$$\mu_5 = 30 \times |F_3| + 120 \times |F_4| + 10 \times |F_6| \tag{1.8}$$

$$\mu_6 = 2 \times |F_2| + 60 \times |F_3| + 480 \times |F_4| + 12 \times |F_5| + 24 \times |F_6| \\ + 6 \times |F_7| + 36 \times |F_8| + 24 \times |F_9| \tag{1.9}$$

where $|F_i|$ denotes the occurrence number of subgraph F_i in molecular graph.

Thus, by substituting spectral moments in the QSPR Equation (1.4) for their expansions (Equations 1.5–1.10) one can obtain the following QSPR equation with fragment descriptors:

$$bp(^\circ C) = -76.719 + 23.992|F_1| + 5.01|F_2| - 13.332|F_3| \\ + 17.880|F_4| + 1.492|F_6| \tag{1.10}$$

Thus, any spectral moment and hence the activities/properties of chemical compounds can be represented by contributions of corresponding fragments. This approach was further extended to molecular graphs containing heteroatoms by weighting the diagonal elements of the bond adjacency matrix.[171]

This methodology has been implemented in TOSS-MODE (TOpological Sub-Structural MOlecular Design) and TOPS-MODE (TOPological Substructural MOlecular DEsign) methods,[173] which were successfully used to assess various physicochemical properties of chemical compounds: retention indices in chromatography,[174] diamagnetic and magnetooptic properties,[175] dipole moments,[176] permeability coefficients through low-density polyethylene,[177] *etc.*), 3D-parameters[178] and a different types of biological activity (sedative/hypnotic activity,[173] anti-cancer activity,[179] anti-HIV activity,[180] skin sensitization,[181] herbicide activity,[182] affinity to A_1 adenosine receptor,[183] inhibition of cyclooxygenase,[184] antibacterial activity,[185] toxicity in *Tetrahymena pyriformis*,[186] mutagenicity,[187–189] *etc.*

1.3.1.9 Mined Subgraphs

The notion of mined subgraphs is closely linked to graph mining (or subgraph mining), a field of searching the graphs (subgraphs) specifically related to some properties or activities.[190–195] The advantage of this approach is that all relevant fragments are available for analysis without the need to consider an almost infinite number of all possible subgraphs, which allows one to select the most "useful" fragments. This methodology[196,197] is based on efficient algorithms for mining the most *frequent fragments* occurring in sets of molecular graphs, such as the AGM (Apriori-based Graph Mining) algorithm by Inokuchi *et al.*,[198] the FSG (Frequent Sub-Graphs) algorithm by Kuramochi and Karypis,[199] the chemical sub-structure discovery algorithm by Borgelt and Berthold,[200] the gSpan (graph-based Substructure pattern mining) algorithm by Yan and Han,[194] the TreeMiner algorithm by Zaki[201] and the HybridTreeMiner and CMTree-Miner algorithms by Chi, Yang and Muntz,[202,203] *etc.* The mined subgraphs approach was originally used to classify chemical structures.[204,205] "Weighted substructure mining, in conjunction with linear programming boosting,[206] allows one to build QSAR regression models involving mined fragment descriptors.[195]

1.3.1.10 Random Subgraphs

The success of different fragmentation schemes in SAR/QSAR studies strongly depends on the initial choice of relevant fragment types. Since it is unrealistic to consider all possible fragments because of their enormous number, one should always select their small subsets. However, any attempt to apply a limited subtype of them (*e.g.*, to use only chains with the user specified length) risks being inefficient because of missing of important fragments. One possible solution is to generate substructural fragments using stochastic techniques. Such an approach has been used by Graham *et al.*, who generated "tape recordings" of chemical structures from atom-bond-atom fragments extracted from molecular graphs by random walks.[207] In the MolBlaster method by Batista, Godden and Bajorath, for each molecule the program generates a "random fragment profile" representing a population of fragments generated by randomly deleting bonds in hydrogen-suppressed molecular graph.[208] This method was successfully applied in similarity-based virtual screening.[209]

1.3.1.11 Library Subgraphs

Many studies employ fixed sets of fragments taken from some libraries containing preliminary selected fragments. Thus, most additive schemes and group contribution methods have been derived using fixed sets of fragments. Some SAR/QSAR/QSPR expert systems also employ fixed sets of selected fragments and often apply an internal language specifically designed for handling the descriptors lists. For example, to describe fragments, the DEREK expert system for assessing toxicity uses the PATRAN language,[210] whereas the ALogP method[86] for predicting the octanol–water partition coefficient log P is based on the SMARTS line notation [as implemented in the MOE (Molecular Operating Environment) software suite[159]].

1.3.2 Fragments Describing Supramolecular Systems and Chemical Reactions

Using "special" bond types, molecular graphs can represent not only individual molecules but also more complex species: supramolecular systems, chemical reactions and polymers with periodic structure. For example, the ISIDA program can recognize a "coordination bond" between central metal atom and donor atoms of the ligand in the metal complexes and "hydrogen bond" in supramolecular assemblies.[32] Varnek *et al.* used fragment descriptors derived from "supramolecular" graphs in QSPR modeling of free energy and enthalpy of formation of 1 : 1 hydrogen bonded complexes.[18]

The concept of molecular graphs can also be expanded to describe chemical reactions by introducing special types of "dynamical" bonds corresponding to formation, modification and breaking of chemical bonds (for a review see ref. 211). The resulting reaction graph contains all necessary information to reconstruct both reactants and products in the corresponding reaction equation. Partial reaction graphs containing only "dynamical" bonds were used to classify and enumerate organic reactions in the framework of Ugi–Dugundji matrix formalism[212] and the Zefirov–Tratch formal-logical approach.[213,214] Vladutz condensed reactants and products of a chemical reaction into a single Superimposed Reaction Skeleton Graph (SRSG)[215] containing both dynamical and conventional (not modified in the reaction) bonds. Similar reaction graphs under the name "imaginary transition state" were also suggested by Fujita[216,217] for classification and enumeration of organic reactions. This approach has been extended recently by Varnek *et al.*[18] in Condensed Graphs of Reactions (CGRs) containing both "dynamical" and conventional bonds (Figure 1.6). Fragment descriptors derived from CGRs were used in similarity search of reactions, in reaction classification and in the development of QSPR models of the rate constant of S_N2 reactions in water.[218]

To encode reaction transformations Borodina *et al.* have developed Reacting Multilevel Neighborhood of Atom (RMNA)[219] descriptors representing an extended version of the MNA descriptors. Unlike CGRs, where reaction information is condensed, in the RMNA approach the information about modified, created or broken bonds is added to the list of the MNA descriptors generated for all products and reactants. The RMNA descriptors were applied to predict metabolic P450-mediated aromatic hydroxylation.[219]

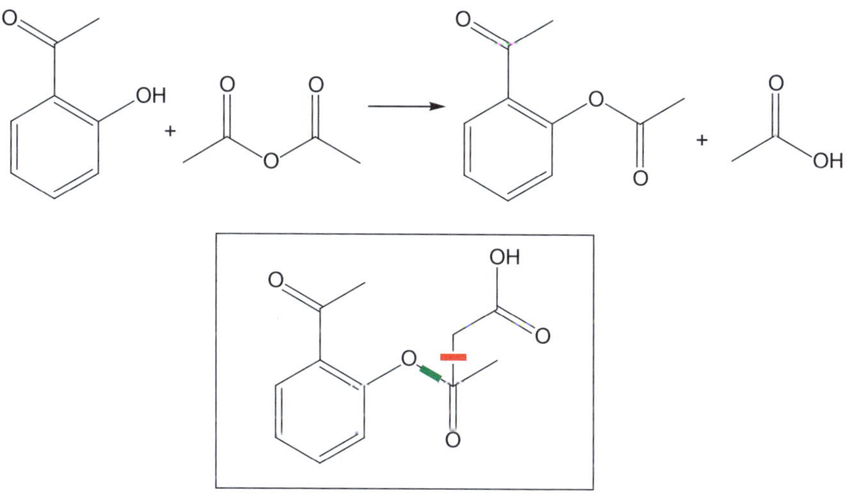

Figure 1.6 Phenol acetylation and related Condensed Graph of Reaction. "Dynamical" bonds marked with green and red correspond, respectively, to formation and breaking a single bond.

1.3.3 Storage of Fragment Information

This section discusses different techniques to store the information about molecular fragments. The most common way is present a given chemical structure as a fixed-size array (vector), in which each element corresponds to the occurrence of a given molecular fragment. Structural keys are descriptor vectors containing binary values indicating presence of absence of fragments. Since structural keys can be kept in computer memory as bit strings they are processed very rapidly, which explains their popularity in chemical database management, similarity search, SAR/QSAR studies and in virtual screening (Figure 1.7).

The composition and length of structural keys always depend on the choice of constituent fragments. Often, structural keys become very sparse, *i.e.*, they contain very few non-zero values. Such highly imbalanced data presentation is rather inefficient for computer processing. As a partial solution to this problem, fragment descriptors can be stored in a list containing the codes (names) of fragments "ON". Although application of lists reduces the storage's size, it is still time consuming to be used for a substructural search in large databases.

Search efficiency can be improved significantly by using hash tables, allowing one to link directly the name of descriptor and location of the descriptor's value. This technology is used in *hashed molecular fingerprints* operating with binary values (Figure 1.8). In contrast to structural keys, in molecular fingerprints each fragment is mapped onto several cells, positions of which are computed from the fragment code. The advantage of hashed fingerprints is a

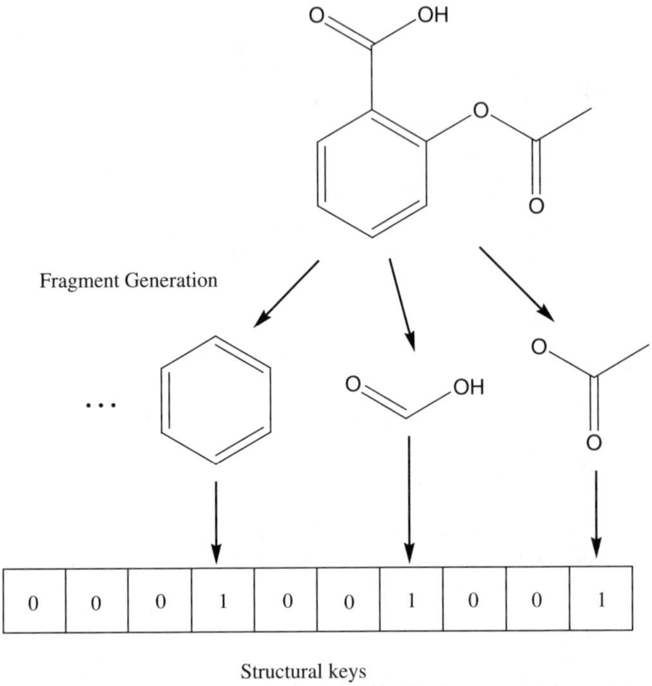

Figure 1.7   Generation of structural keys for a molecule of aspirin.

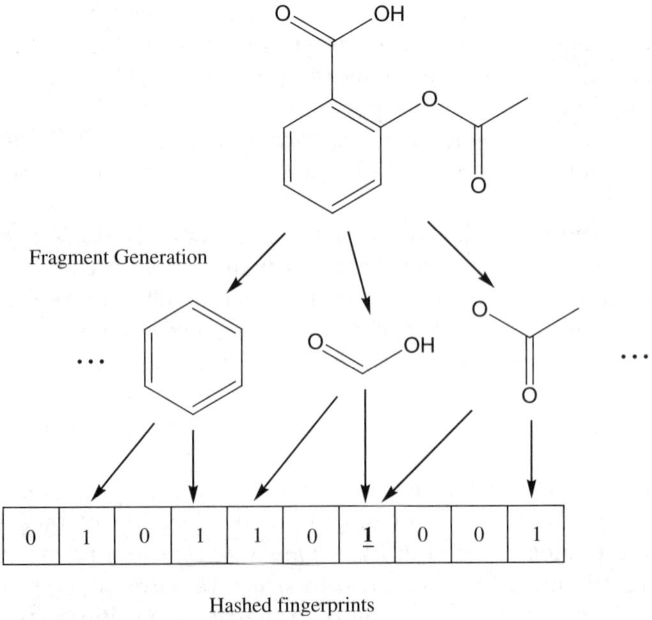

Figure 1.8   Generation of hashed fingerprints. Each fragment leads to "switching on" of several bits. A bit with collisions is underlined and shown in bold.

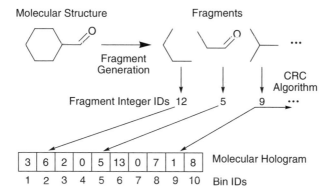

Figure 1.9 Generation of a molecular hologram. A molecule is broken into several structural fragments that are assigned fragment integer identifica tions (IDs) using the CRC algorithm. Each fragment is then placed in a particular bin based on its fragment integer ID corresponding to the bin ID. The bin occupancy numbers are the molecular holo gram descriptors that count structural fragments in each bin. (Adapted from ref. 63.)

possibility to include a big number of fragments in a bit string of reasonable length. Their drawback is related to the existence of collisions when two or more fragments are mapped in the same bit. Nonetheless, this problem could be solved by trade-off between the length of bit string, the number of fragments types and the number of bits allocated for each fragment.

An interesting way of encoding structural information is realized in mole cular holograms, which represent an integer array of bins of predetermined length (hologram length) that contains information about the occurrences of fragments. In the course of generating a molecular hologram, each fragment is coded using the SLN (SYBYL Line Notation).[220] Using the cyclic redun dancy check (CRC) algorithm,[221] this code is transformed into a fragment integer ID, indicating the location of the particular bin in the molecular hologram (Figure 1.9). The occupancy of bins is then incremented by one as soon as the corresponding fragments occur. Since the hologram length I always smaller than the number of fragments, several different fragments map to the same bin in the molecular hologram. The resulting bin occupancy is equal to the sum of occurrence numbers of all these fragments. Molecular holograms were specially designed to be used in the Holographic QSAR (HQSAR) approach.[63]

1.3.4 Fragment Connectivity

Fragments used for building fragment descriptors can be *connected* and *dis connected*. Most applications are based on connected fragments. The point is that the indicators of presence or occurrences of disconnected fragments can always be expressed through the corresponding values obtained for connected fragments.[8] Hence, descriptors based on disconnected fragments are redundant, since they do not carry any additional information compared to their connected counterparts.

Nonetheless, in some cases disconnected fragments descriptors could simplify QSAR/QSPR equations. In particular, nonlinear models involving connected fragments can be replaced with linear models built on disconnected fragments, because the occurrences of disconnected and connected fragments are nonlinearly related. Thus, the use of disconnected fragments may be viewed as an implicit way of introducing nonlinearity into QSARs/QSPRs. If binary descriptor values are used, disconnected fragments implicitly introduce conjunctions (logical .AND.) into logical expressions instead of nonlinear terms for connected fragments. Tarasov *et al.*[222] have shown that the *compound structural descriptors* defined as combinations of unrelated fragments improve significantly the efficiency of mutagenicity predictions. Implicitly, disconnected fragments, as conjugations of binary (logical) connected fragment descriptors, were used to build probabilistic SAR models for some biological activities (see ref. 223 and references therein).

1.3.5   Generic Graphs

In contrast to QSPR studies based on complete (containing all atoms) or hydrogen-suppressed molecular graphs, assessment of biological activity, especially at the qualitative level, often requires greater generalization. In that case, it is convenient to describe chemical structures by *reduced graphs*, in which each vertex – descriptor center or pharmacophoric center – represents an atom or a group of atoms capable of interacting with biological targets, whereas each edge measures the number of bonds between them. Such a biology-oriented representation of chemical structures was invented in 1982 by Avidon *et al.* under the name Descriptor Center Connection Graphs (DCCG)[41] as a generalization of SSFN descriptors (Section 1.3.1.6).

Figure 1.10(b) shows the DCCG for phenothiazine. In this case, the reduced graph consists of 16 edges and 10 vertices corresponding to descriptor centers shown in Figure 1.10(a). Descriptor centers involve four heteroatoms (1–4; see numbering in Figure 1.10a), which can take part in donor–acceptor interaction with biomolecules and in the formation of hydrogen bonds, three methyl groups (5–7), which can take part in hydrophobic interaction with biomolecules, two benzene rings (8, 9) and one heterocycle (10), which can take part in $\pi-\pi$ and $\pi-$cation interactions with biomolecules. Eleven edges in the DCCG labeled with positive numbers indicate the topological distances (counted as the number of bonds) between the atoms included in the corresponding descriptor centers, while the negative labels denote relations between rings within a polycyclic system. Such graphs are very useful not only as a

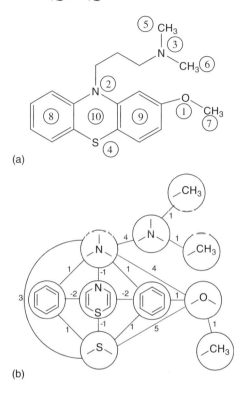

(a)

(b)

Figure 1.10 (a) Structure of phenothiazine with descriptor centers marked on it. (Adapted from ref. 41.) (b) Descriptor center connection graph for phenothiazine. (Adapted from ref. 41.)

source of biology-oriented fragment descriptors but also for pharmacophore based virtual screening.

The atom-pairs proposed by Carhart *et al.*[154] are rather similar to the SSFN descriptors. They can be considered as two-vertex connected fragments of reduced graphs, in which edges correspond to paths between certain atoms. Modifications introduced to the atom-pairs descriptors by Kearsley *et al.*[96] through encoding physicochemical properties of atoms render these fragments even more generic. In 2003 Gillet, Willett and Bradshaw (GWB) introduced another type reduced graphs and proved their high efficiency in a similarity search.[224] A GWB reduced graph consisting of six vertices and five edges is shown in Figure 1.11. Its three vertices R correspond to rings, its two vertices L to linkers, while the vertex F corresponds to a feature – an oxygen atom in this case, which can form hydrogen bonds. In contrast to DCCG, the edges of GWB reduced graphs are not labeled and correspond to ordinary chemical bonds.

An important feature of the GWB reduced graphs is a hierarchical organization of vertex labels. For example, the label Ar_n (non-hydrogen-bonding aromatic cycle) is less general than the label Ar (any aromatic cycle), which, in turn, is less general than R (any ring). Due to this feature, GWB reduced graphs

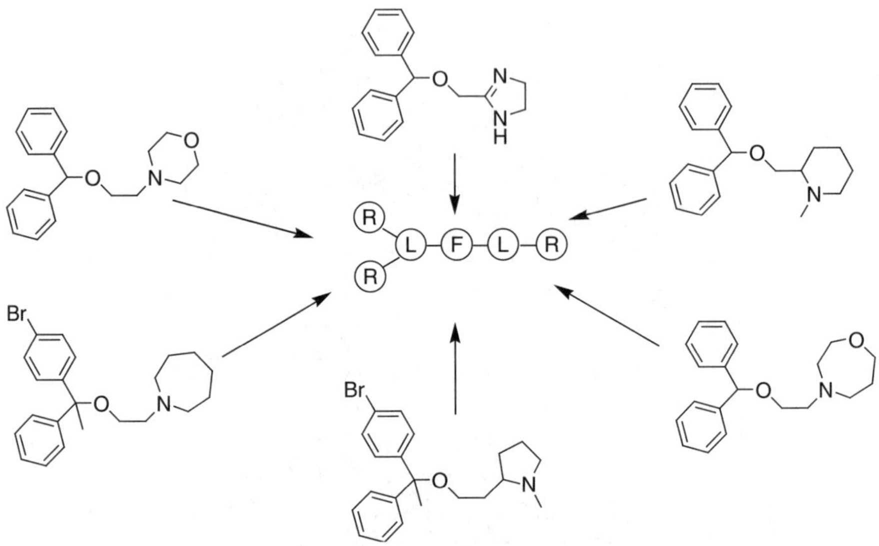

Figure 1.11 Examples of chemical structures corresponding to the same GWB reduced graph of type R/F (shown in center). (Adapted from ref. 224.)

can also be organized hierarchically, and the level of their generalization can be controlled (Figure 1.12). Besides similarity searching, fragment descriptors based on GWB reduced graphs have been applied to derive SAR models using decision trees.[225]

1.3.6 Labeling Atoms

In some cases selected atoms in molecules could be marked with special labels, indicating their particular role in a modeled property. Some examples are (i) local properties, such as atomic charges or NMR chemical shifts, which should always be attributed to a given atom(s), (ii) anchor atoms in the given scaffold to which substituents are attached (Figure 1.13), (iii) atoms forming a main chain in polymers and (iv) reaction centers in a set of reactions. Zefirov *et al.* have applied labeling in QSPR studies of pK_a[226,227] chemical NMR shifts and reaction rate constant for the acid hydrolysis of esters.[226,228] Varnek *et al.*[18] labeled hydrogen bond donor and acceptor centers to model free energies and enthalpies of formation of the 1 : 1 hydrogen-bond complexes.

1.4 Application in Virtual Screening and *In Silico* Design

This section considers the application of fragment descriptors at different stages of virtual screening and *in silico* design.

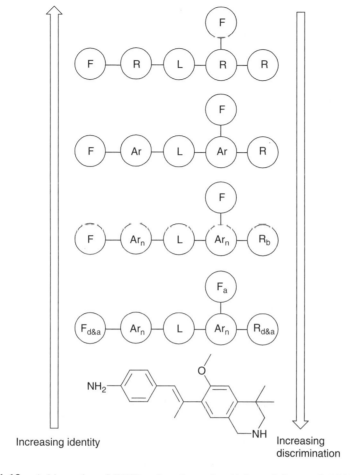

Increasing identity   Increasing discrimination

Figure 1.12 A hierarchy of GWB reduced graphs. (Adapted from ref. 224.)

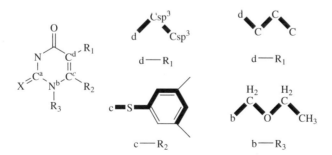

Figure 1.13 Examples of fragments with marked atoms used for modeling inhibitor activity against HIV-I reverse transcriptase for a congeneric set of HEPT derivatives.

1.4.1 Filtering

Filtering is a rule-based approach aimed to perform fast assessment of usefulness of molecules in the given context. In terms of drug design, the filtering is used to eliminate compounds with unfavorable pharmacodynamic or pharmacokinetic properties as well as toxic compounds. Pharmacodynamics considers binding drug-like organic molecules (ligands) to chosen biological target. Since the efficiency of ligand–target interactions depends on spatial complementarity of their binding sites, the filtering is usually performed with 3D-pharmacophores, representing "optimal" spatial arrangements of steric and electronic features of ligands.[229,230] Pharmacokinetics is mostly related to absorption, distribution, metabolism and excretion (ADME) related properties: octanol–water partition coefficients (log P), solubility in water (log S), blood–brain coefficient (log BB), partition coefficient between different tissues, skin penetration coefficient, *etc.*

Fragment descriptors are widely used for early ADME/Tox prediction both explicitly and implicitly. The easiest way to filter large databases concerns detecting undesirable molecular fragments (structural alerts). Appropriate lists of structural alerts are published for toxicity,[231] mutagenicity,[232] and carcinogenicity.[233] Klopman *et al.* were the first to recognize the potency of fragment descriptors for this purpose.[66,67,69] Their programs CASE,[66] MultiCASE,[97,234] as well as more recent MCASE QSAR expert systems,[235] proved to be effective tools to assess the mutagenicity[67,234,235] and carcinogenicity[69,234] of organic compounds. In these programs, sets of biophores (analogs of structural alerts) were identified and used for activity predictions. Several more sophisticated fragment-based expert systems of toxicity assessment – DEREK,[210] TopKat[236] and Rex[237] – have been developed. DEREK is a knowledge-based system operating with human-coded or automatically generated[238] rules concerning toxicophores. Fragments in the DEREK knowledge base are defined by means of the linear notation language PATRAN, which codes the information about atom, bonds and stereochemistry. TopKat uses a large predefined set of fragment descriptors, whereas Rex implements a special kind of atom-pairs descriptors (links). For more information about fragment-based computational assessment of toxicity, including mutagenicity and carcinogenicity, see ref. 239 and references therein.

The most popular filter used in drug design area is the Lipinski "rule of five",[240] which takes into account the molecular weight, the number of hydrogen bond donors and acceptors, along with the octanol–water partition coefficient log P, to assess the bioavailability of oral drugs. Similar rules of "drug-likeness" or "lead-likeness" were later proposed by Oprea,[241] Veber[242] and Hann.[243] Formally, fragment descriptors are not explicitly involved there. However, most computational approaches that assess log P are fragment-based;[244–246] whereas H-donors and acceptor sites are the simplest molecular fragments.

1.4.2 Similarity Search

The notion of molecular similarity (or chemical similarity) is one of the most useful and at the same time one of the most contradictory concepts in chemoinformatics.[247,248] The concept of molecular similarity plays an important role in many modern approaches to predicting the properties of chemical compounds, designing chemicals with a predefined set of properties and, especially, in conducting drug design studies by screening large databases containing structures of available (or potentially available) chemicals. These studies are based on the similar property principle of Johnson and Maggiora, which states: similar compounds have similar properties.[247] The similarity-based virtual screening assumes that all compounds in a database that are similar to a query compound have similar biological activity. Although this hypothesis is not always valid (see discussion in ref. 249), quite often the set of retrieved compounds is considerably enriched with actives.[250]

To achieve high efficacy of similarity-based screening of databases containing millions compounds, molecular structures are usually represented by *screens* (structural keys) or fixed-size or variable-size *fingerprints*. Screens and fingerprints can contain both 2D and 3D-information. However, the 2D-fingerprints, which are a kind of binary fragment descriptors, dominate in this area. Fragment-based structural keys, like MDL keys,[62] are sufficiently good for handling small and medium-sized chemical databases, whereas processing of large databases is performed with fingerprints having much higher information density. Fragment-based Daylight,[251] BCI,[252] and UNITY 2D[253] fingerprints are the best known examples.

The most popular similarity measure for comparing chemical structures represented by means of fingerprints is the Tanimoto (or Jaccard) coefficient T.[254] Two structures are usually considered similar if $T > 0.85$[250] (for Daylight fingerprints[251]). Using this threshold, Taylor estimated a probability to retrieve actives as 0.012–0.50,[255] whereas according to Delancy this probability is even higher, *i.e.*, 0.40–0.60 (ref. 256) (using Daylight fingerprints[251]). These computer experiments confirm the usefulness of the similarity approach as an instrument of virtual screening.

Schneider *et al.* have developed a special technique for performing virtual screening referred to as Chemically Advanced Template Search (CATS).[257] Within its framework, chemical structures are described by means of so-called correlation vectors, each component of which is equal to the occurrence of a given atom pair divided by the total number of non-hydrogen atoms in it. Each atom in the atom pair is specified as belonging to one of five classes (hydrogen-bond donor, hydrogen-bond acceptor, positively charged, negatively charged, and lipophilic), while topological distances of up to ten bonds are also considered in the atom-pair specification. In ref. 257, the similarity is assessed by Euclidean distance between the corresponding correlation vectors. CATS has been shown to outperform the MERLIN program with Daylight fingerprints[251] for retrieving thrombin inhibitors in a virtual screening experiment.[257]

Hull *et al.* have developed the Latent Semantic Structure Indexing (LaSSI) approach to perform similarity search in low-dimensional chemical space.[258,259] To reduce the dimension of initial chemical space, the singular value decomposition method is applied for the descriptor-molecule matrix. Ranking molecules by similarity to a query molecule was performed in the reduced space using the cosine similarity measure,[260] whereas the Carhart's atom pairs[154] and the Nilakantan's topological torsions[95] were used as descriptors. The authors claim that this approach "has several advantages over analogous ranking in the original descriptor space: matching latent structures is more robust than matching discrete descriptors, choosing the number of singular values provides a rational way to vary the 'fuzziness' of the search".[258]

The issue of "fuzzification" of similarity search has been addressed by Horvath *et al.*[155–157] The first fuzzy similarity metric suggested[155] relies on partial similarity scores calculated with respect to the inter-atomic distances distributions for each pharmacophore pair. In this case the "fuzziness" enables comparison of pairs of pharmacophores with different topological or 3D distances. Similar results[156] were achieved using fuzzy and weighted modified Dice similarity metric.[260] Fuzzy pharmacophore triplets (FPT, see Section 1.3.1.6) can be gradually mapped onto related basis triplets, thus minimizing binary classification artifacts.[157] In a new similarity scoring index introduced in ref. 157, the simultaneous absence of a pharmacophore triplet in two molecules is taken into account. However, this is a less-constraining indicator of similarity than simultaneous presence of triplets.

Most similarity search approaches require only a single reference structure. However, in practice several lead compounds are often available. This motivated Hert *et al.*[261] to develop the data fusion method, which allows one to screen a database using all available reference structures. Then, the similarity scores are combined for all retrieved structures using selected fusion rules. Searches conducted on the MDL Drug Data Report database using fragment-based UNITY 2D,[253] BCI,[252] and Daylight[251] fingerprints have proved the effectiveness of this approach.

The main drawback of the conventional similarity search concerns an inability to use experimental information on biological activity to adjust similarity measures. This results in an inability to discriminate relevant and non-relevant fragment descriptors used for computing similarity measures. To tackle this problem, Cramer *et al.*[42] developed substructural analysis, in which each fragment (represented as a bit in a fingerprint) is weighted by taking into account its occurrence in active and in inactive compounds. Subsequently, many similar approaches have been described in the literature.[262]

One more way to conduct a similarity-based virtual screening is to retrieve the structures containing a user-defined set of "pharmacophoric" features. In the Dynamic Mapping of Consensus positions (DMC) algorithm[263] those features are selected by finding common positions in bit strings for all active compounds. The potency-scaled DMC algorithm (POT-DMC)[264] is a modification of DMC in which compounds activities are taken into account. The latter two methods may be considered as intermediate between conventional similarity search and probabilistic SAR approaches.

Batista, Godden and Bajorath have developed the MolBlaster method,[208] in which molecular similarity is assessed by Differential Shannon Entropy[265] computed from populations of randomly generated fragments. For the range $0.64 < T < 0.99$, this similarity measure provides with the same ranking as the

Tanimoto index T. However, for smaller values of T the entropy-based index is more sensitive, since it distinguishes between pairs of molecules having almost identical T. To adapt this methodology for large-scale virtual screening, Proportional Shannon Entropy (PSE) metrics were introduced.[209] A key feature of this approach is that class-specific PSE of random fragment distributions enables the identification of the molecules sharing with known active compounds a significant number of signature substructures.

Similarity search methods developed for individual compounds are difficult to apply directly for chemical reactions involving many species subdivided by two types: reactants and products. To overcome this problem, Varnek *et al.*[18] suggested condensing all participating reaction species in one molecular graph [Condensed Graphs of Reactions (CGR),[18] see Section 1.3.2] followed by its fragmentation and application of developed fingerprints in "classical" similarity search. Besides conventional chemical bonds (simple, double, aromatic, *etc.*), a CGR contains dynamical bonds corresponding to created, broken or transformed bonds. This approach could be efficiently used for screening of large reaction databases.

1.4.3 SAR Classification (Probabilistic) Models

Simplistic and heuristic similarity-based approaches can hardly produce as good predictive models as modern statistical and machine learning methods that are able to assess quantitatively biological or physicochemical properties. QSAR-based virtual screening consists of direct assessment of activity values (numerical or binary) of all compounds in the database followed by selection of hits possessing desirable activity. Mathematical methods used for models preparation can be subdivided into classification and regression approaches. The former decide whether a given compound is active, whereas the latter numerically evaluate the activity values. Classification approaches that assess probability of decisions are called probabilistic.

Various classification approaches have been reported to be used successfully in conjunction with fragment descriptors for building classification SAR models: the Linear Discriminant Analysis (LDA),[266,267] the Partial Least Square Discriminant Analysis (PLS-DA),[268] Soft Independent Modeling by Class Analogy (SIMCA),[269] Artificial Neural Networks (ANN),[270] Support Vector Machines (SVM),[271] Decision Trees (DT),[269,272,273] Spline Fitting with Genetic Algorithm (SFGA),[269] *etc.* Probabilistic methods usually used with fragment descriptors are: Naïve Bayes (NB)[142] and its modification implemented in PASS,[126] Binary Kernel Discrimination,[6] Inductive Logic Programming (ILP),[274] Support Vector Inductive Logic Programming (SVILP),[133] *etc.*

Numerous studies have been devoted to classification (probabilistic) approaches used in conjunction with fragment descriptors for virtual screening. Here we present several examples.

Harper *et al.*[6] have demonstrated a much better performance of probabilistic "binary kernel discrimination" method to screen large databases compared to backpropagation neural networks or conventional similarity search. The Carhart's atom-pairs[154] and Nilakantan's topological torsions[95] were used as descriptors.

Aiming to discover new cognition enhancers, Geronikaki *et al.*[275] applied the PASS program,[126] which implements a probabilistic Bayesian-based approach, and the DEREK rule-based system[210] to screen a database of highly diverse chemical compounds. Eight compounds with the highest probability of cognition-enhancing effect were selected. Experimental tests showed that all of them possess a pronounced antiamnesic effect.

Bender, Glen *et al.* have applied[129–133] several probabilistic machine learning methods (naïve Bayesian classifier, inductive logic programming, and support vector inductive learning programming) in conjunction with circular fingerprints for making classification of bioactive chemical compounds and performing virtual screening on several biological targets. The latter of these three methods (*i.e.*, support vector inductive learning programming) performed significantly better than the other two methods.[133] The advantages of using circular fingerprints were pointed out.[131]

1.4.4 QSAR/QSPR Regression Models

The Multiple Linear Regression (MLR) method was historically the first and to date the most popular method used to develop QSAR/QSPR models with fragment descriptors (Figure 1.14). Linear models involving fragments are built in several program packages: CASE,[66–69] MULTICASE,[97,98] TRAIL,[101,102] ISIDA,[18] EMMA,[276] QSAR Builder from Pharma Algorithms[277] and some others. The Partial Least Squares (PLS) regression,[278,279] an alternative technique for building linear quantitative models, has also been successfully coupled with fragment descriptors.[63,128,280–282] This approach is efficiently used the Holographic QSAR (HQSAR)[63] (implemented in the Sybyl software[253]) and the "Generalized Fragment-Substructure Based Property Prediction Method".[282] The success of treating the fragment descriptors in PLS is explained by efficient handling of multicollinearity, which is a typical problem of fragment descriptors. Two other methods, the Group Method of Data Handling (GMDH)[283] and the more recent Maximal Margin Linear Programming Method (MMLPM),[284,285] also displayed their efficiency in building the linear models from an initial pool of highly correlated fragment descriptors.

Among nonlinear regression methods used in conjunction with fragment descriptors, the Back-Propagation Neural Networks (BPNN)[286–289] occupy a special place. It has been proved[7,8] that any molecular graph invariant can be approximated by an output of a BPNN using fragment descriptors as an input. Indeed, numerous studies have shown that the BPNN models based on fragment descriptors efficiently predict various physicochemical properties[16,290–294] and some biological activities[16,163,295] of organic compounds. A popular ASNN (Associative Neural Networks) approach consists of an ensemble of BPNN coupled with kNN correction in the space of models.[296] This technique,

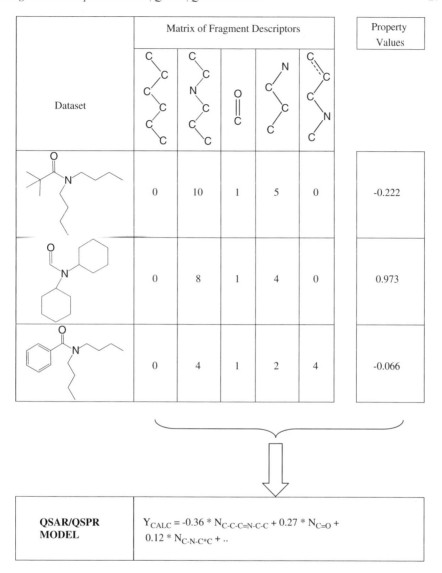

Figure 1.14 General scheme of constructing linear QSAR/QSPR models based on fragment descriptors.

together with fragment descriptors, has been successfully used to model the thermodynamic parameters of metal complexation[285] and melting point of ionic liquids.[297] Besides, the Radial Basis Function Neural Networks[298] (RBFNNs) have also been used with fragment descriptors for predicting the properties of organic compounds.[285,299] The Support Vector Regression (SVR) technique[300–303] is a serious "competitor" of neural networks, as has been demonstrated in QSAR/QSPR studies[285,304] involving fragment descriptors.

In drug design, regression QSAR/QSPR models are often used to assess ADME/Tox properties or to detect "hit" molecules capable of binding a certain biological target. Thus, one could mention fragments based QSAR models for blood–brain barrier,[305] skin permeation rate,[306] blood–air[307] and tissue-air partition coefficients.[307] Many theoretical approaches to calculating the octanol–water partition coefficient log P involve fragment descriptors. In particular, it concerns the methods by Rekker,[308,309] Leo and Hansch (CLOGP),[245,310] Ghose-Crippen (ALOGP),[81–83] Wildman and Crippen,[86] Suzuki and Kudo (CHEMICALC-2),[87] Convard (SMILOGP)[88] and by Wang (XLOGP).[89,90] Fragment-based predictive models for estimation of solubility in water[311] and DMSO[311] are also available.

Benchmarking studies on various biological and physicochemical properties[305–307,312] show that QSAR/QSPR models for involving fragment descriptors in many cases outperform those built on topological, quantum, electrostatic and other types of descriptors.

1.4.5   *In Silico* Design

In this section we consider several examples of virtual screening performed on a database containing only virtual (still non-synthesized or unavailable) compounds. Virtual libraries are usually generated using combinatorial chemistry approaches.[313–315] One of simplest ways is to attach systematically user-defined substituents $R_1, R_2, \ldots, R_N$ to a given scaffold. If the list for the substituent R_i contains n_i candidates, the total number of generated structures is:

$$N = \prod_i n_i \qquad (1.11)$$

although taking symmetry into account could reduce the library's size. The number of substituents R_i (n_i) should be carefully selected to avoid generation of too large a set of structures (combinatorial explosion). The "optimal" substituents could be prepared using fragments selected at the QSAR stage, since their contributions to activity (for linear models) allow one to estimate an impact of combining the fragment into larger species (R_i). In such a way, a focused combinatorial library could be generated.

The technology based on combining QSAR, generation of virtual libraries and screening stages has been implemented in the ISIDA program and applied to computer-aided design of new uranyl binders belonging to two different families of organic molecules: phosphoryl containing podands[316] and mono-amides.[317] QSAR models have been developed using different machine-learning methods (multi-linear regression analysis, associative neural networks[296] and support vector machines[301]) and fragment descriptors (atom/bond sequences and augmented atoms). These models were then used to screen virtual combinatorial libraries containing up to 11000 compounds. Selected hits were synthesized and tested experimentally. Predicted uranyl binding affinity was shown to agree well with the experimental data. Thus, initial data sets were significantly enriched with new efficient uranyl binders, and one of new molecules was found to be more efficient than previously studied compounds. A similar study was conducted for the development of new 1-(2-hydroxyethoxy)methyl)-6-(phenylthio)thymine (HEPT) derivatives potentially possessing high anti-HIV activity.[318] This demonstrates the universality of fragment descriptors and the broad perspectives of their use in virtual screening and *in silico* design.

1.5 Limitations of Fragment Descriptors

Despite the many advantages of fragment descriptors they are not devoid of certain drawbacks, which deserve serious attention. Two main problems should be mentioned: (i) "missing fragments";[319] and (ii) modeling of stereochemically dependent properties.

The term "missing fragments" concerns comparison of the lists of fragments generated for the training and test sets. A test set molecule may contain fragments that, on one hand, belong to the same family of descriptors used for the modeling, and, on the other hand, are different from those in the initial pool calculated for the training set. The question arises whether the model built from that initial pool can be applied to those test set molecules? This is a difficult problem because *a priori* it is not clear if the "missing fragments" are important for the property being predicted. Several possible strategies to treat this problem have been reported. The ALOGPS program,[320] predicting lipophilicity and aqueous solubility of chemical compounds, flags calculations as unreliable if the analyzed molecule contains one or more E-state atom or bond types missed in the training set. In such a way, the program detects about 90% of large prediction errors.[319] The ISIDA program[18] calculates a consensus model as an average over the "best" models developed with different sets of fragment descriptors. Each model corresponds to its "own" initial pool of descriptors. If a new molecule contains fragments different from those in that pool, the corresponding model is ignored. As demonstrated by benchmarking studies,[285] this improves the predictive performance of the method. For each model, the NASAWIN software[99] creates a list of "important" fragments including cycles and all one-atom fragments. The test molecule is rejected if its list of "important" fragments contains those absent in the training set.[321] The LOGP program for lipophilicity predictions[322] uses a set of empirical rules to calculate the contribution of missed fragments.

The second problem of using fragment descriptors deals with accounting for stereochemical information. In fact, its adequate treatment is not possible at the graph-theoretical level and requires explicit consideration of hypergraphs.[323] However, in practice, it is sufficient to introduce special labels indicating stereochemical configuration of chiral centers or (*E/Z*)-isomers around a double bond, and then to use them in the specification of molecular fragments. Such an approach has been used in hologram fragment descriptors[324] as well as in the PARTAN language.[238]

1.6 Conclusion

Fragment descriptors constitute one of the most universal types of molecular descriptors. The scope of their application encompasses almost all existing areas of SAR/QSAR/QSPR studies. Their universality stems from the basic character of structural theory in chemistry as well as from the fundamental possibility of molecular graph invariants being expressed in terms of subgraph occurrence numbers.[8] The main advantages of fragment descriptors lie in the simplicity of their computation, the easiness of their interpretation as well as in efficiency of their applications in similarity searches and SAR/QSAR/QSPR modeling. Progress of their use in virtual screening could be related to the development of new types of fragments and of new mathematical approaches of their processing.

Acknowledgements

The authors thank GDRE SupraChem and ARCUS "Alsace –Russia/ Ukraine" project for support and also Dr V. Solov'ev for fruitful discussions.

References

1. J. Gasteiger and T. Engel, eds., *Chemoinformatics: A Textbook*, Wiley-VCH, Weinheim, 2003.
2. J. Gasteiger, ed., *Handbook of Chemoinformatics: From Data to Knowledge.*, Wiley-VCH, Weinheim, 2003.
3. T. Engel, *J. Chem. Inf. Model.*, 2006, **46**, 2267–2277.
4. W. L. Chen, *J. Chem. Inf. Model.*, 2006, **46**, 2230–2255.
5. N. Brown, *Computing Surveys*, 2006.
6. G. Harper, J. Bradshaw, J. C. Gittins, D. V. S. Green and A. R. Leach, *J. Chem. Inf. Comput. Sci.*, 2001, **41**, 1295–1300.
7. I. I. Baskin, M. I. Skvortsova, I. V. Stankevich and N. S. Zefirov, *Dokl. Chem.*, 1994, **339**, 231–234.
8. I. I. Baskin, M. I. Skvortsova, I. V. Stankevich and N. S. Zefirov, *J. Chem. Inf. Comput. Sci.*, 1995, **35**, 527–531.
9. M. I. Skvortsova, I. I. Baskin, L. A. Skvortsov, V. A. Palyulin, N. S. Zefirov and I. V. Stankevich, *Theochem.*, 1999, **466**, 211–217.
10. M. I. Skvortsova, I. V. Stankevich, I. I. Baskin, V. A. Palyulin and N. A. Zefirov, *Doklady Akademii Nauk*, 1996, **350**, 786–788.
11. M. I. Skvortsova, I. I. Baskin, I. V. Stankevich, V. A. Palyulin and N. S. Zefirov, *J. Chem. Inf. Comput. Sci.*, 1998, **38**, 785–790.
12. M. I. Skvortsova, I. I. Baskin, O. L. Slovokhotova and N. S. Zefirov, *Doklady Akademii Nauk*, 1994, **336**, 496–499.
13. M. I. Skvortsova, I. I. Baskin, I. V. Stankevich and N. S. Zefirov, *Doklady Akademii Nauk*, 1996, **351**, 78–80.
14. N. S. Zefirov and V. A. Palyulin, *J. Chem. Inf. Comput. Sci.*, 2002, **42**, 1112–1122.

15. P. Japertas, R. Didziapetris and A. Petrauskas, *Quant. Struct.-Act. Relat.*, 2002, **21**, 23–37.
16. N. V. Artemenko, I. I. Baskın, V. A. Palyulin and N. S. Zefirov, *Russ. Chem. Bull.*, 2003, **52**, 20–29.
17. C. Merlot, D. Domine and D. J. Church, *Curr. Opin. Drug Discov. Devel.*, 2002, **5**, 391–399.
18. A. Varnek, D. Fourches, F. Hoonakker and V. P. Solov'ev, *J. Comput. Aided Mol. Des.*, 2005, **19**, 693–703.
19. S. Jelfs, P. Ertl and P. Selzer, *J. Chem. Inf. Model.*, 2007, **47**, 450–459.
20. R. Todeschini and V. Consonni, *Handbook of Molecular Descriptors.*, Wiley-VCH Publishers, Weinheim, 2000.
21. A. I. Vogel, *Chemistry & Industry*, 1934, 85.
22. C. T. Zahn, *J. Chem. Phys.*, 1934, **2**, 671–680.
23. M. Souders, C. S. Matthews and C. O. Hurd, *Ind. Eng. Chem.*, 1949, **41**, 1037 1048.
24. M. Souders, C. S. Matthews and C. O. Hurd, *Ind. Eng. Chem.*, 1949, **41**, 1048–1056.
25. J. L. Franklin, *Ind. Eng. Chem.*, 1949, **41**, 1070–1076.
26. J. L. Franklin, *J. Chem. Phys.*, 1953, **21**, 2029–2033.
27. V. M. Tatevskii, *Doklady Akademii Nauk SSSR*, 1950, **75**, 819–822.
28. V. M. Tatevskii, E. A. Mendzheritskii and V. Korobov, *Vestnik Moskovskogo Universiteta*, 1951, **6**, 83–86.
29. H. J. Bernstein, *J. Chem. Phys.*, 1952, **20**, 263–269.
30. K. J. Laidler, *Canadian J. Chem.*, 1956, **34**, 626–648.
31. S. W. Benson and J. H. Buss, *J. Chem. Phys.*, 1958, **29**, 546–572.
32. T. L. Allen, *J. Chem. Phys.*, 1959, **31**, 1039–1049.
33. E. A. Smolenskii, *Zhurnal Fizicheskoi Khimii*, 1964, **38**, 1288–1291.
34. C. Hansch, R. M. Muir, T. Fujita, P. P. Maloney, F. Geiger and M. Streich, *J. Am. Chem. Soc.*, 1963, **85**, 2817–2824.
35. C. Hansch and T. Fujita, *J. Am. Chem. Soc.*, 1964, **86**, 1616–1626.
36. S. M. Free Jr. and J. W. Wilson, *J. Med. Chem.*, 1964, **7**, 395–399.
37. S. A. Hiller, A. B. Glaz, L. A. Rastrigin and A. B. Rosenblit, *Doklady Akademii Nauk SSSR.*, 1971, **199**, 851–853.
38. S. A. Hiller, V. E. Golender, A. B. Rosenblit, L. A. Rastrigin and A. B. Glaz, *Comput. Biomed. Res.*, 1973, **6**, 411–421.
39. V. E. Golender and A. B. Rozenblit, *Avtomatika i Telemekhanika*, 1974, 99–105.
40. V. E. Golender and A. B. Rozenblit, *Med. Chem. (Academic Press)*, 1980, **11**, 299–337.
41. V. V. Avidon, I. A. Pomerantsev, V. E. Golender and A. B. Rozenblit, *J. Chem. Inf. Comput. Sci.*, 1982, **22**, 207–214.
42. R. D. Cramer 3rd, G. Redl and C. E. Berkoff, *J. Med. Chem.*, 1974, **17**, 533–535.
43. W. E. Brugger, A. J. Stuper and P. C. Jurs, *J. Chem. Inf. Model.*, 1976, **16**, 105–110.
44. A. J. Stuper and P. C. Jurs, *J. Chem. Inf. Model.*, 1976, **16**, 99–105.

45. L. Hodes, G. F. Hazard, R. I. Geran and S. Richman, *J. Med. Chem.*, 1977, **20**, 469–475.
46. G. W. Adamson, *Proceedings of the Analytical Division of the Chemical Society*, 1977, **14**, 26–28.
47. G. W. Adamson and J. A. Bush, *Nature*, 1974, **248**, 406–407.
48. G. W. Adamson and D. Bawden, *J. Chem. Inf. Comput. Sci.*, 1975, **15**, 215–220.
49. G. W. Adamson and J. A. Bush, *Journal of the Chemical Society, Perkin Transactions 1*, 1976, 168–172.
50. G. W. Adamson and D. Bawden, *J. Chem. Inf. Comput. Sci.*, 1977, **17**, 164–171.
51. G. W. Adamson and D. Bawden, *J. Chem. Inf. Comput. Sci.*, 1976, **16**, 161–165.
52. M. Milne, D. Lefkovitz, H. Hill and R. Powers, *J. Chem. Doc.*, 1972, **12**, 183–189.
53. G. W. Adamson, J. Cowell, M. F. Lynch, A. H. W. McLure, W. G. Town and A. M. Yapp, *J. Chem. Doc.*, 1973, **13**, 153–157.
54. A. Feldman and L. Hodes, *J. Chem. Inf. Model.*, 1975, **15**, 147–152.
55. P. Willett, *J. Chem. Inf. Model.*, 1979, **19**, 159–162.
56. P. Willett, *J. Chem. Inf. Model.*, 1979, **19**, 253–255.
57. P. Willett, V. Winterman and D. Bawden, *J. Chem. Inf. Model.*, 1986, **26**, 36–41.
58. W. Fisanick, A. H. Lipkus and A. Rusinko, *J. Chem. Inf. Model.*, 1994, **34**, 130–140.
59. L. Hodes, *J. Chem. Inf. Model.*, 1989, **29**, 66–71.
60. M. J. McGregor and P. V. Pallai, *J. Chem. Inf. Model.*, 1997, **37**, 443–448.
61. D. B. Turner, S. M. Tyrrell and P. Willett, *J. Chem. Inf. Model.*, 1997, **37**, 18–22.
62. J. L. Durant, B. A. Leland, D. R. Henry and J. G. Nourse, *J. Chem. Inf. Comput. Sci.*, 2002, **42**, 1273–1280.
63. W. Tong, D. R. Lowis, R. Perkins, Y. Chen, W. J. Welsh, D. W. Goddette, T. W. Heritage and D. M. Sheehan, *J. Chem. Inf. Model.*, 1998, **38**, 669–677.
64. R. D. Cramer, *J. Am. Chem. Soc.*, 1980, **102**, 1837–1849.
65. R. D. Cramer, *J. Am. Chem. Soc.*, 1980, **102**, 1849–1859.
66. G. Klopman, *J. Am. Chem. Soc.*, 1984, **106**, 7315–7321.
67. G. Klopman and H. S. Rosenkranz, *Mutat. Res.*, 1984, **126**, 227–238.
68. G. Klopman and A. N. Kalos, *J. Comput. Chem.*, 1985, **6**, 492–506.
69. H. S. Rosenkranz, C. S. Mitchell and G. Klopman, *Mutat. Res.*, 1985, **150**, 1–11.
70. G. Klopman, M. R. Frierson and H. S. Rosenkranz, *Environmental Mutagenesis*, 1985, **7**, 625–644.
71. H. S. Rosenkranz and G. Klopman, *Progress in Clinical and Biological Research*, 1986, **209A**, 71–104.
72. G. Klopman, K. Namboodiri and A. N. Kalos, *Progress in Clinical and Biological Research*, 1985, **172**, 287–298.

73. G. Klopman, *Environmental Health Perspectives*, 1985, **61**, 269–274.
74. G. Klopman and O. T. Macina, *J. Theor. Biol.*, 1985, **113**, 637–648.
75. G. Klopman and R. Contreras, *Mol. Pharmacol.*, 1985, **27**, 86–93.
76. G. Klopman and R. E. Venegas, *Acta Pharmaceutica Jugoslavica*, 1986, **36**, 189–209.
77. G. Klopman and A. N. Kalos, *J. Theor. Biol.*, 1986, **118**, 199–214.
78. G. Klopman, O. T. Macina, E. J. Simon and J. M. Hiller, *Theochem*, 1986, **27**, 299–308.
79. G. Klopman, O. T. Macina, M. E. Levinson and H. S. Rosenkranz, *Antimicrobial Agents and Chemotherapy*, 1987, **31**, 1831–1840.
80. G. Klopman and O. T. Macina, *Mol. Pharmacol.*, 1987, **31**, 457–476.
81. A. K. Ghose and G. M. Crippen, *J. Comput. Chem.*, 1986, **7**, 565–577.
82. A. K. Ghose and G. M. Crippen, *J. Chem. Inf. Comput. Sci.*, 1987, **27**, 21–35.
83. A. K. Ghose, A. Pritchett and G. M. Crippen, *J. Comput. Chem.*, 1988, **9**, 80–90.
84. V. N. Viswanadhan, A. K. Ghose, G. R. Revankar and R. K. Robins, *J. Chem. Inf. Comput. Sci.*, 1989, **29**, 163–172.
85. A. K. Ghose, V. N. Viswanadhan and J. J. Wendoloski, *Journal of Physical Chemistry A*, 1998, **102**, 3762–3772.
86. S. A. Wildman and G. M. Crippen, *J. Chem. Inf. Comput. Sci.*, 1999, **39**, 868–873.
87. T. Suzuki and Y. Kudo, *J. Comput. Aided. Mol. Des.*, 1990, **4**, 155–198.
88. T. Convard, J.-P. Dubost, H. Le Solleu and E. Kummer, *Quant. Struct.-Act. Relat.*, 1994, **13**, 34–37.
89. R. Wang, Y. Fu and L. Lai, *J. Chem. Inf. Comput. Sci.*, 1997, **37**, 615–621.
90. R. Wang, Y. Gao and L. Lai, *Persp. Drug Discov. Design*, 2000, **19**, 47–66.
91. T. J. Hou, K. Xia, W. Zhang and X. J. Xu, *J. Chem. Inf. Comput. Sci.*, 2004, **44**, 266–275.
92. D. A. Winkler, F. R. Burden and A. J. R. Watkins, *Quantitative Structure-Activity Relationships*, 1998, **17**, 14–19.
93. H. J. Bernstein, *Trans. Faraday Soc.*, 1962, **58**, 2285–2306.
94. A. J. Kalb, A. L. H. Chung and T. L. Allen, *J. Am. Chem. Soc.*, 1966, **88**, 2938–2942.
95. R. Nilakantan, N. Bauman, J. S. Dixon and R. Venkataraghavan, *J. Chem. Inf. Comput. Sci.*, 1987, **27**, 82–85.
96. S. K. Kearsley, S. Sallamack, E. M. Fluder, J. D. Andose, R. T. Mosley and R. P. Sheridan, *J. Chem. Inf. Comput. Sci.*, 1996, **36**, 118–127.
97. G. Klopman, *Quant. Struct.-Act. Relat.*, 1992, **11**, 176–184.
98. G. Klopman, *J. Chem. Inf. Comput. Sci.*, 1998, **38**, 78–81.
99. I. I. Baskin, N. M. Halberstam, N. V. Artemenko, V. A. Palyulin and N. S. Zefirov, in: *EuroQSAR 2002 Designing Drugs and Crop Protectants: processes, problems and solutions.*, M. Ford ed., Blackwell Publishing, 2003, pp. 260–263.
100. M. I. Kumskov, *Zhurnal Organicheskoi Khimii*, 1995, **31**, 1495–1498.

101. V. P. Solov'ev, A. Varnek and G. Wipff, *J. Chem. Inf. Comput. Sci.*, 2000, **40**, 847–858.
102. A. Varnek, G. Wipff and V. P. Solovev, *Solvent Extraction and Ion Exchange*, 2001, **19**, 791–837.
103. A. A. Gakh, E. G. Gakh, B. G. Sumpter and D. W. Noid, *J. Chem. Inf. Comput. Sci.*, 1994, **34**, 832–839.
104. G. Rucker and C. Rucker, *J. Chem. Inf. Comput. Sci.*, 1993, **33**, 683–695.
105. G. W. Adamson, J. Cowell, M. F. Lynch, W. G. Town and A. M. Yapp, *J. Chem. Soc., Perkin Trans. 1*, 1973, 863–865.
106. G. W. Adamson, S. E. Creasey, J. P. Eakins and M. F. Lynch, *J. Chem. Soc., Perkin Trans. 1*, 1973, **1**, 2071–2076.
107. W. J. Wiswesser, *J. Chem. Inf. Comput. Sci.*, 1982, **22**, 88–93.
108. D. Weininger, *J. Chem. Inf. Comput. Sci.*, 1988, **28**, 31–36.
109. D. Weininger, A. Weininger and J. L. Weininger, *J. Chem. Inf. Comput. Sci.*, 1989, **29**, 97–101.
110. G. W. Adamson and D. Bawden, *J. Chem. Inf. Model.*, 1980, **20**, 97–100.
111. G. W. Adamson and D. Bawden, *J. Chem. Inf. Model.*, 1980, **20**, 242–246.
112. G. W. Adamson and D. Bawden, *J. Chem. Inf. Comput. Sci.*, 1981, **21**, 204–209.
113. D. Qu, B. Fu, M. Muraki and T. Hayakawa, *J. Chem. Inf. Model.*, 1992, **32**, 443–447.
114. D. Qu, J. Su, M. Muraki and T. Hayakawa, *J. Chem. Inf. Model.*, 1992, **32**, 448–452.
115. D. Vidal, M. Thormann and M. Pons, *J. Chem. Inf. Model.*, 2005, **45**, 386–393.
116. V. M. Tatevskii, *The Classical Theory of the Structure of Molecules and Quantum Mechanics*, Khimiya, M., 1973.
117. V. M. Tatevskii, *The Theory of Physicochemical Properties of Molecules and Substances.*, MSU Publishing House, 1987.
118. V. M. Tatevskii, *Chemical Structure of Hydrocarbons and Regularities in Their Physicochemical Properties.*, MSU Publishing House, 1953.
119. V. M. Tatevskii, V. A. Benderskii and S. S. Yarovoi, *Methods for Calculating Physicochemical Properties of Paraffin Hydrocarbons.*, MSU Publishing House, M., 1960.
120. G. W. Adamson, M. F. Lynch and W. G. Town, *J. Chem. Soc. C*, 1971, 3702–3706.
121. G. W. Adamson, D. R. Lambourne and M. F. Lynch, *J. Chem. Soc., Perkin Trans. 1*, 1972, 2428–2433.
122. N. V. Artemenko, I. I. Baskin, V. A. Palyulin and N. S. Zefirov, *Dokl. Chem.*, 2001, **381**, 317–320.
123. I. I. Baskin, V. A. Palyulin and N. S. Zefirov, Molecular Graphs in Chemistry Studies, Kalinin, 1990.
124. I. I. Baskin, V. A. Palyulin and N. S. Zefirov, 1st All-Union Conference on Theoretical Organic Chemistry, Volgograd, 1991.
125. O. A. Rayevsky, A. M. Sapegin, V. V. Chistiakov, V. P. Solov'ev and N. S. Zefirov, *Koordinatsionnaya Khimiya*, 1990, **16**, 1175–1184.

126. V. V. Poroikov, D. A. Filimonov, Y. V. Borodina, A. A. Lagunin and A. Kos, *J. Chem. Inf. Comput. Sci.*, 2000, **40**, 1349–1355.
127. D. Filimonov, V. Poroikov, Y. Borodina and T. Gloriozova, *J. Chem. Inf. Comput. Sci.*, 1999, **39**, 666–670.
128. L. Xing and R. C. Glen, *J. Chem. Inf. Comput. Sci.*, 2002, **42**, 796–805.
129. A. Bender, H. Y. Mussa, R. C. Glen and S. Reiling, *J. Chem. Inf. Comput. Sci.*, 2004, **44**, 170–178.
130. A. Bender, H. Y. Mussa, R. C. Glen and S. Reiling, *J. Chem. Inf. Comput. Sci.*, 2004, **44**, 1708–1718.
131. R. C. Glen, A. Bender, C. H. Arnby, L. Carlsson, S. Boyer and J. Smith, *IDrugs*, 2006, **9**, 199–204.
132. S. Rodgers, R. C. Glen and A. Bender, *J. Chem. Inf. Model.*, 2006, **46**, 569–576.
133. E. O. Cannon, A. Amini, A. Bender, M. J. E. Sternberg, S. H. Muggleton, R. C. Glen and J. B. O. Mitchell, *Journal of Computer-Aided Molecular Design*, 2007, **21**, 269–280.
134. J.-L. Faulon, D. P. Visco Jr. and R. S. Pophale, *J. Chem. Inf. Comput. Sci.*, 2003, **43**, 707–720.
135. J.-L. Faulon, C. J. Churchwell and D. P. Visco Jr., *J. Chem. Inf. Comput. Sci.*, 2003, **43**, 721–734.
136. C. J. Churchwell, M. D. Rintoul, S. Martin, D. P. Visco Jr., A. Kotu, R. S. Larson, L. O. Sillerud, D. C. Brown and J. L. Faulon, *J. Mol. Graph. Model.*, 2004, **22**, 263–273.
137. W. Bremser, *Analytica Chimica Acta*, 1978, **103**, 355–365.
138. J.-E. Dubois, A. Panaye and R. Attias, *J. Chem. Inf. Comput. Sci.*, 1987, **27**, 74–82.
139. J. E. Dubois, J. P. Doucet, A. Panaye and B. T. Fan, in *Topological Indices and Related Descriptors in QSAR and QSPR*, eds. J. Devillers and A. T. Balaban, Gordon and Breach Sciences Publishers, Amsterdam, 1999, pp. 613–673.
140. Y. Xiao, Y. Qiao, J. Zhang, S. Lin and W. Zhang, *J. Chem. Inf. Comput. Sci.*, 1997, **37**, 701–704.
141. A. Bender, D. W. Young, J. L. Jenkins, M. Serrano, D. Mikhailov, P. A. Clemons and J. W. Davies, *Comb. Chem. High Throughput Screen.*, 2007, **10**, 719–731.
142. M. G. Nidhi, J. W. Davies and J. L. Jenkins, *J. Chem. Inf. Model.*, 2006, **46**, 1124–1133.
143. G. W. Adamson, J. A. Bush, A. H. W. McLure and M. F. Lynch, *J. Chem. Doc.*, 1974, **14**, 44–48.
144. MDL Information Systems, Inc., www.mdli.com.
145. E. K. F. Ahrensin *Chemical Structures*, ed. W. A. Warr, Springer, London, UK, 1988, pp. 97–111.
146. J. W. Raymond and P. Willett, *J. Comput. Aided Mol. Des.*, 2002, **16**, 521–533.
147. A. B. Rozenblit and V. E. Golender, *Logical-Combinatorial Methods in the Development of Drugs*, Zinatne, Riga, 1983.

148. T. R. Hagadone, *J. Chem. Inf. Model.*, 1992, **32**, 515–521.
149. I. L. Ruiz, C. G. Garcia and M. A. Gomez-Nieto, *J. Chem. Inf. Model.*, 2005, **45**, 1178–1194.
150. M. Stahl and H. Mauser, *J. Chem. Inf. Model.*, 2005, **45**, 542–548.
151. P. A. Bacha, H. S. Gruver, B. K. Den Hartog, S. Y. Tamura and R. F. Nutt, *J. Chem. Inf. Model.*, 2002, **42**, 1104–1111.
152. R. P. Sheridan, *J. Chem. Inf. Comput. Sci.*, 2003, **43**, 1037–1050.
153. V. V. Avidon and L. A. Leksina, *Nauchno.-Tekhn. Inf., Ser. 2*, 1974, 22–25.
154. R. E. Carhart, D. H. Smith and R. Venkataraghavan, *J. Chem. Inf. Comput. Sci.*, 1985, **25**, 64–73.
155. D. Horvath, in *Combinatorial Library Design and Evaluation: Principles, Software Tools and Applications*, A. Ghose and V. Viswanadhan eds., Marcel Dekker, New York, 2001, pp. 429–472.
156. D. Horvath and C. Jeandenans, *J. Chem. Inf. Comput. Sci.*, 2003, **43**, 680–690.
157. F. Bonachera, B. Parent, F. Barbosa, N. Froloff and D. Horvath, *J. Chem. Inf. Model.*, 2006, **46**, 2457–2477.
158. A. Schuffenhauer, P. Floersheim, P. Acklin and E. Jacoby, *J. Chem. Inf. Comput. Sci.*, 2003, **43**, 391–405.
159. MOE, Molecular Operating Environment, Chemical Computing Group Inc., Montreal, www.chemcomp.com.
160. L. Franke, E. Byvatov, O. Werz, D. Steinhilber, P. Schneider and G. Schneider, *J. Med. Chem.*, 2005, **48**, 6997–7004.
161. E. Byvatov, B. C. Sasse, H. Stark and G. Schneider, *ChemBioChem.*, 2005, **6**, 997–999.
162. R. Fleischer, P. Frohberg, A. Büge, P. Nuhn and M. Wiese, *Quant. Struct.-Act. Relat.*, 2000, **19**, 162–172.
163. S. Hatrik and P. Zahradnik, *J. Chem. Inf. Comput. Sci.*, 1996, **36**, 992–995.
164. I. I. Baskin, A. O. Ait, N. M. Halberstam, V. A. Palyulin, M. V. Alfimov and N. S. Zefirov, *Dokl. Akad. Nauk.*, 1997, **357**, 57–59.
165. G. W. Bemis and M. A. Murcko, *J. Med. Chem.*, 1996, **39**, 2887–2893.
166. G. W. Bemis and M. A. Murcko, *J. Med. Chem.*, 1999, **42**, 5095–5099.
167. M. I. Skvortsova, K. S. Fedyaev, I. I. Baskin, V. A. Palyulin and N. S. Zefirov, *Dokl. Chem.*, 2002, **382**, 33–36.
168. M. I. Skvortsova, K. S. Fedyaev, V. A. Palyulin and N. S. Zefirov, *Russian Chemical Bulletin*, 2004, **53**, 1587–1595.
169. V. B. Mnukhin, *in Mathematical Analysis and its Applications*, Rostov-na-Donu, 1983, pp. 55–60.
170. E. Estrada, *J. Chem. Inf. Comput. Sci.*, 1996, **36**, 844–849.
171. E. Estrada, *J. Chem. Inf. Comput. Sci.*, 1997, **37**, 320–328.
172. E. Estrada, *J. Chem. Inf. Comput. Sci.*, 1998, **38**, 23–27.
173. E. Estrada, A. Pena and R. Garcia-Domenech, *J. Comput. Aided Mol. Des.*, 1998, **12**, 583–595.
174. E. Estrada and Y. Gutierrez, *Journal of Chromatography A*, 1999, **858**, 187–199.

175. E. Estrada, Y. Gutierrez and H. Gonzalez, *J. Chem. Inf. Comput. Sci.*, 2000, **40**, 1386–1399.
176. E. Estrada and H. Gonzalez, *J. Chem. Inf. Comput. Sci.*, 2003, **43**, 75–84.
177. M. P. Gonzalez, A. M. Helguera and H. G. Diaz, *Polymer*, 2004, **45**, 2073–2079.
178. E. Estrada, E. Molina and I. Perdomo-Lopez, *J. Chem. Inf. Comput. Sci.*, 2001, **41**, 1015–1021.
179. E. Estrada, E. Uriarte, A. Montero, M. Teijeira, L. Santana and E. De Clercq, *J. Med. Chem.*, 2000, **43**, 1975–1985.
180. E. Estrada, S. Vilar, E. Uriarte and Y. Gutierrez, *J. Chem. Inf. Comput. Sci.*, 2002, **42**, 1194–1203.
181. E. Estrada, G. Patlewicz and Y. Gutierrez, *J. Chem. Inf. Comput. Sci.*, 2004, **44**, 688–698.
182. M. P. Gonzalez, H. G. Diaz, R. M. Ruiz, M. A. Cabrera and R. R. de Armas, *J. Chem. Inf. Comput. Sci.*, 2003, **43**, 1192–1199.
183. M. P. Gonzalez and M. D. T. Moldes, *Bull. Math. Biol.*, 2004, **66**, 907–920.
184. M. P. Gonzalez, L. C. Dias, A. M. Helguera, Y. M. Rodriguez, L. G. de Oliveira, L. T. Gomez and H. G. Diaz, *Bioorganic & Medicinal Chemistry*, 2004, **12**, 4467–4475.
185. E. Molina, H. Gonzales Diaz, M. P. Gonzalez, E. Rodriguez and E. Uriarte, *J. Chem. Inf. Comput. Sci.*, 2004, **44**, 515–521.
186. M. P. Gonzalez, H. G. Diaz, M. A. Cabrera and R. M. Ruiz, *Bioorganic & medicinal chemistry*, 2004, **12**, 735–744.
187. A. M. Helguera, M. P. Gonzalez and J. R. Briones, *Polymer*, 2004, **45**, 2045–2050.
188. M. P. Gonzalez, L. C. Dias and A. M. Helguera, *Polymer*, 2004, **45**, 5353–5359.
189. M. P. Gonzalez, M. d. C. T. Moldes, Y. Fall, L. C. Dias and A. M. Helguera, *Polymer*, 2005, **46**, 2783–2790.
190. S. Kramer, L. De Raedt and C. Helma, Seventh ACM SIGKDD international conference on Knowledge discovery and data mining, San Francisco, California, August 26–29, 2001, 2001.
191. L. De Raedt and S. Kramer, The Seventeenth International Joint Conference on Articial Intelligence, 2001.
192. S. Kramer and L. De Raedt, The eighteenth International Conference on Machine Learning, 2001.
193. A. Inokuchi, in *Proceedings of the Fourth IEEE International Conference on Data Mining (ICDM'04)* IEEE Computer Society, 2004, pp. 415–418.
194. X. Yan and J. Han, in *Proceedings of the 2002 IEEE International Conference on Data Mining*, IEEE Computer Society, Washington DC, USA, 2002, pp. 721–724.
195. H. Saigo, T. Kadowaki and K. Tsuda, International Workshop on Mining and Learning with Graphs 2006, 2006.
196. T. Asai, K. Abe, S. Kawasoe, H. Arimura, H. Satamoto and S. Arikawa, in *SIAM SDM'02*, 2002.

197. Y. Chi, R. R. Muntz, S. Nijssen and J. N. Kok, *Fundamenta Informaticae*, 2005, **66**, 161–198.
198. A. Inokuchi, T. Washio and H. Motoda, 4th European Conf. on Principles and Practice of Knowledge Discovery in Databases (PKDD), Lyon, France, September 2000, 2000.
199. M. Kuramochi and G. Karypis, 1st IEEE Conference on Data Mining, 2001.
200. C. Borgelt, T. Meinl and M. Berthold, in *Proceedings of the 1st international Workshop on Open Source Data Mining: Frequent Pattern Mining Implementations* ACM Press, New York, NY, Chicago, Illinois, August 21–21, 2005, 2005, pp. 6–15.
201. M. J. Zaki, in*Proceedings of the eighth ACM SIGKDD international conference on Knowledge discovery and data mining*, ACM Press, Edmonton, Alberta, 2002, pp. 71–80.
202. Y. Chi, Y. Yang and R. R. Muntz, in *The 16th International Conference on Scientific and Statistical Database Management (SSDBM'04), June 2004*, Editon edn., 2004.
203. Y. Chi, Y. Yang, Y. Xia and R. R. Muntz, in *The Eighth Pacific Asia Conference on Knowledge Discovery and Data Mining (PAKDD'04)*, May 2004 Editon, Springer, London, UK, 2004.
204. L. Dehaspe, H. Toivonen and R. D. King, in *4th International Conference on Knowledge Discovery and Data Mining*, R. Agrawal, P. Stolorz and G. Piatetsky-Shapiro eds., AAAI Press, 1998, pp. 30–36.
205. M. Deshpande, M. Kuramochi and G. Karypis, in *Proceedings of the Third IEEE international Conference on Data Mining (November 19–22, 2003). ICDM.*, IEEE Computer Society, Washington, DC, 2003, pp. 35–49.
206. A. Demiriz, K. P. Bennett and J. Shawe-Taylor, *Mach. Learn.*, 2002, **46**, 225–254.
207. D. J. Graham, C. Malarkey and M. V. Schulmerich, *J. Chem. Inf. Comput. Sci.*, 2004, **44**, 1601–1611.
208. J. Batista, J. W. Godden and J. Bajorath, *J. Chem. Inf. Model.*, 2006, **46**, 1937–1944.
209. J. Batista and J. Bajorath, *J. Chem. Inf. Model*, 2007, **47**, 59–68.
210. D. M. Sanderson and C. G. Earnshaw, *Hum. Exp. Toxicol.*, 1991, **10**, 261–273.
211. L. Chen, in *Handbook of Chemoinformatics*, ed. J. Gasteiger, Wiley-VCH, Weinheim, 2003, vol. 1, pp. 348–388.
212. J. Dugundji and I. Ugi, *Topics Curr. Chem.*, 1973, **39**, 19–64.
213. N. S. Zefirov and S. S. Trach, *Chemica Scripta*, 1980, **15**, 4–12.
214. N. S. Zefirov, *Accounts of Chemical Research*, 1987, **20**, 237–243.
215. G. Vladutz, in *Approaches to Chemical Reaction Searching*, ed. P. Willett, Gower, London, 1986, pp. 202–220.
216. S. Fujita, *J. Chem. Inf. Comput. Sci.*, 1986, **26**, 205–212.
217. S. Fujita, *J. Chem. Inf. Comput. Sci.*, 1987, **27**, 120–126.
218. F. Hoonakker, PhD Thesis, ULP, Strasbourg, 2008.

219. Y. Borodina, A. Rudik, D. Filimonov, N. Kharchevnikova, A. Dmitriev, V. Blinova and V. Poroikov, *J. Chem. Inf. Comput. Sci.*, 2004, **44**, 1998–2009.

220. S. Ash, M. A. Cline, R. W. Homer, T. Hurst and G. B. Smith, *J. Chem. Inf. Comput. Sci.*, 1997, **37**, 71–79.

221. D. E. Knuth, *Sorting and searching*, Addison-Wesley, Reading, MA, 1988.

222. V. A. Tarasov, O. N. Mustafaev, S. K. Abilev and V. A. Mel'nik, *Russian Journal of Genetics*, 2005, **41**, 814–821.

223. C. S. Kadyrov, L. A. Tjurina, V. D. Simonov and V. A. Semenov, *Machine Search for Chemicals with Specified Properies*, FAN, Tashkent, 1989.

224. V. J. Gillet, P. Willett and J. Bradshaw, *J. Chem. Inf. Comput. Sci.*, 2003, **43**, 338–345.

225. E. J. Barker, E. J. Gardiner, V. J. Gillet, P. Kitts and J. Morris, *J. Chem. Inf. Comput. Sci.*, 2003, **43**, 346–356.

226. I. I. Baskin, N. I. Zhokhova, V. A. Palyulin, A. A. Ivanova, A. N. Zefirov and N. S. Zefirov, in *Book of Abstracts of the XVI European Symposium on Quantitative Structure-Activity Relationships and Molecular Modelling, 10–17 September 2006, Mediterranean Sea, Italy*, 2006, p. 206.

227. A. A. Ivanona, I. I. Baskin, V. A. Palyulin and N. S. Zefirov, *Dokl. Chem.*, 2007, **413**, 90–94.

228. N. I. Zhokhova, I. I. Baskin, V. A. Palyulin, A. N. Zefirov and N. S. Zefirov, *Dokl. Chem.*, 2007, **417**, 282–284.

229. O. F. Guener, *Pharmacophore Perception, Development, and Use in Drug Design.*, Wiley-VCH Publishers, Weinheim, 2000.

230. T. Langer and R. D. Hoffman, *Pharmacophores and Pharmacophore Searches.*, Wiley-VCH Publishers, Weinheim, 2000.

231. J. Wang, L. Lai and Y. Tang, *J. Chem. Inf. Comput. Sci.*, 1999, **39**, 1173–1189.

232. J. Kazius, R. McGuire and R. Bursi, *J. Med. Chem.*, 2005, **48**, 312–320.

233. A. R. Cunningham, H. S. Rosenkranz, Y. P. Zhang and G. Klopman, *Mutat. Res.*, 1998, **398**, 1–17.

234. G. Klopman and H. S. Rosenkranz, *Mutat. Res.*, 1994, **305**, 33–46.

235. G. Klopman, S. K. Chakravarti, N. Harris, J. Ivanov and R. D. Saiakhov, *SAR QSAR Environ. Res.*, 2003, **14**, 165–180.

236. V. K. Gombar, K. Enslein, J. B. Hart, B. W. Blake and H. H. Borgstedt, *Risk Anal.*, 1991, **11**, 509–517.

237. P. N. Judson, *Pestic. Sci.*, 1992, **36**, 155–160.

238. P. N. Judson, *J. Chem. Inf. Comput. Sci.*, 1994, **34**, 148–153.

239. M. D. Barratt and R. A. Rodford, *Curr. Opin. Chem. Biol.*, 2001, **5**, 383–388.

240. C. A. Lipinski, F. Lombardo, B. W. Dominy and P. J. Feeney, *Adv. Drug Deliv. Rev.*, 2001, **46**, 3–26.

241. T. I. Oprea, *J. Comput. Aided Mol. Des.*, 2000, **14**, 251–264.

242. D. F. Veber, S. R. Johnson, H. Y. Cheng, B. R. Smith, K. W. Ward and K. D. Kopple, *J. Med. Chem.*, 2002, **45**, 2615–2623.

243. M. M. Hann and T. I. Oprea, *Curr. Opin. Chem. Biol.*, 2004, **8**, 255–263.
244. A. A. Petrauskas and E. A. Kolovanov, *Perspectives in Drug Discovery and Design*, 2000, **19**, 99–116.
245. A. J. Leo, *Chem. Rev.*, 1993, **93**, 1281–1306.
246. I. V. Tetko and D. J. Livingstone, in *Comprehensive Medicinal Chemistry II: In silico tools in ADMET*, B. Testa and H. van de Waterbeemd eds., Elsevier, 2006, vol. 5, pp. 649–668.
247. A. M. Johnson and G. M. Maggiora eds., *Concepts and Applications of Molecular Similarity*, John Willey & Sons, New York, 1990.
248. N. Nikolova and J. Jaworska, *QSAR & Combinatorial Science*, 2003, **22**, 1006–1026.
249. H. Kubinyi, *Persp. Drug Discov. Design*, 1998, **9–11**, 225–252.
250. Y. C. Martin, J. L. Kofron and L. M. Traphagen, *J. Med. Chem.*, 2002, **45**, 4350–4358.
251. Daylight Chemical Information Systems Inc., http://www.daylight.com (accessed May 2008).
252. Barnard Chemical Information Ltd., http://www.bci.gb.com/ (accessed May 2008).
253. Tripos Inc., http://www.tripos.com (accessed May 2008).
254. P. Jaccard, *Bull. Soc. Vaud. Sci. Nat.*, 1901, **37**, 241–272.
255. R. Taylor, *J. Chem. Inf. Comput. Sci.*, 1995, **35**, 59–67.
256. J. S. Delaney, *Mol. Divers.*, 1996, **1**, 217–222.
257. G. Schneider, W. Neidhart, T. Giller and G. Schmid, *Angew. Chem. Int. Ed.*, 1999, **38**, 2894–2896.
258. R. D. Hull, S. B. Singh, R. B. Nachbar, R. P. Sheridan, S. K. Kearsley and E. M. Fluder, *J. Med. Chem.*, 2001, **44**, 1177–1184.
259. R. D. Hull, E. M. Fluder, S. B. Singh, R. B. Nachbar, S. K. Kearsley and R. P. Sheridan, *J. Med. Chem.*, 2001, **44**, 1185–1191.
260. P. Willett, J. M. Barnard and G. M. Downs, *J. Chem. Inf. Comput. Sci.*, 1998, **38**, 983–996.
261. J. Hert, P. Willett, D. J. Wilton, P. Acklin, K. Azzaoui, E. Jacoby and A. Schuffenhauer, *J. Chem. Inf. Comput. Sci.*, 2004, **44**, 1177–1185.
262. A. Ormerod, P. Willett and D. Bawden, *Quant. Struct.-Act. Relat.*, 1989, **8**, 115–129.
263. J. W. Godden, J. R. Furr, L. Xue, F. L. Stahura and J. Bajorath, *J. Chem. Inf. Comput. Sci.*, 2004, **44**, 21–29.
264. J. W. Godden, F. L. Stahura and J. Bajorath, *J. Med. Chem.*, 2004, **47**, 5608–5611.
265. J. W. Godden and J. Bajorath, *J. Chem. Inf. Comput. Sci.*, 2001, **41**, 1060–1066.
266. P. C. Jurs, T. R. Stouch, M. Czerwinski and J. N. Narvaez, *J. Chem. Inf. Comput. Sci.*, 1985, **25**, 296–308.
267. R. I. Zalewski and J. Jasiczak, *J. Chem. Inf. Model.*, 1994, **34**, 179–183.
268. H. Sun, *J. Chem. Inf. Comput. Sci.*, 2004, **44**, 1506–1514.
269. J. J. Sutherland, L. A. O'Brien and D. F. Weaver, *J. Chem. Inf. Comput. Sci.*, 2003, **43**, 1906–1915.

270. M. Brinn, P. T. Walsh, M. P. Payne and B. Bott, *SAR and QSAR in Environmental Research*, 1993, **1**, 169–210.
271. C. Helma, T. Cramer, S. Kramer and L. De Raedt, *J. Chem. Inf. Comput. Sci.*, 2004, **44**, 1402–1411.
272. A. Rusinko III, M. W. Farmen, C. G. Lambert, P. L. Brown and S. S. Young, *J. Chem. Inf. Comput. Sci.*, 1999, **39**, 1017–1026.
273. M. Wagener and V. J. Van Geerestein, *J. Chem. Inf. Comput. Sci.*, 2000, **40**, 280–292.
274. R. D. King, A. Srinivasan and L. Dehaspe, *Journal of Computer-Aided Molecular Design*, 2001, **15**, 173–181.
275. A. A. Geronikaki, J. C. Dearden, D. Filimonov, I. Galaeva, T. L. Garibova, T. Gloriozova, V. Krajneva, A. Lagunin, F. Z. Macaev, G. Molodavkin, V. V. Poroikov, S. I. Pogrebnoi, F. Shepeli, T. A. Voronina, M. Tsitlakidou and L. Vlad, *J. Med. Chem.*, 2004, **47**, 2870–2876.
276. D. E. Petelin, D. V. Sukhachev, I. I. Baskin, V. A. Palyulin and N. S. Zefirov, 9th All-Union Conference on Chemical Informatics, 1992.
277. Pharma Algorithms, http://pharma-algorithms.com/qsar_builder.htm (accessed May 2008).
278. H. Martens and T. Naes, *Multivariate Calibration*, Wiley, Chichester, etc, 1989.
279. A. Höskuldsson, *J. Chemometrics*, 1988, **2**, 211–228.
280. L. Xing, R. C. Glen and R. D. Clark, *J. Chem. Inf. Comput. Sci.*, 2003, **43**, 870–879.
281. D. Butina and J. M. R. Gola, *J. Chem. Inf. Comput. Sci.*, 2003, **43**, 837 841.
282. M. Clark, *J. Chem. Inf. Comput. Sci.*, 2005, **45**, 30–38.
283. H. R. Madala and A. G. Ivakhnenko, *Inductive Learning Algorithms for Complex System Modeling*, CRC Press, Boca Raton, Ann Arbor, London, Tokyo, 1994.
284. A. V. Antonov, I. V. Tetko, M. T. Mader, J. Budczies and H. W. Mewes, *Bioinformatics*, 2004, **20**, 644–652.
285. I. V. Tetko, V. P. Solov'ev, A. V. Antonov, X. Yao, J. P. Doucet, B. Fan, F. Hoonakker, D. Fourches, P. Jost, N. Lachiche and A. Varnek, *J. Chem. Inf. Model.*, 2006, **46**, 808–819.
286. D. E. Rumelhart, G. E. Hinton and R. J. Williams, in *Parallel Distributed Processing: Explorations in the Microstructure of Cognition, Volume 1: Foundations.*, D. E. Rumelhart and J. L. McClelland eds., MIT Press, Cambridge, MA, 1986, pp. 318–362.
287. J. Zupan and J. Gasteiger, *Neural Networks in Chemistry*, Wiley-VCH, Weinheim, 1999.
288. D. A. Winkler and F. R. Burden, *Methods Mol. Biol.*, 2002, **201**, 325–367.
289. N. M. Halberstam, I. I. Baskin, A. Palyulin Vladimir and N. S. Zefirof, *Russian Chemical Reviews*, 2003, **72**, 629–649.
290. I. I. Baskin, V. A. Palyulin and N. S. Zefirov, *Doklady Akademii Nauk*, 1993, **332**, 713–716.

291. N. V. Artemenko, V. A. Palyulin and N. S. Zefirov, *Doklady Chemistry (Translation of the chemistry section of Doklady Akademii Nauk)*, 2002, **383**, 114–116.
292. N. I. Zhokhova, I. I. Baskin, V. A. Palyulin, A. N. Zefirov and N. S. Zefirov, *Russian Chemical Bulletin (Translation of Izvestiya Akademii Nauk, Seriya Khimicheskaya)*, 2003, **52**, 1885–1892.
293. N. I. Zhokhova, I. I. Baskin, V. A. Palyulin, A. N. Zefirov and N. S. Zefirov, *Russian Journal of Applied Chemistry (Translation of Zhurnal Prikladnoi Khimii)*, 2003, **76**, 1914–1919.
294. N. I. Zhokhova, I. I. Baskin, V. A. Palyulin, A. N. Zefirov and N. S. Zefirov, *Journal of Structural Chemistry*, 2004, **45**, 626–635.
295. T. M. Martin and D. M. Young, *Chem Res Toxicol*, 2001, **14**, 1378–1385.
296. I. V. Tetko, *J. Chem. Inf. Comput. Sci.*, 2002, **42**, 717–728.
297. A. Varnek, N. Kireeva, I. V. Tetko, I. I. Baskin and V. P. Solov'ev, *J. Chem. Inf. Model.*, 2007, **47**, 1111–1122.
298. E. Hartman, D. Keeler and J. Kawalski, *Neural Computation*, 1990, **2**, 210–215.
299. J. Tetteh, T. Suzuki, E. Metcalfe and S. Howells, *J. Chem. Inf. Comput. Sci.*, 1999, **39**, 491–507.
300. B. Schölkopf and A. J. Smola, *Learning with Kernels: Support Vector Machines, Regularization, Optimization, and Beyond*, MIT Press, Cambridge, MA, London, England, 2002.
301. V. N. Vapnik, *The Nature of Statistical Learning Theory*, Springer, 1995.
302. N. Christianini and J. Shawe-Taylor, *An introduction to Support Vector Machines and Other Kernel-Based Learning Methods*, Cambridge University Press, 2000.
303. R. Herbrich, *Learning Kernel Classifiers: Theory and Algorithms*, MIT Press, 2002.
304. P. Lind and T. Maltseva, *J. Chem. Inf. Comput. Sci.*, 2003, **43**, 1855–1859.
305. A. R. Katritzky, M. Kuanar, S. Slavov, D. A. Dobchev, D. C. Fara, M. Karelson, W. E. Acree Jr., V. P. Solov'ev and A. Varnek, *Bioorg. Med. Chem.*, 2006, **14**, 4888–4917.
306. A. R. Katritzky, D. A. Dobchev, D. C. Fara, E. Hur, K. Tamm, L. Kurunczi, M. Karelson, A. Varnek and V. P. Solov'ev, *J. Med. Chem.*, 2006, **49**, 3305–3314.
307. A. R. Katritzky, M. Kuanar, D. C. Fara, M. Karelson, W. E. Acree Jr, V. P. Solov'ev and A. Varnek, *Bioorg. Med. Chem.*, 2005, **13**, 6450–6463.
308. R. Mannhold, R. F. Rekker, C. Sonntag, A. M. ter Laak, K. Dross and E. E. Polymeropoulos, *J. Pharm. Sci.*, 1995, **84**, 1410–1419.
309. G. G. Nys and R. F. Rekker, *Eur. J. Med. Chem.*, 1973, **8**, 521–535.
310. A. Leo, P. Y. C. Jow, C. Silipo and C. Hansch, *J. Med. Chem.*, 1975, **18**, 865–868.
311. K. V. Balakin, N. P. Savchuk and I. V. Tetko, *Curr. Med. Chem.*, 2006, **13**, 223–241.
312. A. Varnek, N. Kireeva, I. V. Tetko, Baskin II and V. P. Solov'ev, *J. Chem. Inf. Model.*, 2007, **47**, 1111–1122.

313. B. P. Feuston, S. J. Chakravorty, J. F. Conway, J. C. Culberson, J. Forbes, B. Kraker, P. A. Lennon, C. Lindsley, G. B. McGaughey, R. Mosley, R. P. Sheridan, M. Valenciano and S. K. Kearsley, *Curr. Top. Med. Chem.*, 2005, **5**, 773–783.
314. D. V. Green and S. D. Pickett, *Mini Rev. Med. Chem.*, 2004, **4**, 1067–1076.
315. D. V. Green, *Prog. Med. Chem.*, 2003, **41**, 61–97.
316. A. Varnek, D. Fourches, V. P. Solov'ev, V. E. Baulin, A. N. Turanov, V. K. Karandashev, D. Fara and A. R. Katritzky, *J. Chem. Inf. Comput. Sci.*, 2004, **44**, 1365–1382.
317. A. Varnek, D. Fourches, V. Solov'ev, O. Klimchuk, A. Ouadi and I. Billard, *Solvent Extraction and Ion Exchange*, 2007, **25**, 433–462.
318. V. P. Solov'ev and A. Varnek, *J. Chem. Inf. Comput. Sci.*, 2003, **43**, 1703–1719.
319. I. V. Tetko, P. Bruneau, H.-W. Mewes, D. C. Rohrer and G. I. Poda, *Drug Discovery Today*, 2006, **11**, 700–707.
320. I. V. Tetko, V. Y. Tanchuk and A. E. P. Villa, *J. Chem. Inf. Comput. Sci.*, 2001, **41**, 1407–1421.
321. N. M. Halberstam, Ph.D. Thesis, Moscow State University, 2001.
322. A. J. Leo and D. Hoekman, *Persp. Drug Discov. Des.*, 2000, **18**, 19–38.
323. E. V. Konstantinova and V. A. Skorobogatov, *Discrete Mathematics*, 2001, **235**, 365–383.
324. K. M. Honorio, R. C. Garratt and A. D. Andricopulo, *Bioorganic & Medicinal Chemistry Letters*, 2005, **15**, 3119–3125.
325. L. Hodes, *J. Chem. Inf. Comput. Sci.*, 1981, **21**, 132–136.
326. M. Randič, *J. Chem. Inf. Comput. Sci.*, 1992, **32**, 57–69.

CHAPTER 2

Topological Pharmacophores

DRAGOS HORVATH

UMR 7177 CNRS – Laboratoire d'Infochimie, Université Louis Pasteur, 4, rue Blaise Pascal, 67000 Strasbourg, France

2.1 Introduction

Pharmacophores, defined[1] as "the ensemble of steric and electronic features that is necessary to ensure the optimal supramolecular interactions with a specific biological target structure and to trigger (or to block) its biological response" conceptually emerged as an attempt by chemists to empirically rationalize structure–property relationships. Following the understanding of the three-dimensional nature of molecules and of the stereochemical rules determining the preferred conformations, ligand binding to macromolecules was explained by the (oversimplified) key-and-lock paradigm[2] of shape complementarity. The nature of the non-covalent binding forces – electrostatic, hydrogen bonding and dispersive contributions, including solvation/hydrophobic effects[3] – is, however, prohibitively complex. Affinity predictions based on an in-depth study of the physicochemical ligand–target–solvent interactions – flexible docking[4] or free energy perturbation simulations[5] – are typically far too time-consuming to be of large-scale practical use (even though they are based on severe approximations of the physical reality, using empirical force field energy calculations). Instead, the principle of functional group complementarity (cations interact favorably with anions, donors with acceptors and hydrophobes among themselves) was coined as "ligand-site physical chemistry in a nutshell," to become the second pillar of the pharmacophore concept, next to shape complementarity.

In medicinal chemistry, the pharmacophore is often viewed as being complementary to the molecular scaffold, *i.e.*, the molecular topology. Scaffold

Chemoinformatics Approaches to Virtual Screening
Edited by Alexandre Varnek and Alex Tropsha
© Royal Society of Chemistry, 2008
Published by the Royal Society of Chemistry, www.rsc.org

hopping[6] – the quest of bioisosteric, topologically different structures, which nevertheless orient their interacting groups in space in a similar way to the starting compound and therefore display similar interactions with the biological targets – became a central paradigm in drug design. Its importance stems from its ability to open up new synthetic routes once all the analogs around a given scaffold have been explored, to escape the chemical space covered by scaffold-based patent applications or to discover molecules with different pharmacokinetic properties but similar binding affinities with respect to the aimed target. Lead optimization is therefore alternatively oriented along two conceptually orthogonal research directions:[7] the sampling for various scaffolds compatible with a given pharmacophore pattern and the sampling of various pharmacophore patterns that can be supported by a given scaffold.

2.1.1 3D Pharmacophore Models and Descriptors

With the advent of computer-aided drug design,[8] the intuitive pharmacophore concept was rapidly adopted by chemoinformaticians.[9] Modern substructure search tools can be easily adapted to recognize specific functional groups and categorize them into complementary pharmacophore features: hydrogen bond donors (HD) interact with acceptors (HA), cations (PC – positive charges) form salt bridges with anions (NC), while hydrophobes (Hp) are complementary to themselves. Often, aromatics (Ar) are considered as a specific category, complementary to both themselves and to hydrophobes. In the following, N_T represents the number of considered types ($N_T = 6$ according to the enumeration above). Formally, pharmacophore-type information can be represented under the form of a binary pharmacophore flag matrix $F(a,T)$, with $F(a,T) = 1$ if atom a is of type T and $F(a,T) = 0$ otherwise. An alleged binding pharmacophore can then be derived by generating a three-dimensional alignment of actives and picking the space regions in which all of the training set compounds chose to place pharmacophorically equivalent groups. Pharmacophore hypotheses[10,11] are a list of conserved, overlapping features found in the alignment model. Any candidate compound possessing such groups, and able to orient each within the sphere circumscribing the equivalent overlapping training set features, is then assumed to match the "binding pharmacophore" and to be, therefore, active. A large variety of tools gravitating around the same general idea have been developed, testifying to the great popularity of the pharmacophore paradigm. Pharmacophore spheres may be replaced by fuzzy "pharmacophore fields" (ComPharm[12]) to be matched by the aligned candidate molecules, or pharmacophore typing may be skipped altogether, in favor of straightforward monitoring of steric and electrostatic fields (CoMFA[13]). This notwithstanding, the principle is the same: given the common alignment, space zones featuring occupancies/field values that correlate[14–16] with experimental activities over the whole series of training set examples enter the "pharmacophore". At the next stage, test molecules and external candidates in virtual screening need to prove their ability to align to the considered template such as to appropriately occupy (or generate the appropriate field intensities in) these important zones. Claiming that the important zones actually correspond to the space regions in which ligand-site interactions actually happen is only a small conceptual step away. Such a step should, however, be taken with extreme caution,[17] for the observed field/activity correlations are not proof of any causal relationship. Alternatively, binding pharmacophores can be extracted as the "negative image" of a protein binding site,[18] by programs searching for the most appropriate locations of hydrophobic, polar and charged probes within the site, and then combined into one or several pharmacophore hypotheses.

Overlay models are, however, time-consuming and, more important, limited to data sets that share a significant common substructure, in the absence of which no meaningful alignment can be achieved. To circumvent these drawbacks, alignment-invariant pharmacophore fingerprints, representing the pharmacophore pattern of the molecule, have been introduced. The pharmacophore pattern can be defined as the relative spatial arrangement of all the present pharmacophore features – whether involved in actual site-ligand interactions or not. A simple way to characterize the pharmacophore pattern is to generate density distribution histograms of the atom pairs corresponding to each pharmacophore feature combinations,[19,20] with respect to the distance separating them. The pharmacophore pattern may then be characterized by a vector in which every element T_i–$T_j\Delta_k$ stands for the number of pairs of atoms in which the first is of pharmacophore type T_i, the second represents the feature T_j and the distance separating them falls within the binning range Δ_k. If N_B represents the number of considered distance bins, so that binning ranges uniformly span a distance interval between some d_{min} and d_{max}, then $\Delta_k = [d_{min} + (k-1).\varepsilon, d_{min} + k.\varepsilon]$ with $\varepsilon = (d_{max}-d_{min})/N_B$, and the dimension of the fingerprint will equal $N_B N_T (N_T-1)/2$. The space points between which distances are measured and binned must not be actual atoms but rather space zones[21,22] where a given (hydrophobic or polar) probe reaches optimal energy levels: in these cases, the type of the considered probes count as features F. Pharmacophore triplets[23] or quadruplets[24] may be monitored instead of pairs. In binary three-point fingerprints, basis triangles i are fully specified by a list of three pharmacophore types $T_j(i)$ – each type of T_j being associated to a corner $j = 1, 2, 3$ of the triangle – plus a set of three tolerance ranges $[d_{kj}^{min}(i), d_{kj}^{max}(i)]$ specifying constraints for triangle edge lengths. Basis triangles should thus be understood as the meshes of a grid onto which a molecule is being mapped. Considering an atom triplet (a_1, a_2, a_3) in a molecule, this triplet is said to match a basis triangle i if:

1. Each atom a_j is of pharmacophore type $T_j(i)$, e.g., $F[a_j, T_j(i)] > 0$ for each corner j.
2. The calculated – geometric or other – interatomic distances dist(a_j, a_k) each fall within the respective tolerance ranges: $d_{kj}^{min}(i) \leq$ dist(a_j, a_k) $< d_{kj}^{max}(i)$.

If in a molecule M an atom triplet simultaneously fulfilling the above-mentioned conditions can be found, then the fingerprint of M will highlight the bit i corresponding to this basis triangle.

Such fingerprints are static descriptors of the global pharmacophore pattern in the molecule – they describe how the existing representatives of each pharmacophore type are oriented with respect to each other, but make no statement whatsoever about the actual subset of functional groups that actually participate in (or block) ligand binding. Therefore, such fingerprints were mainly – and successfully – used in molecular similarity calculations, within the conceptual framework of the molecular similarity principle,[25,26] which is briefly stated as "similar molecules have similar properties", or, more precisely, "Similar molecules are more likely to share similar properties than any pairs of randomly chosen compounds." Pharmacophore fingerprints are the exponents of a stronger version of the similarity principle: "Molecules with similar overall pharmacophore patterns are likely to share similar reversible non-covalent binding behavior to biological targets."

Beyond similarity-based applications, machine learning techniques[27–29] may pick the specific descriptor elements that appear to correlate with the observed activity trends throughout a training set. Unlike in overlay models, where there is an obvious link between pharmacophore spheres or "fields" in space and their source atoms, the actual pairs (triplets, *etc.*) of atoms in molecules that incarnate the picked descriptor elements must be first established, to gain any potential insights into the binding mechanisms.

2.1.2 Topological Pharmacophores

Pharmacophores are intrinsically three-dimensional – what, then, is "topological pharmacophore" supposed to mean? This chapter highlights the key aspects of this topic along with some published studies. Its goal is to convey a general introduction to the main concepts and issues in 2D pharmacophore modeling, and was not conceived as an exhaustive literature review. This section briefly introduces key topics that are then detailed later on.

2.1.2.1 Topological Pharmacophores from 2D-alignments

Certainly, molecular geometry is an implicit function of connectivity. However, to build pharmacophore hypotheses as outlined in Section 2.1.1, stable conformer(s) of the compounds concerned have first to be explicitly generated. Nevertheless, avoiding the conformational sampling step – time-consuming and notoriously noisy, in particular for flexible molecules, where stochastic procedures return different and incomplete sets of conformers at every run – is worth every effort. Encouragingly, the success of topological indices in QSAR studies hint that the bypassing of the explicit 3D modeling step may prove feasible, and the pharmacophore feature detection is, *per se*, purely based on molecular connectivity considerations. The literature-wide debate[25] on the relative performances of 2D *vs.* 3D descriptors, opposing simplicity and robustness of the former against the higher information content of the latter, is ongoing. Overlay-based 3D pharmacophore models may perform excellently if

the bioactive conformations of the training set actives used for model calibration is known. Otherwise, what geometries should one use? The most stable
(provided the structures are rigid enough to allow a reproducible sampling of
their phase space)? Would some randomly picked conformer within the most
stable ones returned by a stochastic search do? What happens if the geometry
returned for two very close analogues turn out to be radically different?

Although overlapping 2D molecular sketches leads to nothing but unreadable
patches of atom symbols and crossing bonds, making little sense from a graphical point of view, methods aimed at establishing a mapping of the groups of a
molecule onto equivalent groups in another do exist, and may be used to
highlight conserved patterns in actives. Typically, compounds in a training set
are merged to form a hypermolecule, with their chemically and topologically
equivalent atoms fused into unique hypermolecule vertices. Eventually, vertices
specifically populated in actives and inactives can be learned. Subsets that
(almost) exclusively occur among actives are termed "pharmacophores" and are
thought to be responsible for activity, whereas subsets specifically seen to occur
in inactives are termed "antipharmacophores" and are claimed to prevent the
ligand from binding to an active site. These subsets do not have to represent
contiguous graphs – therefore, such methods should be in principle capable of
scaffold hopping. Unfortunately, hypermolecule-based approaches are most
often seen to exploit 3D information as well – there appears to be no explicit
study addressing the advantages and pitfalls of topology-based *vs.* geometry-
based overlay models in QSAR build-up and pharmacophore elucidations.

2.1.2.2   Topological Pharmacophores from 2D
Pharmacophore Fingerprints

The analogy between 3D and topological pharmacophore pattern descriptors[30,31] is quite obvious: it suffices to replace the Euclidean distances in 3D
fingerprints with shortest-path topological interatomic distances. These are
integers and thus simplify the distance-based binning scheme (all pairs at a
given topological distance enter a same bin). 2D pharmacophore fingerprints
can then be used exactly like their 3D counterparts – either for similarity-based
screening or for machine learning of pharmacophore models based on the
selected descriptor elements. If actual interatomic 3D distances correlate well
with their topological separations, then 2D and 3D pharmacophore descriptors
should behave similarly. Certainly, molecules may fold such as to close up
atoms that are separated by many bonds – a situation in which 2D descriptors
are bound to fail. However, in practice the bioactive conformer may be
unknown – therefore, 3D fingerprints often rely on an ensemble of conformers
rather than on a single geometry. It has been shown that single geometry-based
3D descriptors often behave erratically, because geometry build-up programs
may return different structures for topologically similar compounds (and different programs may return different conformers of the same compound) so
that close analogs may be coded by very different fingerprints. Alternatively,

interatomic distance averages over a set of geometries are much stronger correlated with topological distances, if the conformational diversity is sufficient (*i.e.*, the set enumerates both folded and unfolded structures). In this sense, 3D fingerprints are actually not as "3D" as their authors claim.[23,32] Using geometry ensembles for fingerprint generation means diluting the 3D information based on the bioactive structure with topology-related information: "2.5D" fingerprints would be a more fitting designation for these descriptors. 2D pharmacophore fingerprints, therefore, may be strong performers in both similarity calculations and machine learning. Even at lesser performance with respect to their 3D counterparts they might still rank as the method of choice, in view of their much lower computational cost.

After discussing 2D pharmacophore fingerprints and their applications in similarity searches, this chapter focuses on the issue of 2D pharmacophore models obtained by machine learning. In principle, any machine learning method can be used to mine for a correlation between the (binary) presence/ absence or the (continuous) population levels of certain elements of the 2D pharmacophore fingerprint. Since each element stands for a particular pattern (pair, triplet), selected elements can be traced back to the matching atom subsets in each molecule. Therefore, "pharmacophores" and "antipharmacophores" in the sense of Section 1.2.1 may in principle be extracted from pharmacophore fingerprint-based QSAR models. A central question to be addressed here is the quest for the actual meaning of such "pharmacophores" and "antipharmacophores" – is the claim that these stand for actually interacting ligand groups justified? If so, then these patterns must be scaffold-independent, *i.e.*, models must apply to various families of bioisosteric structures and not only to the series of differently substituted variants of a same scaffold. A classical pitfall likely to occur with pharmacophore fingerprints is learning from a homogeneous series of actives, which leads to the selection of fingerprint elements that are specific to the scaffold, so that the alleged pharmacophore model actually behaves like a fragment based approach and turns out to be unable of scaffold hopping.

2.1.2.3 Topological Index-based "Pharmacophores"?

Last but not least, the pharmacophore concept was often (mis)used in the literature in a much larger sense, *i.e.*, as the "recipe" to make an active molecule. In SAR studies, claims such as "the 3-hydroxyphenyl substituent in position 4 of the scaffold is a key element of the binding pharmacophore" or "the pharmacophore requests a substituent with high Balaban[33] index value at position 5 of the scaffold" abound. In this sense, every QSAR model based on classical 2D indices may arguably be regarded as a "topological pharmacophore" model. In the first case, however, the statement lacks generality. In the second, the statement is mechanistically obscure: how does the Balaban index relate to the steric and physicochemical properties and enable the favorable interaction between the group and the site? Replacing a $-CH_3$ by an $-NH_3^+$ group in the substituent has no impact whatsoever on the Balaban index, but is not likely to leave affinity unchanged. The correlation between activity and the topological index is likely to owe its existence to limited diversity in model training or validation sets. The final goal of this chapter is to issue a warning against overhasty claims of "pharmacophore" extractions from topological index-based QSARs. Topological indices may meaningfully correlate with various activities, but the obtained models are either (i) whole molecule property (log *P*, boiling point, *etc.*) predictors or (ii) they specifically apply to a limited structural family. Neither of these cases is compatible with the common definition of pharmacophores. In case (ii), if atom type sensitive topological indices such as E-state keys are used, the model may describe the activity-enabling substitution patterns of a given scaffold in terms of substituent polarity, very much like in a classical pharmacophore approach. However, scaffold-bound approaches obviously lack scaffold-hopping ability, are mechanistically obscure and strongly prone to over-interpretation.

2.2   Topological Pharmacophores from 2D-Aligments

Generating a 2D-aligment, the precursor step of 2D pharmacophore elucidation consists in first establishing, for each atom of the aligned molecule, a list of possible matches[34] (equivalent atoms) in the target compound (subsequently referred as the "template"). Unless the aligned compound is a close analogue of the template, this problem is not trivial, for certain atoms may not have any appropriate equivalents in the partner compound. Putatively matching atom pairs are vertices of similar "chromatism" (nature) located in similar neighborhoods (with similar successive coordination spheres). This information (atom nature, neighborhood) can be synthetically rendered by means of specific atomic Topological Indices (TI). The most likely matched template atoms for a given vertex of the aligned graph can be pinpointed easily as the ones with the closest TI values. Next, a unique equivalence map needs to be established, linking each atom of the aligned molecule to at most one template atom. The various mapping alternatives need to be scored, to enable searching for the optimal one: "bonuses" are considered each time atoms were successfully mapped onto an equivalent of very similar TI value, and "gap penalties" subtracted for unmatched atoms. Unlike in bioinformatics, where specific sequence alignment tools[35] have emerged as industry standards, there is no universally accepted 2D alignment tool for organic compounds. First, there is no consensus on the "coloring" of the molecular graphs (the default being by atom symbols, although a coloring by pharmacophore feature could be meaningful – to our knowledge, such an attempt has not been reported). In addition, the 2D nature of organic ligands complicates the construction of the equivalence map, by contrast to the straightforward 1D sequence matching procedures. 2D alignment-based approaches emerged in the 1980s,[36] introducing the concept of "hypermolecule", a graph obtained by fusion of individual molecular graphs, by matching equivalent atoms as outlined above. Although the alignment criterion is of topological nature (MTD – Minimal Topological Difference, *i.e.*, matching of topologically equivalent vertices), the approach was typically used[37,38] to pilot the overlay of 3D molecular models to build 3D pharmacophores or pseudo-receptor models. Hypermolecule vertices can be classified in "cavity" points (specifically present in active training set compounds, therefore allegedly harboring favorable site-ligand interactions), "site wall" points (present in inactives, therefore allegedly representing space points that should not be occupied by the ligand) and "indifferent" points. A purely topological[39] version of MTD relies on the alignment of each candidate molecule on the hypermolecule, to calculate specific topological indices with respect to the atom subsets that map onto "cavity" and "wall" points, respectively. No 3D aspects are taken into account, and significant QSAR models are obtained within a series of close trimethoprime analogues. The models are, however, far from intuitive and cannot be used to interpret the mechanism of reversible binding to dihydrofolate reductase. In an additional QSAR study concerning carbonic anhydrase inhibitors (2-substituted 1,2,3-thiadiazol-5-sulfonamides), the authors found the topological approach to be competitive with the classical 3D-based MTD model.

More recently, a purely topological alignment-based strategy relying on subtree matching, the MTree approach,[40] has been reported to lead to meaningful topological pharmacophores. First, molecules are reduced to "feature trees" in which every vertex stands for interconnected functional groups and is colored according to the pharmacophore type of the representing group. A new pairwise alignment algorithm (dynamic match search)[41] leads to a consistent topological molecular alignment based on chemically reasonable matching of corresponding functional groups. On the basis of such alignment, a new tree (MTree model) combining the information from several input feature trees can be created. The nodes represent the matches containing the features of the mapped subtrees, while edges are formed by following the topologies of the input feature trees. Each MTree node can be color-coded by the degree of its conservation in active molecules. The authors found that, in a study of α_{1A} antagonists featuring two aromatic moieties connected by a central linker featuring a cationic secondary ammonium group, the most heavily conserved features are, indeed, the cation and the flanking phenyls. This perfectly coincides with a Catalyst-built pharmacophore model – however, the considered actives are all obvious representatives of this $\Phi-NH_2^+-\Phi$, so that the outcome could hardly have been different, irrespective of the 2D or 3D nature of the alignment tool. Virtual screening for angiotensin converting enzyme (ACE) inhibitors, either based on the feature trees of individual well-known inhibitors or on their consensus tree, employed the similarity score between the feature tree of a candidate compound and the reference and returned excellent enrichments and allowed the retrieval of topologically diverse "seeded" hits. In some cases, enrichments better than the ones obtained by Catalyst-based 3D pharmacophore screening were obtained – probably a consequence of either active compounds not having the hypothesis-fitting conformer included in their representative conformer families or inactive compounds fortuitously folded such as to match the hypothesis. Also, note that the Catalyst pharmacophore hypotheses did not incorporate excluded volume information. The MTree approach does not explicitly account for exclusion zones either, but, since it is based on similarity scoring, candidates much larger than the template hypotheses, thus at risk of clashing against the site wall, will be implicitly discarded as dissimilar even if they include the template as a substructure.

MTree models are *per se* not "conceptually equivalent to topological pharmacophores", as the authors claim, but rather conceptually equivalent to the above-mentioned hypermolecules. They represent a fusion of nodes seen in training set compounds – some conserved in many molecules, other occurring less often. To generate a topological pharmacophore out of an MTree model, node weights should be introduced to account for the relative occurrence of each node within actives, by contrast to inactives – as pointed out by the authors in the perspectives section.

2.3 Topological Pharmacophores from Pharmacophore Fingerprints

Recently, a plethora of topological pharmacophore fingerprints, binary presence indicators or fuzzy population level counts of pharmacophore-typed atom pairs[20,25,31,42,43] or triplets[23,30,44] have been developed. Typically, topological pharmacophore build-up consists of several steps: molecule import, standardization (counterion deletion, ambiguous bond order fixing, hydrogen atom adding/deletion, *etc.*) and topological analysis (calculation of the shortest-path topological distance matrix). Eventually, pharmacophore typing of the atoms/functional group is undertaken and the feature pairs or triplets are detected and classified with respect to topological separations and pharmacophore types. Notably, although topological pharmacophore fingerprints are being thought of as conceptually different from fragment counts, the borderline between these two key categories is not as sharp as it seems. In fact, 2D pharmacophore fingerprints can be obtained[45] by means of a SMARTS[46]-driven generic substructure search procedure and the pharmacophore patterns counted by each element of the 2D pharmacophore fingerprint are generic "wildcard-matching" fragments. For example, the expression "Hp" $\sim$ * $\sim$ * $\sim$ * $\sim$ "HA", where "Hp" and "HA" would be generic SMARTS definitions for "hydrophobes" and "acceptors," respectively, matches any hydrophobe-acceptor pair separated by four bonds – a typical CATS[42] term. Technically, the choice of SMARTS-driven pharmacophore detection is quite powerful, for it may implicitly allow for any arbitrary degree of refinement of pharmacophore type definitions (*e.g.*, donors and acceptors categories may be split into finer subtypes) without having to modify the software.

2.3.1 Topological Pharmacophore Pair Fingerprints

One of the most widely used pairwise descriptors, CATS[42] (Chemically Advanced Template Search), represents counts of the 150 different atom pair types defined as the combinations of five considered pharmacophore features (HA, HD, PC, NC and lipophilic = Hp + Ar), times ten monitored shortest path distance values (1–10 separating bonds). The same principle holds for all pairwise pharmacophore fingerprints: differences occur only in terms of the explicitly considered pharmacophore types (a distinction between Hp and Ar may be considered[30,32]), in terms of applied pharmacophore typing rules (in ref. 43, for example, tertiary amino groups are, by contrast to primary or secondary ones, not considered under cationic form) and in terms of monitored distance ranges. Although fuzzy counting of pharmacophore patterns was initially introduced as a means to smooth out noise due to conformational sampling-dependent 3D distances in 3D fingerprints,[25] blurring the borders between the clear-cut distance categories defined in terms of integer topological distances appears to be nevertheless beneficial.[25,30] It may indirectly account for the implicit tolerance of receptors that may typically tolerate an insertion/ deletion of a –CH_2– group in linkers without dramatic shifts of affinity. The use of reduced graphs[47] represents a further step towards fuzziness, as structural details are being merged into generic functional groups. Interestingly, the authors find that in addition to this implicit "fuzzifying" an explicit fuzzy pair counting strategy may, up to a certain point, still enhance retrieval rates in virtual screening.

2.3.2 Topological Pharmacophore Triplets

To generate topological triplets, a basis set of reference pharmacophore triplets is chosen, enumerating all possible combinations of pharmacophore features of the corners, times all the considered integer edge lengths obeying triangle inequalities, within a finite range [E_{min}, E_{max}]. A basic example for such fin- gerprints are Typed Graph Triangles (TGT),[44] a binary fingerprint monitoring the presence/absence of each of the considered pharmacophore triangles. A series of improvements,[48] such as allowing for overlapping pharmacophore types (*i.e.*, allowing functional groups to represent several types – carboxylates counting both as acceptors and anions, for example) and monitoring triplet counts rather than binary presence/absence, led to improved performances in similarity screening. Eventually, the recently developed topological Fuzzy Pharmacophore Triplets 2D-FPT adopted these improvements and introduced two more innovations: fuzzy mapping of atom triplets onto basis triplets and pK_a-dependent pharmacophore typing. As a consequence of the former, the grid of basis triplets may now be spaced, in considering only triangles with edge lengths being multiples of a user-defined E_{step} value, controlling the graininess of 2D-FPT. At $E_{step} = 2$, for example, triplets associated to the elements of the 2D- FPT vectors are exclusively triangles of even edge lengths. Odd edge length triplets in the molecules are, however, not ignored: they will partially map onto several basis triplets of neighboring even edge lengths. In general, each triangle in the molecule will contribute to the population levels of several similar basis triplets, allowing for recognition of analogues with spacer group insertions or deletions. Total population levels of basis triplets form a sparse vector, the 2D-FPT descriptor, with non-zero elements corresponding to the basis triangles that are either present *per se* or are represented by similar triplets in the molecule. The second major improvement in 2D-FPT, pK_a-dependent pharmacophore typing, consists of the enumeration of all the significantly contributing proto-nation states at a given pH, according to predicted[49] pK_a values of ionizable groups. The molecular 2D-FPT is then returned as a population-weighed average of fingerprints of considered protonation states. Rule-based flagging considers the protonation states of functional groups taken out of the molecular context (therefore, due to the mechanical application of the rule "secondary amines are protonated" to an ethylenediamino moiety, R_1-NH-CH_2-CH_2-NH-R_2 will be flagged as doubly protonated R_1-NH_2^+-CH_2-CH_2-NH_2^+-R_2). By con-trast, pK_a-dependent flagging returns an average fingerprint of the two domi-nant species R_1-NH_2^+-CH_2-CH_2-NH-R_2 and R_1-NH-CH_2-CH_2-NH_2^+-R_2, with two key differences over the rule-based. First, the pK_a-sensitive triplet will not display any populated triplets featuring a cation–cation edge, as the two putative cations are never simultaneously present in a same species. It will therefore be quite similar to the one of R_1-NH-CH_2-CH_2-O-R_2 or even R_1-NH-CH_2-CH_2-CH_2-R_2, which is good news for medicinal chemists: a piperazine ring, for example, behaves very much like morpholine or even cyclohexylamine, while rule-based flagging would suggest significant differences due to an addi-tional charge.[i] Furthermore, changes in R_1 and R_2, including substitutions that would not make any difference at all in rule-based pharmacophore typing (*e.g.*, replacement of a –CH_3 group by –Cl, both being hydrophobes), may now significantly affect the obtained fingerprint if they affect the pK_a of the amino groups, which would trigger a change in relative population levels of the two main protonated states and, therefore, alter the participations of the respective fingerprints when calculating the molecular average 2D-FPT.

2.3.3   Similarity Searching with Pharmacophore Fingerprints – Technical Issues

Prior to a brief discussion of specific applications of topological pharmaco-phore fingerprints in similarity-based virtual screening, some general remarks can be made.

2.3.3.1   The Fundamental Flaw of Similarity Searches

Similarity-driven retrieval of (hopefully) active analogs has nothing to do with actual pharmacophore elucidation. In a certain sense, it assumes that the query

---

[i] This is the reason for the somewhat peculiar decision of ChemAxon to consider tertiary amines, a group often seen in drugs, as not protonated in their default pharmacophore flagging scheme: similarity scoring is better off when ignoring charges altogether instead of considering doubly charged piperazines.

structure *per se* is the pharmacophore, as the aim of similarity searching is to find analogues entirely matching the query pharmacophore pattern. For example, if the reference contains a carboxylate that played an essential role during its synthesis, but does not interact at all with the biological target, pharmacophore-based similarity searching will nonetheless insist on retrieving negatively charged species. These may contain a tetrazole ring, an elegant bioisosteric replacement for the useless carboxylate, but miss some unapparent hydrophobe that is actually responsible for most of the binding free energy of the reference. The similarity search results will be disastrous, but neither descriptors nor search metric should be blamed – the problem is having used a "holistic" approach, not making the distinction between the overall pharmacophore pattern and the actual binding pharmacophore. Certainly, in the absence on any additional SAR data, overall similarity is the only rational choice, and its intrinsic risks need to be accepted. If analogues having both a carboxylate and the key hydrophobe would have been present in the database, they would have been ranked at the top of the list. Analogues containing the hydrophobe but not the –COOH may well be returned, but rank lower than the inactive acids missing the hydrophobe (negative charges are much rarer than ubiquitous hydrophobes, so that their insertion/deletion in a structure will typically have much more impact on the fingerprints than the insertion/deletion of one ethyl group among many others).

2.3.3.2 Knowledge-enhanced Similarity Searching: A Workaround

It is possible[50] to perform knowledge-enhanced pharmacophore fingerprint-driven similarity searching, by assigning higher weights to the fingerprint elements that specifically code for the patterns involving the functional groups actually interacting with the site. In the cited publication, these groups were pinpointed by looking at experimental X-ray structures, but in principle they could also be inferred by machine-learning driven pharmacophore elucidation (see below). The weight-biased similarity search still behaves much more like a traditional similarity search (NB this still is a topological pharmacophore based similarity search – the only 3D information being used to determine the weights of fingerprint elements), in the sense that the holistic matching of the entire pharmacophore pattern of the query is being aimed at. However, failure to match the highlighted groups would result in higher penalty scores – in terms of the example mentioned in the previous paragraph, analogues containing the key hydrophobe but not the carboxylate would now be made to rank higher than the inactives with carboxylate and no hydrophobe. Unsurprisingly, the authors achieved significant increases in performance upon the introduction of knowledge-based weighing. In fact, such a hybrid method regroups both advantages of pharmacophore hot spot matching and of similarity searching. Discarding the reference compound moieties that are not directly interacting with the site, *i.e.*, going for "pure" pharmacophore hot-spot matching might not necessarily be a good idea, for it may open the door to many false positives that are too large or structurally too different to actually bind. Keeping an intrinsic holistic similarity component in the score implicitly amounts to checking whether the candidates matching the hot spots are within the applicability domain[51] of the method. Of course, the X-ray structure of the active site with a co-crystallized ligand is mandatory for the approach, so the direct use of docking[52] methods could be preferred, in view of the fact that the similarity search would discard any actives adopting different binding modes, whereas docking has at least a theoretical chance of picking them up. The authors, however, showed that quick similarity-based filtering still retrieves enough chemically diverse candidates, so that it can be used as high-throughput filtering before actual docking of passing candidates.

2.3.3.3 Similarity Metrics Should take Descriptor Nature into Account

Typically, the metrics[53] used for pharmacophore fingerprint-based virtual screening are the same universal distance functions used in conjunction with any arbitrary descriptor space. However, pharmacophore descriptors (or, for that matter, fragment descriptors,[54] to which the present discussion applies as well) are not real numbers on an arbitrary scale, but presence indicators or, even better, counts of actual patterns in molecules. In a descriptor space defined, for example, on hand of topological and 3D whole molecule indices, two molecules m and M with low Balaban index values β are, in fact, as "similar" as two molecules having high Balaban index values. It therefore makes sense to rely on the difference $[\beta(m) - \beta(M)]$ in similarity scoring – to estimate an Euclidean distance, for example. With pharmacophore descriptors π, the situation is different: a pharmacophore pair or triplet that is seen to occur in both m and M $- \pi(m) = \pi(M) > 0$ – is, indeed, an indicator of similarity. However, $\pi(m) = \pi(M) = 0$ is clearly a less stringent indicator of similarity – all that can be said is that m and M are not different in terms of π, but it would be an overstretch to claim that shared absence of something should count as a common point. Furthermore, $\pi(m) \neq \pi(M)$ is synonymous for dissimilarity, but $[\pi(m) = 0, \pi(M) = 1]$ is a different situation than $[\pi(m) = 4, \pi(M) = 3]$ – one pattern more or less with respect to the three already present is not on equal footing with the fundamental transition from absence to presence. Therefore, it makes little sense to base pharmacophore dissimilarity calculations on $\pi(m)$–$\pi(M)$. Euclidean distances[25] were found to be weak performers in Neighborhood Behavior (NB) studies with pharmacophore descriptors. They are prone to size artifacts: a random pair of small molecules is guaranteed to have low Euclidean distances because $\pi(m) = \pi(M) = 0$ for all but very few of the features π in the fingerprint. Complex molecules have, in contrast, many populated features. Therefore, in a pair of similar complex molecules most of these will be present in both, but those few that happen not to be shared may yet be more numerous than the sum of features (overlapping or not) found in a pair of small compounds. In consequence, a similar pair of complex compounds may score higher Euclidean distances than a pair of completely unrelated small molecules. Correlation coefficient-based scores seem to perform much better as far as pair descriptors are concerned – most of the studies that will be mentioned below bear witness to this. Triplets, however, may form a much too large and sparse fingerprint to be properly handled by classical metrics: their NB can be significantly enhanced upon introduction of tailor-made metrics[30] that acknowledge the different impact of shared absence *vs.* shared presence of triangles. This topic was never directly addressed in the context of pairwise 2D pharmacophore fingerprints, or fragment descriptors. Recently, pharmacophore fingerprints have been increasingly used as more or less sophisticated kernels in Support Vector Machine[55,56] learning. Pharmacophore kernels were introduced as classical kernels based on standard similarity scoring formulas, to be used in connection with pharmacophore fingerprints. However, kernels specifically tuned to account for the peculiarities of pharmacophore pattern information are yet to be developed. Self-organizing maps are another interesting alternative[57,58] to classical similarity searching with 2D pharmacophore fingerprints.

2.3.3.4   What can be Really Learned from Virtual Screening Simulations of Seeded Databases?

All articles introducing novel topological pharmacophore fingerprints validate their methodologies by typical virtual screening studies such as similarity-driven retrieval of hidden actives in seeded databases. It is, however, for various reasons, difficult to extract any strong general statements concerning an absolute utility scale of topological pharmacophore *vs.* fragment-based fingerprints. A key problem therein is the nature of the involved compound sets: used references (query compounds, in virtual screening terminology), hidden analogues (sometimes referred as "the needles") and the decoy compounds ("the haystack"). If the queries and the "needles" belong to a same series of analogues, whereas the "haystack" is a large diverse more or less drug-like database, virtual screening will be successful with both pharmacophore and fragment-based fingerprints. Take a series of 200 analogues around a common scaffold, half of which are actives, and a haystack of 10000 random compounds. Depending on the part of the molecule represented by the scaffold (*i.e.*, the fraction of functional groups tied to the scaffold, and therefore conserved throughout the series, by contrast to the varying "ornaments" seen in specific compounds only), all these molecules will display some potentially significant degree of overlap of both their fragment thesaurus and their pharmacophore pattern. If the substituents tend to be relatively small with respect to, say, a common benzodiazepine ring system, there may be an upper bound concerning the fragment-based dissimilarity scores within the series (no matter what small "ornaments" are entered, the scaffold is so large that it always contributes say 80% of fragments, and those are shared throughout the series). However, there is no upper bound in terms of the pharmacophore-based dissimilarities within the set: pharmacophore features carried by the scaffold are here to stay, but small "ornaments" may significantly alter the pharmacophore landscape. Insertion of ammonium groups and/or carboxylates would not dramatically disturb the fragmentation scheme, but highlight many of the previously absent cation and anion-involving feature combinations. Pharmacophore fingerprints are combinatorial, *i.e.*, adding a novel feature to a context already including *t* other types will highlight *t* novel pairs (grafting a carboxylate onto a hydrophobic scaffold will only highlight Hp-NC pairs, whereas many more – HA-NC, HD-NC, PC-NC – will spring into life if the same –COOH is grafted on a highly functionalized compound containing acceptors, donors and cations). Supposing that the "haystack" does not contain any representatives of the considered scaffold, the 200 actives and inactives of the seeding set will be ranked at the top of the list by any fragment-based screening. This means that the selected top 200 would include 100 actives, whereas in the unsorted haystack 10200 compounds would have to be tested before retrieving the same 100 actives. This amounts to a huge enrichment score of $(100/200) \times (100/10200)^{-1} = 51$, which is, however, not attributable to any "predictive power" of the approach (a simple substructure query of the nude scaffold would have lead to exactly the same results – therefore, beware of taking high enrichment factors for intrinsic proofs of the predictive power). With pharmacophore fingerprints, results may be less straightforward if the haystack contains original scaffolds that are nevertheless compatible with the query pharmacophore. The caveat here is that these unexpected "hits" are usually of unknown activities and are by default considered as inactives, thus unfavorably biasing the enrichment statistics. Scaffold hopping, which is a landmark of pharmacophore fingerprints, may actually be penalizing in carelessly designed study cases. Furthermore, another imponderable factor of the virtual screening success is the ability of the method to discriminate between actives and inactives from the set. It should be kept in mind that such sets were developed under real-life constraints in the drug discovery laboratory, not aimed at providing the ideal set for virtual screening validation studies. If, for example, medicinal chemists have picked an activity-neutral position to insert various polar groups to modulate the pharmacokinetic properties, the set will contain many actives with quite different overall pharmacophore patterns. These will be ranked rather towards the bottom of the similarity-sorted list. For all these reasons, it is not astonishing that no absolute goodness score can be defined in Neighborhood Behavior studies. Virtual screening of "seeded haystacks" should be interpreted with extreme precaution, and the expectation that in benchmarking the un-avoidable data set artifacts will similarly skew the results of all the methods – so that relative comparisons may yet hold – is not automatically granted.

2.3.4 Similarity Searching with Pharmacophore Fingerprints – Some Examples

Going beyond typical seeded haystack analyses, neighborhood behavior studies conducted with respect to the set of drugs on the market,[59] and based on an entire activity panel rather than on a single property,[30,60] form a much more solid basis for benchmarking, but are still not fully exempt of artifacts – note, for example, the dominance of G-Protein Coupled Receptor ligands and the associated dominance of the aromatic–cation pharmacophore pattern among the nowadays marketed drugs. In this context,[25] pure topological fuzzy pairwise pharmacophore descriptors performed only slightly worse than their 3D counterparts (all things being equal, except for the average 3D distances from multiconformational models being replaced by topological distances). This is consistent with the expectation to see properly (fuzzily) binned 3D distances to win over topological information. Practically, however, the achieved improvement in NB may not compensate for the additional effort of 3D buildup and conformational sampling. Both 3D and 2D fuzzy pharmacophore pairs dramatically outperform 3D bitwise 3-point descriptors, which are sensitive to geometry artifacts to the point of failing to recognize obviously similar pairs of compounds because of 3D build-up differences.

Alternatively, topological pharmacophores were thought to be potentially helpful in classification studies, aimed at discriminating between some more or less well-defined compound category and the rest of the world (drugs *vs.* non-drugs,[61] natural *vs.* synthetic compounds[62] or G-Protein Coupled Receptor GPCR binders *vs.* non-binders[58]). Such studies unavoidably lead to fuzzy discussions and conclusions – the more nebulous the employed classification criterion, the less meaningful the study. While the difference between synthetic and natural products is mainly a question of complexity (but also, to some extent, a matter of latest "fashionable" organic synthesis routes), the discrimination between GPCR binders and non-binders is expected to rely on some objective structural patterns. The cited[58] study found a slight advantage of the in-house dictionary-based fingerprint Acte-lionFP over topological two-point pharmacophore fingerprints, but sheds little light on the underlying reasons. Nothing is told about the actual composition of the GPCR binder sets: what percentage of bioactive amines did it contain? A significant excess thereof can be safely assumed, as these dominate the pool of currently marketed drugs, while inhibitors of peptide-binding GPCRs are sparse. Are we then simply witnessing a correct recognition of the bioactive amine pattern, and are the more "exotic" GPCR ligands the misclassified ones? If so, it is not astonishing to see the dictionary-based approach scoring better than the topological pharmacophore: the dictionary terms were well chosen – but how? The pharmacophore-based method may be confused by the presence of the typical aromatic–cation pair of bioactive amines in decoy structures (which were labeled inactive because they were not designed as GPCR ligands – but not strictly disproved not to have some affinity for such targets). Certainly, the practical importance of the work should not be minimized – the main message is that supplier libraries can be analyzed in terms of their propensities to contain GPCR hits. It should, however, not be regarded as a benchmarking study and its conclusions should not be presented out of context, as general statements about relative descriptor quality.

More general studies might require access to large big-pharma corporate databases (which are still far from ideal environments, given their fair share of historical bias). The truth is that incontestably unbiased benchmarking studies would need to involve a significant fraction of the "universe" of drug-like molecules, and it is impossible to set up any training set that represents a significant fraction out of allegedly 10^{56} compounds. Without denying the importance of benchmarking, all sources of bias notwithstanding, the chemical meaningfulness of fingerprints may sometimes intuitively be evidenced from specific observations, rather than overall performance scores. For example,[30] specific pairs of strikingly similar compound pairs with nevertheless diverging biological activities were proven not to be as similar as typical rule-based pharmacophore descriptors might suggest (or, for that matter, not as similar as medicinal chemists may guess, at a first glance). These "activity gaps" found in many classical SAR landscapes no longer show up as discontinuities in the 2D-FPT mapping of activity space, due to the explicit accounting for pK_a shifts in pharmacophore typing (Section 2.3.2).

Eventually, the ultimate judge of molecular descriptors is not being the best in benchmarking but having practically contributed to the discovery of new compounds. Logically, 2D pharmacophore fingerprints were typically employed as quick and effective means to perform scaffold hopping[32,63,64] and/or bio-isosteric replacements[65] in virtual screening or *de novo* drug design.[66] A typical similarity screening example[42] based on CATS is depicted in Figure 2.1. The two isofunctional compounds (T-channel blockers) both feature the same global pharmacophore pattern: two aromatic systems (one including some polar groups) separated by a flexible aminodialkyl linker. Although there is significant variance within the aromatic terminal groups, the overall pharmacophore pattern – and the activity – was conserved. Notably, however, chemotypes with flexible linkers spanning two rigid ends are most likely to perform well in topological pharmacophore-based similarity screening. These fingerprints would also return flexible linker analogs by "opening" rigid cyclic systems in

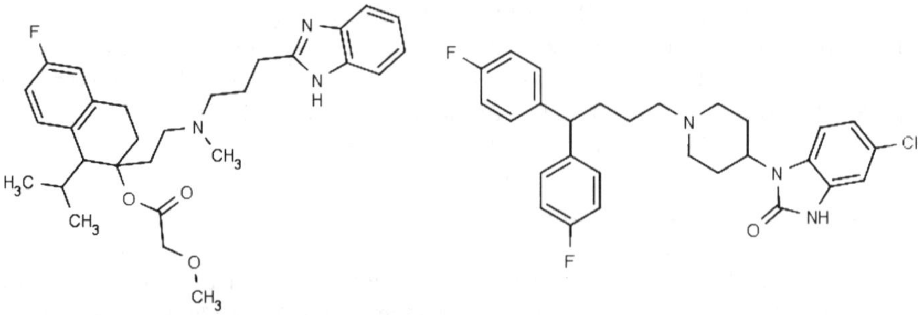

Figure 2.1   Typical example of scaffold hopping, obtained by similarity screening with CATS descriptors. Starting from the left-hand reference T-channel blocker mibefradil, the right-hand compound clopimozid, a submicromolar T-channel blocker, is found among the 12 top ranking analogs.

active references – and probably lose the scaffold-hopping "bet", for the initial rigidity is likely to be a paramount contributor to activity. Reversely, they may come up with rigid analogs of the flexible query compounds – with potential benefits, but significant risks: if the substituents in the rigid analogs are oriented compatibly with the active site, activity will dramatically increase. Otherwise, analogs will be plainly inactive. Starting from a flexible central linker and sticking to that pattern (introducing modifications of the extremes) is certainly the failsafest scenario in "scaffold hopping". 2D pharmacophore pattern similarity of topologically different structures is necessary, but not sufficient, to ensure activity similarity. Therefore, "scaffold hopping" is always a risky undertaking – however, topological pharmacophore similarity is nevertheless the most rational manner to address the problem. Perhaps (that is certainly, except that negative results, not being published, are not at hand to substantiate this claim) most of the scaffold hopping attempts using 2D pharmacophore fingerprint similarity are bound to fail due to wrong stereochemistry, despite conserved pharmacophore pattern. Considering 3D models of rigid scaffolds is very likely to improve the success rate of virtual screening (if the 3D modeling is performed properly – the likelihood of which is inversely proportional to the sizes of the compound libraries to be screened). Conversely, the random alternative of blindly choosing scaffolds with neither correct stereochemistry nor conserved pharmacophore pattern is certainly bound to fail. Therefore, using 2D pharmacophore fingerprints for scaffold hopping certainly represents an advantageous performance/cost tradeoff – one excellent example being the retrieval of completely original 5-lipoxigenase inhibitors.[67]

Interestingly, but not surprisingly, the Euclidean metric is found to be the top performer in the quest for bioisosteric replacements[65] based on a fuzzy binary 2D pharmacophore fingerprint. In this specific context, focused on the description and comparing of pharmacophore patterns within functional groups rather than in whole molecules (a specific "attachment point" dummy feature being considered to situate the free valences of the fragments in the context of their pharmacophore features), size artifacts as mentioned in Section 2.3.3 will not occur. Bioisosteric substitutions suggested by this approach were shown to make chemical sense (*e.g.*, aminoisoquinolines as a replacement for benzamidines in thrombin inhibitors).

2.3.5 Machine-learning of Topological Pharmacophores from Fingerprints

Machine learning can be used to select or weigh specific pharmacophore elements from a fingerprint, to improve the predictive power of the model, in letting it focus on the actually important patterns in the ligands and ignore unbound ligand moieties pending out into solvent (recall discussion in Section 2.3.3).

An intermediate between similarity searching and descriptor-selection based QSAR, self-organizing maps[68] (SOM), will be mentioned first. SOMs try to classify a population of individuals (each described by a fingerprint) into a fixed number of final categories, by assigning to each such category (or "neuron") a characteristic fingerprint. Individuals will be assigned to the category that is most similar in terms of its characteristic fingerprint. Since these characteristic fingerprints evolve during the training step, their final states may be thought as "consensus fingerprints" of the individuals in each category. As training is completely unbiased with respect to the properties of the objects to be categorized, the utility of a SOM only becomes apparent after it has been built: if objects sharing a certain property are assigned to a same neuron (more often than randomly expectable) then it may be assumed that any external objects assigned to that neuron are likely to share that property as well. To resume, a topological pharmacophore fingerprint-driven SOM performs a similarity-based screening operation, except that the known actives are not used straight ahead as queries but are first classified, together with the inactives, into pharmacophore pattern categories. The resulting categories are assessed with respect to enhanced likelihood of harboring actives. For those passing this test, the fingerprints – consensus fingerprints of the predominantly active members assigned to that neuron, in which activity-favorable features are becoming enforced and unimportant features down-weighted – will practically serve as queries to retrieve candidate compounds. In principle, an investigation on what specific features are being enforced and what others were downscaled within active neurons may indirectly highlight activity-specific features in molecules. Although such analysis was (unfortunately) not undertaken by the authors, a CATS-driven SOM was successfully employed to discover purinergic antagonists.[57]

There are relatively few QSAR build-up attempts with 2D pharmacophore fingerprints in the literature. Some of those focus on the question whether 2D pharmacophore fingerprints can yield statistically valid models (*i.e.*, whether they encode the proper chemical information required to explain reversible binding), and find that this is actually the case.[15,45,69] However, analysis of the pharmacophore pairs or triplets selected in regressions (or entering the model with high weighing factors, in PLS) may, at least in principle, shed some light on the actual ligand functional groups involved in direct interactions with the site. Indeed, it was shown that the (binary topological) pharmacophore triplets appearing to play an important role in recognition of Cox-2[15] and respectively thrombin[69] inhibitors closely match functional groups seen to effectively participate in binding. Unlike in linear regression-based QSAR, where the coefficients fitted for the entering descriptors are a directly accessible measure of their relative importance, SVM models[15] are "black boxes" that do not allow such a straightforward analysis. This notwithstanding, the importance of each triplet in a molecule was calculated by comparing the predicted activity in the real compound, with the triplet "on" to the prediction that would have be obtained if the triplet were set to "off". Then, individual atom weights are determined as the average of the above-determined importance of all the triplets containing the atom. The procedure raises a series of problems, since bluntly turning a triplet "off" may not necessarily make physical sense: to assess the role of an atom on the activity, one should perhaps simultaneously switch off all the triplets involving that atom – in a nonlinear SVM model, this is not the same as switching the triplets off one at a time. Furthermore, fingerprint degeneracy (which "haunts" regression-based pharmacophore fingerprints models as well, see below) may complicate a lot of the interpretation of the models. In this binary context, the issue is particularly sensitive: if a pharmacophore triplet is embodied by several distinct atom triplets in a molecule, what exactly is switching off supposed to mean? Despite these numerous issues related to SVM interpretability, the fact is that apparently correct site-ligand anchoring points could be successfully highlighted. As the authors point out, the SVM-derived information is far from accounting for all the subtleties of Cox-2 binding (in fact, the extraction of a Cox-2 pharmacophore is a relatively easy challenge, since Celecoxib analogues are tightly bound and virtually completely buried into the active site – so it is difficult to find a moiety that actually does *not* interact[17]). Also, some bias induced by the peculiar make-up of the training set can be seen to affect the highlighted pharmacophore. Furthermore, the authors have also successfully employed the same methodology for thrombin inhibitor pharmacophore extraction, which is a more difficult problem than Cox-2.

In a recent[69] benchmarking exercise involving 13 typical QSAR training sets from the literature, fuzzy topological pharmacophore triplets (2D-FPT) were shown to fare extremely well, outperforming not only 2D and 3D-index-based models, but also the elaborate, overlay-based CoMFA approaches. The biological property less well handled by pharmacophore triplet models is, unsurprisingly, the heme alkylating activity of artemisinin analogues, the only studied property not reflecting a reversible non-covalent target inhibition process. Topological pharmacophore triplets are thus information-rich and relevant descriptors of site-ligand recognition processes. The study of optimal 2D-FPT fuzziness highlighted the problem of 2D-FPT degeneracy, which may be of serious concern in descriptor selection-based QSARs (much more so than in similarity scoring), although pharmacophore triplets suffer much less from this problem than pair-wise descriptors.

Nevertheless, the topological pharmacophores defined by triplets entering 2D-FPT models are not necessarily representatives of ligand-site anchoring points. This work highlighted the very limited scope of typical QSAR training and validation sets, showing many situations where the successful QSAR fitting and validation relied on family-specific idiosyncrasies. A clear symptom of training set limitations consists in set-specific artifacts gaining the upper hand over pK_a-related effects: the best performing pharmacophore flagging scheme was often the one best exploiting some set-specific coincidence, not necessarily the one based on physicochemically sound ionization states. Furthermore, the work depicted antipharmacophores (selections of unwanted features) to be one more warning signal of poor training set diversity. Thrombin models with excellent validation performance (against an external set, issued, however, from the same structural family) turned out to be pIC_{50} (bigger values mean stronger affinity) predictor equations with very high intercept values and penalties for the presence of certain triplets specifically encountered in inactives. This makes sense from a machine learning point of view: presented with a homogeneous series of inhibitors based on a common scaffold, the system may assume inactivity "by default" and pick features that are positively correlated. Equivalently, it may as well assume all molecules to be active (all the molecules the learning agent knows about contain the activity-relevant scaffold!), and pick features likely to specifically decrease activity scores for those that are not. The latter approach, however, produces meaningless models from a chemical point of view: they return the plain intercept value (of up to 9 pIC_{50} units) for the alleged nanomolar thrombin inhibitor H_2O, which obviously contains none of the antipharmacophore triplets contributing activity decrements. In principle, the mere presence of a pharmacophore element or fragment somewhere in the molecule cannot be held responsible for a drop in activity, unless it is intimately interwoven with the binding moieties. Alternatively, a fingerprint descriptor merely states whether a given pair or triplet is populated, not where it lies.[ii] It is in principle possible to design an active compound core to which the alleged antipharmacophore is connected through a spacer, harmlessly pending in the solvent (unless it includes charged groups, while the target generates a long-range electrostatic potential blocking its access to the site neighborhood). Therefore, equations based on pharmacophore (or, for the matter, fragment) pattern counts should avoid including negative contributions. This can be achieved by adding chemically diverse alleged inactives to the training set, which also have the merit of forcing the system to learn structural elements of the conserved scaffold (otherwise, essential scaffold-typical features, present in all training molecules, actives and inactives alike, might finish up as a fully useless constant fingerprint column in the activity-descriptor table). Under these circumstances, topological pharmacophores extracted from the models could be successfully matched with reported thrombin site–ligand interaction hot spots. All in all, the question of whether selected triplets match actual binding pharmacophores turned out to be, primarily, a matter of training set diversity. 2D-FPT may lead to valuable QSAR models, provided the training set diversity is sufficient to force the learning of key features, not of secondary pharmacophore signatures that serendipitously reflect subsets locally enriched in actives. If this is the case, the applicability range of such models may extend over several chemotypes – and may even go beyond expectations if the targeted active site offers alternative models to accommodate a topological triplet.

2.4   Topological Index-based "Pharmacophores"?

Arguments in favor of QSAR studies systematically included the claim that the models may help in elucidating ligand binding mechanisms. Unfortunately, recent results reveal that this is unlikely to be the case, unless draconian standards in terms of training set diversity and size are being fulfilled. Even in the presence of enlarged data sets, and benefiting from:

- inclusion of a vast majority of published actives and thousands of experimentally certified inactives;

[ii] That information may be implicitly coded by other pairs or triplets that involve the end points of the monitored element, but is difficult to extract and analyze in fingerprint-based QSARs.

- a reputedly easy-to-model target (Cox-2);
- full-blown three-dimensional pharmacophore descriptors and model building tools that lead to excellent statistical validation criteria (marginally better than training values) and an outstanding ratio of 6 explaining variables for two thousand explained data points.

the extracted pharmacophore models made sense only for Cox-2 (-coxib-like) inhibitors, but did not fit at all the Cox-1/Cox-2 unspecific inhibitors (despite the fact that their activity values were properly predicted).[17] Therefore, claims that QSAR-derived conclusions are mechanistically relevant should be endorsed with extreme caution. With this in mind, a skeptical stance is strongly advised as to whether relevant pharmacophores can be obtained from QSAR modeling with standard topological descriptors (TIs), bypassing the pharmacophore fingerprint-specific graph coloring step by pharmacophore features. We failed to find any convincing example thereof in the literature, but retrieved many counter-examples of heavily overinterpreted models instead. For example, a QSAR study[70] of 24 benzodiazepine derivatives with anti-Alzheimer activity allegedly solved the pharmacophore responsible for this activity – thought to consist of the two phenyl rings A and B in Figure 2.2.

This conclusion has been derived on the basis of QSAR models using electrotopological state atom indices (ETSA) for each of the numbered atoms of the common scaffold of the series. Atoms from the substituents are being implicitly accounted for, through the perturbing influences they exercise on the ETSA values of the numbered atoms. There are several obvious reasons why

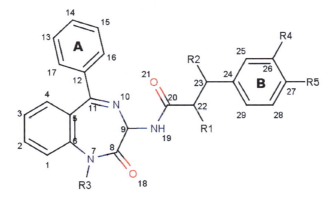

Figure 2.2 Common scaffold, with standardized numbering of atoms for which electrotopological state atom indices (ETSA), sensitive to the perturbations induced by the substituents R1–R5, were calculated and used in QSAR modeling. The study overhastily concluded that phenyl rings A and B must be part of the pharmacophore, explaining anti-Alzheimer activity, because the QSAR study picked the average ETSA values of the atoms within the rings as key descriptors.

the pharmacophore elucidation claim can, to put it mildly, be suspected to be an overstatement:

- The authors have found alternative equations of comparable quality, involving either ring A or ring B, but never both rings at the same time. If both rings were mechanistically involved they should both enter the model.
- The authors basically claim that the substituents involved modulate the π electron densities of the considered phenyl rings, which in response form more or less strong stacking interactions with some aromatic protein residues. That might sound plausible as far as ring B is concerned, but no inductive effect of reasonable strength should be felt by ring A, from any of the considered substituents. Of course, ETSA calculations implicitly capture some very weak "inductive effects," propagated as inversely proportional to the square of the topological distances. QSAR model build-up is, *per se*, not at all concerned with the strength of the perturbations – it is enough if ETSA scores show some non-zero variance, for this can be arbitrarily amplified by multiplication with a large fitted coefficient. The biological receptor, however, is not working as an amplifier of minute electron density fluctuations, to achieve expected modulations in activity. Note that, on an absolute scale, the coefficients fitted for ring A contributions are one order of magnitude greater than the one associated to contributions from B. It may be that the pocket binding A has a preference for electron-enriched rings, while the B-binding one prefers electron-depleted phenyls (hence explaining the different signs of the respective coefficients), but it is difficult to believe that their relative sensitivity to ring electron density fluctuations may differ by one order of magnitude.
- Note that any numbered atom within the scaffold contributes a constant ETSA increment with respect to all the other scaffold atoms, throughout the series. As a consequence, their direct involvement in binding cannot be addressed by the method. The importance of rings A and B has not been deduced from first-order evidence – there is no example showing what happens if either of them is deleted or replaced. They were singled out because they reflect the perturbations seen elsewhere in the structure in a way that happens to correlate with activity. The numbered reference nodes used to locally probe ETSA perturbations feel differently weighted contributions from the varying substituents R1–R5, because of differing sets of separating topological distances. With such a small data set of only 24 members, at least some of the scaffold atoms appear to be "properly" placed, in the sense that the peculiar interplay of substituent group electronegativities and separating topological distances lead to final ETSA values that match activities. Data scrambling tests should have been performed to discard the hypothesis of such chance correlations.
- Conversely, suppose that the R groups themselves interact with the site. Unfortunately, the model could not monitor this – only indirectly (if at all) by means of the perturbations the R groups cause at the level of the reference atoms. In other words, the model is "prejudiced" in the sense that

it will compulsory design the "culprits" for the incriminated activity among the numbered reference vertices – even if the actual responsibles are elsewhere in the molecule.

2.5 Conclusions

Binding pharmacophores are three-dimensional models of ligand–site key interactions, and represent an arguably simplistic, but intuitive and effective, way to rationalize the search for novel active organic compounds in drug design. Since, however, modeling of ligand geometries is, in many respects, a bottleneck in modern day computer-aided drug design (computational effort, accuracy, uncertainty regarding the bioactive conformer, reproducibility of geometries obtained by stochastic methods, *etc.*) various chemoinformatics tools and applications emerged, approaching the pharmacophore problem from a purely topological perspective. These are the main topic of this chapter.

2.5.1 How Important is 3D Modeling for Pharmacophore Characterization?

In many respects, these methods rely on the central working hypothesis that actual geometrical distances used by their 3D counterparts can be replaced by the interatomic topological distances (*i.e.*, the number of interposed bond on the shortest path connecting two atoms). In many situations, this working hypothesis is remarkably fruitful and the performance of topological pharmacophore fingerprints in various benchmarking studies was shown to be comparable or sometimes superior to the that of their 3D counterparts. It is, however, not easy to come up with some general statements about how much information is lost when choosing a topological approach over its three-dimensional counterpart. There no questioning whether a loss of information does occur – if the 2D method performs better then it may only be because of specific shortcomings of the competing 3D approach. To our knowledge there is only one report[25] of a direct comparison of 3D multiconformational models (with their implicit drawbacks) against topological ones, all other things being strictly equal in the fingerprint setup. Most of the literature is concerned with comparisons of 2D approaches with fully unrelated 3D methods (docking, Catalyst hypotheses, *etc.*), so there can be no interpolation of the exact impact imputable to approximating geometry by topology. Furthermore, the multiple and unavoidable sources of bias in the benchmarking data sets seriously limit the generality of the conclusions that can be drawn from such comparative tests. However, topological pharmacophore-based approaches are likely to be state-of-the-art performers in drug design, unless applied to problems where stereochemistry is known to be a primordial issue (rigid ring systems with strictly oriented terminal groups, chiral centers, *etc.*).

2.5.2   2D Pharmacophore Fingerprints are Mainstream Chemoinformatics Tools, whereas 2D Pharmacophore Elucidation has been Rarely Attempted

Following the classical distinction between overlay-based and overlay-independent (fingerprint-based) 3D modeling techniques, this chapter has followed the same classification to monitor the typical topological pharmacophore-derived techniques. It has been shown that purely topological considerations (atom-to-atom mapping techniques) can be successfully applied to pilot molecular superimposition processes, but actual 3D alignment remains by far the one preferred by medicinal chemists, over the computationally feasible but much less readable 2D alignment of putative ligands. 2D alignments of pharmacophore feature trees were shown to yield a chemically reasonable estimation of molecular similarity, but were not (yet) used for pharmacophore elucidation.

Clearly, overlay-independent approaches are the dominant trend in topological pharmacophore modeling, with similarity search-based applications largely outnumbering machine learning-based modeling studies. Of these latter, very few address the issue of the relevance of the highlighted key features (topological pharmacophore) compared to known 3D binding modes. While 2D pharmacophore fingerprints are now main-stream technology, attempts to highlight binding pharmacophores on a purely topological basis are rare. Although the pertinent papers were largely successful at pinpointing mechanistically relevant groups in inhibitors of known binding modes, and clearly showed that training set richness and diversity may be much more important for successful elucidation than the actual use of 3D geometries instead of topological distances, binding pharmacophore extraction is still perceived as *the* typical 3D chemoinformatics application.

Various implementations of topological pharmacophore fingerprints have been proposed in terms of considered multiplets (pairs, triplets), pharmacophore flagging strategies (based on various chemical common-sense rules or, alternatively, on predicted ionization states at given pH) or population level monitoring (binary, multiplet counts, fuzzy multiplet counts). None of these may claim absolute advantages in terms of relative neighborhood behavior (*i.e.*, the extent to which they verify the similarity principle), for no exhaustive benchmarking study has ever compared all of them, under identical conditions. Reports of various authors seem to agree that a certain amount of fuzziness enhances scaffold hopping performance, while too much tends to "blur" the query structures to the point where almost any candidate may match. Also, an accurate assignment of ionization states when flagging the donors, acceptors, cations and anions may be able to explain otherwise surprising activity cliffs, involving pairs of almost identical compounds (except for an intuitively unobvious shift of a ionization constant, leading to different proteolytic behaviors that may well explain the unexpected activity differences).

Machine learning (using both linear and nonlinear techniques) with topological pharmacophore fingerprints was shown to be largely successful in explaining binding affinities for various QSAR series – at least as successful as elaborate 3D modeling techniques. However, training set-related bias was seen to play a paramount role in QSAR build-up based on congeneric series, so that none of these studies can be used to derive any general scale of QSAR-building propensities for the various 2D pharmacophore fingerprints, in the context of other molecular descriptors. A good point for 2D pharmacophore fingerprints is their straightforward interpretability, which allowed the particular flaws of the studied training sets to be pinpointed easily. For example, activity models largely based on "antipharmacophore" contributions (penalties for the presence of allegedly unwanted pharmacophore features) should be rejected unless one has good physicochemical reasons to explain why the presence of these features is incompatible with activity.

2.5.3 Each QSAR Problem should be Allowed to Choose its Descriptors of Predilection

The choice of descriptors may be, in itself, relevant to pharmacophore elucidation – if founded on an information-rich training set.

A general conclusion emerging from this study is that all the exploratory work concerning descriptor-based structure–activity relationships is split over a plethora of more or less biased data sets and therefore does not lead to any objective, general recommendations concerning the rational choice of descriptors for a given context. It has been proven that giving up 3D modeling in favor of topological description is certainly not synonymous with a dramatic loss in predictive power, but we cannot advance any solid estimate of the likelihood of seeing 2D modeling performing either as well or even better. Chemoinformaticians must thus rely on common sense in choosing the proper descriptors to address a problem. Similarly, are 2D pharmacophore fingerprints better than fragment-based descriptors? Studies are split on the issue, and typically do not go into a detailed analysis of the reasons for the different behaviors (which would likely lead to the insight that the winning descriptors claimed victory due to a peculiar set-specific artifact). On closer inspection, however, the question as formulated above makes no sense, since pharmacophore descriptors are nothing but wild-card matching fragments that may be counted using SMARTS-driven substructure search. The real question is "How precise or fuzzy should fragment definition be?" Or, this is obviously a target-specific problem. Suppose that a target accepts an acceptor-substituted phenyl ring in one of its pockets. Depending on the exact nature of the interaction and the flexibility of the side-chain interacting with the donor, there are two extreme scenarios: if the site donor is a backbone amide $>$N–H, there will probably be little tolerance with respect to steric effects near the acceptor lone pair. Both Φ-OH and Φ-OMe fragments may therefore display different activities and fragment-based modeling could effortlessly learn the rule. From a topological pharmacophore point of view, the two analogues are very probably undistinguishable,[iii] and fragment-based modeling wins. The drawback of well-defined fragments is that they do not allow any generalizations at all: the learnt rule "Φ-OH is favorable" does not allow any extrapolation to Φ-NH$_2$, Φ-CH=O, *etc.* On the contrary, the pharmacophore model, having learnt form Φ-OH that Ar-HA patterns are favorable, will – rightly or erroneously – adventure to predict that Φ-NH$_2$ and Φ-CH=O are actives too. Suppose now the opposite scenario, where interaction occurs with a flexible lysine side-chain cation: a diverse set of acceptor-decorated aromatics may be accepted and even variable length spacers between the ring and its acceptor may work. In this case, different species, a pyridyl or Φ-CH$_2$-COO$^-$, may count among the actives as well. Now fuzzy pharmacophore fingerprints will have a significant advantage, for the characteristic pharmacophore signature of all these fragments is now being condensed into very few specific fingerprint elements (say, a fuzzy aromatic-acceptor counts at three bonds). Certainly, the fragment-based model may, in principle, learn by allowing all the specific fragment counts to enter, but the variable-rich resulting model may run into cross-validation issues. If, for example, there is only one training set example of an active Φ-CH$_2$-COO$^-$, that is technically enough to fit a coefficient for the participation of that fragment, but "leaving-the-only-example-out cross-validation" makes no sense. In the pharmacophore model, this compound will be assimilated to one of the many examples with Ar-HA pairs and will effort-lessly pass leave-one-out prediction tests. The pharmacophore model may erroneously conclude that Φ-NO$_2$ derivatives are strongly active as well (if the flagging rules considered nitro groups as acceptors), whereas in practice the hydrogen bond accepting propensity of the nitro derivatives only suffices for a modest affinity level. In this case, a pharmacophore-based general term com-bined with a nitro fragment-specific penalty would be the ideal answer.

To sum up the preceding discussion, topological pharmacophore models are, unlike specific fragment-based approaches, able to make chemotype-transcending generalizations: the acceptor-aromatic pattern can be learned from phenols alone, and extrapolated to anilines. For fragment-based approaches, anilines would be outside the applicability domain of the phenol-trained model. The great success of fragment-based descriptors is very much due to their conservative "no risk – no errors" character. The pharmacophore model suggesting the "bioisosteric hop-ping" from Φ-OH to Φ-NH$_2$ may be commended for its enhanced extrapolability if it is right – otherwise, false positives wasting resources on synthesis and testing are typically a stronger issue than false negatives, representing only a virtual "missed opportunity" out of the 10^{56} drug-like ligand candidates. In any case, *poor training sets* (including only Φ-OH) *will lead to poor models*, which are either

---

[iii] Except when learning from highly biased training sets: if, for example, the inactive –OMe deri-vatives happen to be the only training set compounds that do not contain any donor at all, machine learning may pick any donor-related fingerprint element and give it a positive weight, concluding "donor is needed (phenyl –OH acts as a donor!)" instead of "do not sterically crowd the acceptor."

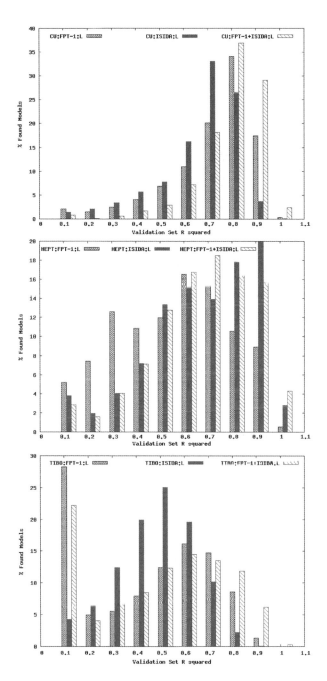

Figure 2.3 Comparative density distribution histograms (on X, bins with respect to the validation set correlation coefficients R^2_V, on Y – percentage of models scoring R^2_V values within each bin) of linear QSAR model sets obtained with fuzzy pharmacophore triplets FPT-1 (grid filling), ISIDA fragments (solid gray) and composite descriptors: FPT and ISIDA mixture, CD (hashed). Descriptors, the employed Stochastic QSAR Sampler and the three anti-HIV compound data sets (labeled CU, HEPT and TIBO) have been described elsewhere.[27] The synergy effect of combined fragments and pharmacophores (histogram shift to the right) is particularly strong within the CU series.

unable to extrapolate (with fragments) or may be technically able to extrapolate, at the user's own risk (with pharmacophore patterns). Model mining of rich training sets, simultaneously including both specific fragments and fuzzy fingerprints, may, however, allow the procedure to settle for the best target-specific choices. In this sense, the nature of the chosen descriptors may carry an implicit message concerning the nature of site–ligand interactions: if data mining selected fuzzy pharmacophore fingerprint components, then the interaction site is probably flexible. Conversely, if only certain fragments entered the equations, whereas others of their isopharmacophoric analogues do not, binding at that specific anchoring point may be more of a rigid key-and-lock scenario. Potential synergies between pharmacophore and fragment descriptors are furthermore likely to ensure improved validation propensities of the QSAR models (see, for example, Figure 2.3). While ceaselessly repeating the call for high training set diversity standards, it should also be pointed out that model mining in the context of several thousands of candidate descriptors, if fragment and pharmacophore fingerprints are to be merged, does call for some methodological progress as well. To confirm its credibility, the quest for meaningful binding model extraction from structure–activity data needs to mature, in moving from artisan regressions on 30 congeneric compounds to aggressive, massively parallel mining of models with (tens of) thousands of compounds.

Abbreviations

2D-FPT	Topological fuzzy pharmacophore triplets
Ar	Aromatic
HA/HD	Hydrogen bond acceptor/donor
Hp	Hydrophobe
NB	Neighborhood behavior
PC/NC	Positive/negative charge
QSAR	Quantitative structure–activity relationships
TI	Topological index

References

1. IUPAC, *Glossary of Terms used in Medicinal Chemistry*, http://www.chem.qmul.ac.uk/iupac/medchem.
2. W. L. Jorgensen, *Science*, 1991, **254**, 954–955.
3. N. Choudhury and B. Montgomery-Pettitt, *J. Am. Chem. Soc.*, 2007, **129**, 4847–4852.
4. C. Wang, P. Bradley and D. Baker, *J. Mol. Biol.*, 2007, **373**, 503–519.
5. V. De Grandis, A. R. Bizzarri and S. Cannistraro, *J. Mol. Recognition*, 2007, **20**, 215–226.
6. R. Bergmann, A. Linusson and I. Zamora, *J. Med. Chem.*, 2007, **50**, 2708–2717.

7. R. Poulain, D. Horvath, B. Bonnet, C. Eckoff, B. Chapelain, M.-C. Bodinier and B. Deprez, *J. Med. Chem.*, 2001, **44**, 3378–3390.
8. H.-J. Bohm and G. Schneider, *Virtual Screening for Bioactive Molecules.*, Wiley-VCH, Weinheim, 2000.
9. O. F. Güner, *Pharmacophore Perception, Use and Development in Drug Design.*, International University Line, La Jolla, CA, 2000.
10. Y. C. Martin, M. G. Bures, E. A. Danaher, J. Delazzer, I. Lico and P. A. Pavlik, *J. Comput.-Aided Mol. Des.*, 1993, **7**, 83–102.
11. G. Jones, P. Willet and R. C. Glen, *J. Comput.-Aided Mol. Des.*, 1995, **9**, 532–549.
12. D. Horvath, in *QSPR/QSAR Studies by Molecular Descriptors*, ed. M. V. Diudea, Nova Science Publishers, Inc, New York, 2001, pp. 395-439.
13. R. D. Cramer, D. E. Patterson and J. E. Bunce, *J. Am. Chem. Soc.*, 1988, **110**, 5959–5967.
14. M. L. Barreca, S. Ferro, A. Rao, L. De Luca, M. Zappala, A. M. Monforte, Z. Debyser, M. Witvrouw and A. Chimirri, *J. Med. Chem.*, 2005, **48**, 7084–7088.
15. L. Franke, E. Byvatov, O. Werz, D. Steinhilber, P. Schneider and G. Schneider, *J. Med. Chem.*, 2005, **48**.
16. C. M. Low, I. M. Buck, T. Cooke, J. R. Cushnir, S. B. Kalindjian, A. Kotecha, M. J. Pether, N. P. Shankley, J. G. Vinter and L. Wright, *J. Med. Chem.*, 2005, **48**, 6790–6802.
17. D. Horvath, B. Mao, R. Gozalbes, F. Barbosa and S. L. Rogalski, in *Chemoinformatics in Drug Discovery*, ed. T. I. Oprea, Wiley-VCH, Weinheim, 2004, pp. 117–137.
18. T. L. Blundell, *Nature*, 1996, **384**, 23–36.
19. U. Fechner, L. Franke, S. Renner, P. Schneider and G. Schneider, *J. Comput.-Aided Mol. Des.*, 2003, **17**, 687–698.
20. U. Fechner, J. Paetz and G. Schneider, *QSAR Comb. Sci.*, 2005, **24**, 961–967.
21. F. Fontaine, M. Pastor, I. Zamora and F. Sanz, *J. Med. Chem.*, 2005, **48**, 2687–2694.
22. S. Sciabola, I. Morao and M. J. de Groot, *J. Chem. Inf. Model.*, 2007, **47**, 76–84.
23. S. D. Pickett, J. S. Mason and I. M. McLay, *J. Chem. Inf. Comput. Sci.*, 1996, **36**.
24. J. S. Mason, I. Morize, P. R. Menard, D. L. Cheney, C. Hulme and R. F. Labaudiniere, *J. Med. Chem.*, 1998, **38**, 144–150.
25. D. Horvath and C. Jeandenans, *J. Chem. Inf. Comput. Sci.*, 2003, **43**, 691–698.
26. D. Horvath and F. Barbosa, *Curr. Trends Med. Chem.*, 2004, **4**, 589–600.
27. D. Horvath, F. Bonachera, V. Solov'ev, C. Gaudin and A. Varnek, *J. Chem. Inf. Model.*, 2007, **47**, 927–939.
28. S. Oloff, R. B. Mailman and A. Tropsha, *J. Med. Chem.*, 2005, **48**, 7322–7332.

29. C. Rolland, R. Gozalbes, E. Nicolai, M. F. Paugam, L. Coussy, F. Barbosa, D. Horvath and F. Revah, *J. Med. Chem.*, 2005, **48**, 6563–6574.
30. F. Bonachera, B. Parent, F. Barbosa, N. Froloff and D. Horvath, *J. Chem. Inf. Model.*, 2006, **46**, 2457–2477.
31. E. R. Carhart, D. H. Smith and R. Venkataraghavan, *J. Chem. Inf. Comput. Sci.*, 1985, **25**, 64–73.
32. D. Horvath, in *Combinatorial Library Design and Evaluation. Principles, Software Tools, and Applications in Drug Discovery.*, ed. A. K. Ghose, Viswanadhan V.N Marcel Dekker, Inc., New York, 2001, pp. 429–472.
33. A. T. Balaban, *Chem. Phys. Lett.*, 1982, **89**, 399–404.
34. T. Laidboeur, D. CabrolBass and O. Ivanciuc, *J. Chem. Inf. Comput. Sci.*, 1997, **37**, 87–91.
35. Y. Qi, R. I. Sadreyev, Y. Wang, B.-H. Kim, N. V. Grishin, *BMC BIOINFORMATICS*, 2007, **8**, Art. No. 314.
36. Z. Simon, A. Chiriac, S. Holban, D. Ciubotariu and G. I. Mihalas, *Minimum Steric Difference. The MTD Method for QSAR Studies.*, Letchworth and Wiley, New York, 1984.
37. L. Kurunczi, E. Seclaman, T. I. Oprea, L. Crisan and L. Simonx, *J. Chem. Inf. Model.*, 2005, **45**, 1275–1281.
38. C. D. Chiriac, S. Funar-Timofei, L. Kurunczi, M. Mracec, M. Mracec, Z. Szabadai, E. Seclaman and Z. Simon, *Rev. Roum. Chim.*, 2006, **51**, 71–99.
39. O. M. Minailiuc and M. V. Diudea, in *QSPR/QSAR Studies by Molecular Descriptors.*, ed. M. V. Diudea, Nova Science Publishers, Inc., New York, 2001, pp. 363–388.
40. G. Hessler, M. Zimmermann, H. Matter, A. Evers, T. Naumann, T. Lengauer and M. Rarey, *J. Med. Chem.*, 2005, **48**, 6575–6584.
41. M. Zimmermann, *Rechnerunterstütze Analyse von HTS-Daten.*, Mathematisch Natur-wissenschaftliche Fakultät; Rheinische Friedrich-Wilhelms-Universität Bonn, Bonn, 2003.
42. G. Schneider, W. Neidhart, T. Giller and G. Schmid, *Angew. Chem. Int. Ed.*, 1999, **38**, 2894–2896.
43. ChemAxon, *Screen User Guide*, http://www.chemaxon.com/jchem/index.html?content=doc/user/Screen.html.
44. MOE (Molecular Operating Environment), (2005) Chemical Computing Group, Inc., Montreal.
45. M. Olah, C. Bologa and T. I. Oprea, *J. Comput.-Aided Mol. Des.*, 2004, **18**, 437–439.
46. DayLight, *SMARTS*, http://www.daylight.com/dayhtml/doc/theory.smarts.html.
47. N. Stiefl, I. A. Watson, K. Baumann and A. Zaliani, *J. Chem. Inf. Model.*, 2006, **46**, 208–220.
48. T. Ewing, C. Baber and M. Feher, *J. Chem. Inf. Model.*, 2006, **46**, 2423–2431.
49. ChemAxon, *pKa Calculator Plugin*, http://www.chemaxon.com/marvin/chemaxon/marvin/help/calculator-plugins.html#pka.
50. N. Stiefl and A. Zaliani, *J. Chem. Inf. Model.*, 2006, **46**, 587–596.

51. R. W. Stanforth, E. Kolossov and B. Mirkin, *QSAR Comb. Sci.*, 2007, **26**, 837–844.

52. G. M. Morris, D. S. Goodsell, R. S. Halliday, R. Huey, W. E. Hart, R. E. Belew and A. J. Olson, *J. Comp. Chem.*, 1998, **19**, 1639–1662.

53. P. Willett, J. M. Barnard and G. M. Downs, *J. Chem. Inf. Model.*, 1998, **38**, 983–996.

54. V. P. Solov'ev, A. Varnek and G. Wipff, *J. Chem. Inf. Comput. Sci.*, 2000, **40**, 847–858.

55. C.-A. Azencott, A. Ksikes, J. Swamidass, J. H. Chen, L. Ralaivola and P. Baldi, *J. Chem. Inf. Model.*, 2007, **47**, 965–974.

56. P. Mahé, L. Ralaivola, V. Stoven and J.-P. Vert, *J. Chem. Inf. Model.*, 2006, **46**, 2003–2014.

57. G. Schneider and M. Nettekoven, *J. Comb. Chem.*, 2003, **5**, 233–237.

58. M. von Korff and K. Hilpert, *J. Chem. Inf. Model.*, 2006, **46**, 1580–1587.

59. Cerep, *BioPrint Database*, http://www.cerep.fr/cerep/users/pages/Collaborations/Bioprint.asp, 2007.

60. D. Horvath and C. Jeandenans, *J. Chem. Inf. Comput. Sci.*, 2003, **43**, 680–690.

61. E. Byvatov, U. Fechner, J. Sadowski and G. Schneider, *J. Chem. Inf. Comput. Sci.*, 2003, **43**, 1182–1189.

62. M.-L. Lee and G. Schneider, *J. Comb. Chem.*, 2001, **3**, 284–289.

63. G. Schneider, P. Schneider and S. Renner, *QSAR Comb. Sci.*, 2006, **25**, 1162–1171.

64. S. Renner and G. Schneider, *ChemMedChem*, 2006, **1**, 181.

65. M. Wagener and J. P. M. Lommerse, *J. Chem. Inf. Model.*, 2006, **46**, 677–685.

66. G. Schneider, M.-L. Lee, M. Stahl and P. Schneider, *J. Comput.-Aided Mol. Des.*, 2000, **14**, 487–494.

67. L. Franke, O. Schwarz, L. Müller-Kuhrt, C. Hoernig, L. Fischer, S. George, Y. Tanrikulu, P. Schneider, O. Werz, D. Steinhilber and G. Schneider, *J. Med. Chem.*, 2007, **50**, 2040–2046.

68. T. Kohonen, *Self-Organization and Associative Memory.*, Springer, Heidelberg, 1984.

69. F. Bonachera and D. Horvath, *J. Chem. Inf. Model.*, 2008, **48**(2), 409.

70. B. Debnath, S. Gayen, A. Basu, K. Srikanth and T. Jha, *J. Mol. Model.*, 2004, **10**, 328–334.

CHAPTER 3

Pharmacophore-based Virtual Screening in Drug Discovery

CHRISTIAN LAGGNER,[a] GERHARD WOLBER,[b]
JOHANNES KIRCHMAIR,[b] DANIELA SCHUSTER[a] AND
THIERRY LANGER[a, b]

[a] Department of Pharmaceutical Chemistry, Faculty of Chemistry and
Pharmacy and Center for Molecular Biosciences (CMBI), University of
Innsbruck, Innrain 52, A-6020 Innsbruck, Austria; [b] Inte:Ligand Software-
Entwicklungs und Consulting GmbH, Clemens Maria Hofbauer-Gasse 6,
A-2344 Maria Enzersdorf, Austria

3.1 Introduction

In past years, considerable efforts have been devoted to compacting the early
phase of hit-to-lead development in the drug discovery process within the pharmaceutical industry. In particular, as combinatorial chemistry and high-
throughput screening failed to show the expected success, *in silico*-based virtual
screening approaches emerged and largely evolved.[1] Several issues related to
efficient search algorithms, but also to library design, to diversity selection, to
drug and/or to lead-likeness assessment arose that were addressed in numerous
papers and some reviews.[2–8] In this chapter, we focus on work in the context of
generation and use of pharmacophore models and related methods for mining
virtual compound libraries and on new developments in the field of pharmacophore usage for *in silico* screening. All these procedures in general aim at
obtaining hits (or leads) that have enhanced likelihoods of leading to successful
clinical candidates by medicinal chemistry efforts. The goal is to reduce the overall

Chemoinformatics Approaches to Virtual Screening
Edited by Alexandre Varnek and Alex Tropsha
© Royal Society of Chemistry, 2008
Published by the Royal Society of Chemistry, www.rsc.org

cost associated with the discovery and development of a new drug, by identifying the most promising candidates to focus the experimental efforts on.

In many cases, drug discovery projects have reached an already well-advanced stage before detailed structural data on the protein target has become available. Even though it has been shown that novel methods of molecular biology together with biophysics and computational approaches enhance the likelihood of successfully obtaining detailed atomic structure information, full elucidation of the target's 3D structure often lags behind the first results of screening experiments. A consequence is that medicinal chemists have to develop novel compounds for a target using preliminary structure–activity information, together with theoretical models of interaction. In such a case, responses that are consistent with the working hypotheses will contribute to an evolution of the used models. In this context, the chemical feature-based pharmacophore approach has proven to be extremely successful, allowing the perception and understanding of key interactions between a receptor and a ligand on a generalized level. Such feature-based pharmacophore models together with large 3D structure databases originating either from commercial vendors, from in-house compound collections, or from virtual combinatorial chemistry have proven to be of great value for performing *in silico* database mining. Successful application to virtual screening described in recent papers is summarized within this chapter.

3.2 Virtual Screening Methods

In parallel with the widely used high-throughput screening (HTS) technology, virtual screening (VS) has become an indispensable tool for identifying possible lead structures.[9] VS is now established as one of the most important computational techniques used to discriminate between wanted (presumably active) and unwanted (presumably inactive) molecules within compound libraries. This task has to be done as early as possible in the drug discovery pipeline to reduce drug discovery costs. Hristozov *et al.*[10] have recently presented a classification of four different main scenarios for VS: (i) prioritizing compounds for subsequent HTS, (ii) selecting a predefined (small) number of potentially active compounds from a large chemical database, (iii) assessing the probability that a given structure will exhibit a given activity and (iv) selecting the most active structure(s) for a biological assay. The different scenarios may require different amounts of data available for model generation, favor different approaches and require different benchmarking values.

When the three-dimensional structure of the target is unknown, pharmacophore approaches play a predominant role among the compound selection filters that have been designed for retrieving bio-active molecules. Additional pre-filtering based upon favorable physicochemical properties necessary for, *e.g.*, oral bioavailability,[11] aqueous solubility, metabolic clearance, and chemical reactivity or the presence of toxic chemical groups,[12,13] clearly will enhance the success rate of finding possible candidates for further optimization.

If the target 3D structure is known, docking turned out to be a valuable structure-based VS method to be applied for successful identification of novel bioactive molecules.[14-17] However, the biggest and so far unsolved problem of docking methods remains the scoring of the docked compounds. In fact, the major weakness of these procedures currently does not lie in the docking algorithms themselves, *i.e.*, in the ability to find the correct ligand-poses, but rather in the inaccuracy of the scoring functions that are used to estimate the binding affinity between ligand and target – including a rough separation between actives and inactives. Such scoring functions for VS have been analyzed,[18] giving insight into weaknesses and strengths of currently used models for affinity estimation. A review highlighting the most relevant advances in docking and scoring has been published recently,[19] also indicating that the major drawbacks still exist. When using pharmacophore models as screening filters instead of protein 3D structures, affinity estimation is based on a geometric fit of structures to the model. In such a case, the values calculated are often also far from reality; however, they are useful for filtering possible hits from non-binding molecules. Moreover, in the pharmacophore fitting procedures, computational demands are considerably lower than when docking algorithms are applied for VS. This allows the number of compounds to be processed in a comparable time to be by far higher than even in so-called high-throughput docking. This advantage becomes even more important considering the fact that for compound profiling the use of parallel screening on different targets will become indispensable in the near future.

Apart from the afore-mentioned pharmacophore and docking methods, other approaches have been successfully applied to VS scenarios: especially at the early stages of a drug discovery program, where little is known about both target and ligands, similarity searching is a useful method.[20,21] To analyze the results from a high-throughput screening campaign, consisting of both active and inactive compounds, various machine learning techniques that can differentiate between the two classes have proven succesfull.[22,23]

Parallel screening for targets and anti-targets, enabling the prediction of unfavorable side-effects and therefore allowing some risk assessment in the early stage of drug discovery, based on different methods, has been proposed recently.[24-26] Finally, VS is not only useful for retrieving hits from existing compound libraries, but additionally represents a valuable tool for the design of a combinatorial library with a given target.

3.3   Chemical Feature-based Pharmacophores

In the past few years an increasing influence on rational drug design has been exerted by several software programs from major software companies (CATALYST[2,4,27,28] – now integrated into DiscoveryStudio[29] by Accelrys; DISCOtech,[30] GALAHAD,[31] GASP,[32-35] and UNITY[36,37] by Tripos; MOE[38] by Chemical Computing Group; and Phase[39-42] by Schrödinger) that rely on the concept of chemical feature-based pharmacophore models. The following section explains the basic concepts of 3D pharmacophore modeling and highlights differences and similarities between some state-of the art software-packages.

3.3.1 The Term "3D Pharmacophore"

The term "pharmacophore" has become increasingly used in medicinal chemistry in recent years and has had different meanings attributed to it. "Pharmacophores" are often regarded as structural fragments or functional groups being related to a chemical compound. However, the official IUPAC definition[43] from 1998 is more precise: "A pharmacophore is the ensemble of steric and electronic features that is necessary to ensure the optimal supra-molecular interactions with a specific biological target structure and to trigger (or to block) its biological response".[44] This definition clearly emphasizes the abstraction of common steric and electronic interactions of bio-active compounds exhibiting comparable biological effects within the same binding site in a comparable situation. This abstract model, containing chemical functionalities (such as "positive ionizable" instead of "primary amine") can serve as an effective search filter for VS. This concept is not new in medicinal chemistry and has already been successfully applied before computers were used in chemistry.[45,46] Still, in recent years an increasing influence on rational drug design has been exerted by several software programs from major software companies (Accelrys, Chemical Computing Group, Schrödinger, Tripos) that rely on the concept of chemical feature-based pharmacophore models.

To be a useful tool for drug design, a pharmacophore model[47] has to provide valid information for medicinal chemists investigating structure–activity relationships. First, the pharmacophore model has to describe the nature of the functional groups involved in ligand–target interactions, as well as the type of the non-covalent bonding and intercharge distances. The model also has to show predictive power which, at its best, enables the design of novel chemical structures that are not evidently derived by the translation of structural features from one active series to the other, or even allows effective scaffold hopping.[48]

3.3.2 Feature Definitions and Pharmacophore Representation

Selecting the right chemical feature types is a first crucial step for the development of a high quality pharmacophore model. In early pharmacophore modeling techniques, such as the active analog approach described by Marshall *et al.*,[49] features constituting a pharmacophore could contain any fragment or atom type. More recent techniques, such as the software package Catalyst[50] use a more general way for building pharmacophore queries, *e.g.*, a single geometric entity for all negative ionizable groups. The discussion below will show that several arguments exist for continuing with this trend and even to further extend the generalization of chemical functionalities. In real-life applications, however, built-in features are often tailored to achieve a desired filtering restrictivity level,[51] *i.e.*, the ability to restrict a model to identify a specified set of compounds that the model was created for.

General definitions may result in models that are universal, at the cost of restrictivity. However, restrictivity is an important issue in pharmacophore searching and, therefore, feature descriptions that are too general need to be

Table 3.1 Abstraction layers of pharmacophoric feature contraints.

Layer	Classification	Universality	Specifity
4	Chemical functionality without geometric constraint, e.g., an H-bond acceptor without a projected point or a lipophilic group	+++	–
3	Chemical functionality (H-bond acceptor, H-bond donor, positive ionizable, negative ionizable, hydrophobic) with geometric constraint, e.g., an H-bond acceptor vector including an acceptor point as well as a projected donor point; aromatic ring including a ring plane	++	+
2	Molecular graph descriptor (atom, bond) without geometric constraint, e.g., a geometrically unconstrained phenol group	–	++
1	Molecular graph descriptor (atom, bond) with geometric constraint, e.g., a phenol group facing a parallel benzenoid system within a distance of 2–4 Å	–	+++

changed from reflecting universal chemical functionality to representing distinct functional groups. To describe the levels of universality and specificity of chemical features, a simple layer model is used in the following discussion to allow referral to these properties more easily. Table 3.1 shows a proposed classification of abstraction layers of the most important chemical features.[52] A lower level corresponds to higher specificity and, therefore, to lower universality.

The most frequent reason for creating features on the low universality levels 1 and 2 is that the definitions of the higher levels are not sufficient to describe the features occurring within the collection of known active ligands (see ref. 51 for an example). Even if customization results in a layer 1 or layer 2 feature, there should be a possibility of including layer 3 or 4 information to categorize and, thus, increase comparability (e.g., a carboxylic acid as a layer 2 feature is a subcategory of "negative ionizable", which is a layer 4 feature).

Software packages for pharmacophore modeling always have to face a trade-off in the design of a generally applicable feature set that is universal and, at the same time, still selective enough to reflect all relevant types of ligand–receptor interactions. The most relevant interactions and their geometric representation in some current software packages, Catalyst,[53] Phase,[54] MOE,[55] and LigandScout,[56,57] are described in the following section, with Table 3.2 providing a summary.

3.3.2.1 Hydrogen Bonding Interactions

Hydrogen bonding occurs when covalently bound hydrogen atoms with a positive partial charge interact with another atom with a negative partial charge. To capture the characteristics of hydrogen bonding, Catalyst and LigandScout model H-bond donor and acceptor features as a position for the

Table 3.2 Summary of pharmacophoric feature abstraction, representation and customization in different pharmacophore modeling applications.

Feature	Phase	MOE	Catalyst	LigandScout
Hydrogen bonding interaction	H-donors located at hydrogen atom, acceptors at heavy atom. Modeled as layer 3 feature with direction constraint and position tolerance	Geometric constraints depend on selected pharmacophore scheme. Features located at heavy atoms with tolerance sphere.	Acceptor and donor features positioned on heavy atom with tolerance sphere. (Layer 3 feature) Max one donor or acceptor feature per atom	Acceptor and donor features positioned on heavy atom with tolerance sphere. Represented as layer 3 features with feature position and projected point
Lipophilic area	Represented as tolerance spheres. Aromatic rings not recognized as hydrophobic areas	Represented as tolerance spheres. Aromatic rings not recognized as hydrophobic areas	Represented as tolerance spheres	Represented as tolerance spheres
Aromatic interaction	Represented as position with tolerance and ring plane orientation	Modeling of aromatic features depends on the selected pharmacophore scheme	Represented as position with tolerance and ring plane orientation	Represented as position with tolerance and ring plane orientation
Electrostatic interaction	Represented as tolerance spheres, no explicit charges necessary	Represented as tolerance spheres, require charges (cationic and anionic)	Represented as tolerance spheres, no explicit charges necessary	Represented as tolerance spheres, no explicit charges necessary
Definition and customization	Definition and customization through SMARTS patterns with associated geometry attributes	Implementation of new pharmacophore schemes possible with scripting language (SVL)	Definition of new features and customization *via* graphical interface	Feature definitions as SMARTS or boxed algorithms, adjustment of feature specific parameters and geometric constraints

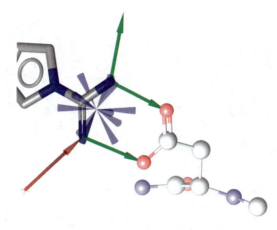

Figure 3.1   Thrombin inhibitor SN3401 in complex with Thrombin (PDB entry
2GDE): three hydrogen bond donors (green), one acceptor (red), and one
charge transfer feature (blue) are recognized by LigandScout. In order to
show more detail of the interaction, only a portion of the structure of
SN3401 has been shown.

heavy atom and a projected point representing the position from which the
participating hydrogen will extend. These two positions form a vector that
indicates the direction from the heavy atom to the projected point of the
hydrogen bond. However, in Catalyst, only a single hydrogen bonding feature
is permitted per heavy atom, whereas LigandScout allows an acceptor or donor
atom to be involved in more than one H-bonding interaction. In the situation
shown in Figure 3.1 (PDB code 2GDE) either all hydrogen bonds or two
hydrogen bonds and the ionizable feature must be omitted in Catalyst, or
several different models must be created to reflect all interactions.

   In a similar manner, Phase positions H-bond acceptor features on heavy
atoms that carry one or more lone pairs, and, depending on the hybridization of
the acceptor atoms, assigns vector attributes to each idealized hydrogen bond
axis. Hydrogen bond donor features are centered on donor hydrogen atoms with
a single vector constraint directed along the hydrogen bond axis. An alternative
to this ligand-centric convention is to represent acceptors and donors as pure
projected points, located at complementary positions on theoretical binding
sites. The projected point approach does not incorporate vector character into
the site definition and permits situations where two ligands form hydrogen
bonds to the same receptor atom, but from different locations and directions.

   In MOE the modeling of H-donor and acceptor features depends on the
selected pharmacophore scheme. There are six schemes supplied with MOE:
PCH, PCHD, PCH_All, PPCH, PPCH_All, and CHD. The PCH scheme
(which is the default) defines H-bond acceptors and donors as layer 4 point
features. In contrast, the PCHD scheme also includes putative points from
hydrogen bond donors and acceptors that are projected in the approximate
direction of the hydrogen bond.

3.3.2.2 Lipophilic Areas

Lipophilic contacts represent layer 4 features with no geometric constraints and are generally represented as tolerance spheres located in the centre of hydrophobic atom chains, branches or groups. Although the perception of hydrophobic areas in Catalyst, Phase and LigandScout is based on the same algorithm described during Catalyst development by Greene *et al.*,[50] subtle deviations seem to exist and the results differ considerably, which makes an otherwise possible program interoperability hard to accomplish.

3.3.2.3 Aromatic Interactions

Aromatic features can be modeled as layer 4 point features or as layer 3 features. In Catalyst, Phase, and LigandScout these features are also attributed with a ring plane normal defining the spatial orientation (layer 3 feature). In MOE, the selected pharmacophore scheme determines whether a ring plane orientation constraint is included or if aromaticity is modeled as a pure point feature with a tolerance sphere.

3.3.2.4 Electrostatic Interactions

Positive or negative ionizable areas are single atoms or groups of atoms that are likely to be protonated or deprotonated at physiological pH. Ionizable features are commonly implemented as spheres with a certain tolerance radius for pharmacophore matching. While Catalyst, LigandScout and Phase are insensitive to the protonation state from the input molecules, MOE requires preprocessing of the molecules and the assignment of explicit charges. Additionally, in MOE, positive and negative ionizable areas are limited to single atoms (including resonance anions and cations) carrying a corresponding charge, causing limits for groups like carboxylic acids or guanidines.

3.3.2.5 Customization and Definition of New Features

Although feature definitions should be general and describe all possible interactions that are observed in ligand binding, some models still do not fulfill all requirements for restrictivity. For this reason, most applications provide means for customization or extension of the predefined feature set. Catalyst, *e.g.*, allows customization and extension of the built-in features *via* the graphical user interface, allowing specification of one or more chemical groups that satisfy a particular feature (OR logic). Similarly, Phase allows specification of matching chemical groups and exclusions for a particular feature as a list of SMARTS[58] patterns. Each pattern can be associated with a geometric representation (point, group, or vector) and additional flags for hydrogen bond acceptors and donors. Although the Phase approach is very flexible and user friendly, only a maximum of three additional custom feature definitions can be added. In MOE, the user has the possibility of implementing a new or modifying an existing pharmacophore scheme, using the scripting language SVL.

LigandScout defines all chemical features as SMARTS patterns in a single configuration file and additionally provides a means for defining geometric constraints, *e.g.*, for hydrogen bonding or stacking of aromatic ring systems. No graphical user interface is provided for modifying feature definitions, but angle constraints and distance ranges for feature recognition in a macromolecule ligand complex can be fine-tuned in a graphical user interface.

Some recent examples of user-customized features include extended negative ionizable features including sulfonamides,[51] Michael acceptor features for covalent addition,[59] zinc ion binding features,[60] and extended positive ionizable features that can map carbocation intermediates.[61]

Once chemical feature points are detected, they can be used as input for the computationally more challenging part: aligning a molecule to a pharmacophore. The algorithms described below can all be used to superposition two molecules according to their pharmacophoric annotation points, to overlay a molecule to a pharmacophore, or to overlay two 3D pharmacophores. These algorithms form the basis for both the elucidation of common pharmacophore patterns amongst different ligands with similar known biological activity and for VS. Both tasks are time-consuming, even on modern hardware, and therefore efficiency remains important.

3.3.2.6   Current Super-positioning Techniques for Aligning 3D Pharmacophores and Molecules

In the broad field of possible pharmacophoric alignment techniques one can distinguish between either point- or property-based approaches.[62] With point-based approaches, atoms or chemical feature point distances are minimized, while property-based approaches generate a pharmacophore by assessing the molecular interaction potential (MIP) similarity, based on Goodford's GRID[63] method, to generate alignments. Programs representing both approaches have been applied for the generation of pharmacophore models of neuronal nicotine acetylcholine receptor (nAChR), resulting in good agreement between the two methods.[64] Considering recent trends and examining currently-available commercial software packages, most of the programs use point-based alignment algorithms, superposing pairs of points by minimizing distances. As already described, an important issue is the positioning of pharmacophoric anchor points, since this is the only chemical representation of the molecule for the algorithms. To discover the relevant chemical feature points, Dror *et al.* differentiate between points being either atoms, fragments or chemical features.[65] Compared to property-based techniques, this abstraction is one of the greatest limitations of all point-based methods, because aligning dissimilar ligands can become problematic in either case. Nevertheless, the feature-based approach – in the sense of pharmacophoric features – has become widely accepted and is used in nearly all drug discovery toolkits nowadays. If advanced feature point algorithms are used, the points represent bioisosterically comparable molecule parts, and the geometric sensitivity becomes an advantage of these algorithms.

Three-dimensional alignment incorporates the problem of conformational flexibility. One possibility to address the problem is to pre-generate conformations (like in DISCO[66] or Catalyst[50]), which makes the actual alignment less time-consuming, but the user has to ensure that all relevant conformations are included. On the other hand, there are in-process approaches that perform the pattern identification and conformational search simultaneously.

A well-known pharmacophore elucidation program using such an approach is GASP[33,67] that was developed by Jones and co-workers in the mid-1990s and is marketed by Tripos. The program is based on a genetic algorithm, *i.e.*, a non-deterministic method that simulates evolution by randomly mutating chromosomes of a certain population. In terms of pharmacophore pattern matching, each chromosome represents a potential flexible pharmacophore by encoding all bond angles and by listing all feature mappings to a manually-selected reference compound. In each run, chromosomes are selected that score best, according to some crucial fitness function, and those are then mutated by applying random torsional rotations to cover conformational space on-the-fly. GALAHAD,[32,35,68] developed at the University of Sheffield, Novo Nordisk, and Biovitrum, and also marketed by the company Tripos, uses a modified genetic algorithm reducing bias towards a single template (base) molecule and introduces partial matching and an improved multi-objective scoring function. Searching is faster than with GASP, since GALAHAD allows the use of pre-generated conformations. For pharmacophore elucidation, GALAHAD uses a very efficient atom-based alignment technique,[31] but lacks important feature definitions like the flexible placement of hydrophobic features which is possible with the other programs.

In contrast to the in-process techniques providing fully flexible models, the other very different class of algorithms relies on rigid-body techniques for aligning molecular structures.[69] Either these methods are completely structure-based approaches or consider conformational flexibility in terms of handling pre-generated conformations sequentially. The advantage of these techniques is that the time-consuming process of generating conformations is out-sourced and the conformations can be stored persistently – because multi-conformational generators actually have the ability to provide generally applicable conformational ensembles that sufficiently sample most small organic drug-like molecules.[70]

Nearly all commercial software using rigid-body alignment techniques is based on maximum common substructure search. One of the first programs that go along with this and which had a considerable influence on modern techniques[62] is DISCO.[66] It is based on distance geometry[71] and the alignment is implemented using the Bron–Kerbosh clique-detection algorithm[72] in terms of inter-distance comparisons. Because of exploring the complete conformational space, the technique is limited to a small number of input compounds of preferably limited flexibility – the main drawback when looking for an optimal solution.

A further development of the exhaustive search is Catalyst's HipHop[27] algorithm, just relaxing the GASP requirement that each feature in the pharmacophore must be present in each of the input compounds. Furthermore, HipHop starts by finding all two-feature models and expands the model until no more configurations can be found. All of the numerous results are listed and ranked according to their rarity-based score.

Another example of an exhaustive search is employed by Schrödinger's program Phase.[40] The algorithm enhances performance by narrowing the search space. The conformational search is dedicated to find a pharmacophore containing a user-defined number of features that are shared by a user-defined number of input molecules regarding a user-defined tolerance. All possible pharmacophores are grouped in a tree according to their inter-site distance, *i.e.*, a vector containing distances of all feature pairs. The tree is traversed and if a node fails to contain pharmacophores from the minimum number of actives, the complete sub-tree is eliminated from further investigation. Phase places high emphasis on user interaction, providing user intervention possibilities at each step of the pharmacophore elucidation process.

The most recent development regarding pharmacophore alignment technique is LigandScout's pattern matching approach.[73] In a first step, feature pairs are formed based on feature types and distance characteristics, encoding the whole pharmacophore for each feature in a rotationally and translationally independent manner. For each feature type and feature, a distance shell contains several bins counting neighboring features, with each bin representing space at a certain distance interval. The core part of the algorithm for identifying pairs is a fast maximum weighted bipartite matching algorithm that scales polynomially with the number of features involved and thus allows its application to larger molecules like peptides. Through its performance, this algorithm has the possibility to perform rigid alignments within less than 100 ms on a modern single CPU, allowing for interactive usage within the graphical user interface of LigandScout.

3.4 Generation and Use of Pharmacophore Models

In the generation of pharmacophore models, one can generally distinguish two approaches: the ligand-based approach and the structure-based approach. This separation is based on whether structural information (*e.g.*, from an X-ray crystal structure of a protein–ligand complex) is available and can be used for the generation of the model, or whether only the structures, but not the bioactive conformations, of a group of ligands are known. Both approaches are highly valuable for the retrieval of novel lead structure scaffolds and there are studies available that demonstrate the efficiency of both techniques in direct comparison[74,75] as well as synergism by using both approaches in combination.[76,77]

3.4.1 Ligand-based Pharmacophore Modeling

In the ligand-based approach, a pharmacophore is deduced from the arranged key interactions of active ligands – having the same binding mode to the target – with respect to conformational flexibility.[78] As described in the previous section, known active compounds are usually aligned by aligning the pharmacophore features present in the compounds, while keeping the conformational energy of the ligands within reasonable limits. Ideally, the compounds used for deriving these models cover a broad structural space, showing a different distribution of flexible and rigid parts, thus minimizing the number of possible overlays.

While several ligand-based pharmacophore building methods are already implemented in commercial software products (Catalyst by Accelrys, now included in DiscoveryStudio[29] Sybyl/Unity by Tripos,[79] MOE by CCG,[55] Phase by Schrödinger[54]) and widely distributed in the drug discovery community, still new promising methods have recently been described in the literature. Among these, the approach by Feng *et al.*[80] seems rather promising. Fingerprints of 3D features and a modification of Gibbs sampling to align a set of known flexible ligands, where all compounds are active, are used to discern possible pharmacophores. A clique detection method is used to map the features back onto the binding conformations. The complete algorithm is described in detail, and it is shown that the method can find common superimpositions. The method reproduces answers very close to the crystal structure and literature pharmacophores in the examples presented. The basic algorithm is relatively fast and can easily deal with up to 100 compounds and tens of thousands of conformations, as shown in test sets of D2 and D4 ligands.[80] In a recent review, Renner *et al.*[81] describe the application of alignment-free pharmacophore pattern using a correlation vector approach for several purposes in bio-active compound selection, similarity searching, and virtual library design. Feature-trees as described by Rarey *et al.*[82,83] differ from most other descriptors in structure. The node-labeled tree structure is more closely related to the molecular than linear structures and it implies, however, more complex comparison algorithms. The descriptor combines conformation independence with alignment dependence, which makes it somehow unique. Alignment dependence is often seen as a disadvantage due to the more time-consuming comparison and the potential bias resulting from heuristic alignment schemes. Both arguments do not hold in the case of Feature Trees (FTrees available from BioSolveIT[84]) since the optimal alignment can be computed within milliseconds by employing dynamic programming techniques.

Starting from a model composed of features placed on one conformation of a rather rigid cyclic peptide, Jia *et al.* have shown how genetic algorithms can be applied to successively improve their model for melanocortin type 4 receptor agonists by optimizing the classification of a large training set of both known agonists and agonists. The optimized model was able to correctly retrieve 37 out of 55 agonists, and none out of 51 nonagonists, while the initial model started off with a ratio of 37/32. Similarly, the retrieval rate for a test set of 55 agonists and 50 nonagonists improved from 40/31 to 33/8.[85] Frequently, ligand-based models are used to guide the synthesis of novel compounds, as has been shown, *e.g.*, for T-type calcium channel blockers,[86] HIV integrase strand transfer inhibitors,[87] metabotropic glutamate receptor 5 inhibitors,[88] and sigma-1 receptor ligands.[89]

3.4.2   Structure-based Pharmacophore Modeling

During the last decade, structure-based methods (also called target-based or direct approaches), which rely on the availability of structural data of the target, have gained significant interest, since the number of experimentally determined three-dimensional structure of targets has grown constantly. Information on the target structure is preferentially taken from experimental investigations, but can also be taken from homology modeling.[90] Today, the largest database of X-ray and NMR structures of protein and protein–ligand complexes in public domain, the Protein Data Bank (PDB),[91] is just approaching the 50 000 entries mark, which is about eight times more than in 1997. This number does not include the large amount of propriety structural data. Despite being originally conceived as primarily a ligand-based method, pharmacophore modeling approaches have been successfully used in a wide range of structure-based VS applications. They aim at being complementary to docking procedures, including the same level of information, but are less demanding with respect to computational demands and therefore much more efficient.

There are different ways to generate structure-based pharmacophore models, both in the presence or absence of a ligand. Additionally, while structure-based models derived from a single protein–ligand complex may be highly specific at retrieving compounds similar to the complexed ligand, they may be too restrictive for retrieving other active compounds that do not share all of the detected pharmacophore interactions. This may be one of the reasons why some authors still prefer ligand-based methods that identify common features among their ligands, despite the availability of X-ray structures for their targets.[92–94] The strategies for the identification of all features present in a given complex structure, as well as of those features that are common to most or all ligands, are discussed in the following.

The traditional way of identifying pharmacophore features in a given protein–ligand complex was to just look at the binding site, analyze what interactions can be detected, and then add the corresponding features manually. Nowadays, all major pharmacophore modeling software packages allow for the construction of the pharmacophore model on top of the imported complex structure. We have automated this task of detecting protein–ligand interactions with our software package LigandScout.[52,56,73] LigandScout is a novel tool for structure-based pharmacophore model generation, comparison and manipulation. The PDB, which is the largest available public repository of biologically relevant proteins complexed with small organic molecules, serves as a starting point. The major focus of this work has been put on the ligands with the aim of extracting relevant information about the respective binding mode. Owing to poor data quality of the ligands in some complexes resulting from historic growth, existing algorithms were adopted and new strategies were developed to interpret ligand topology adequately. A step-by-step interpretation is performed on the PDB ligand entries: planar ring detection, assignment of functional group patterns, hybridization state determination, and Kekulé pattern assignment. The interpretation procedure forms the basis for the next step, the fully automated creation of pharmacophore models, implementing a rule set that automatically detects and classifies protein–ligand interactions into H bond interactions, charge transfers and lipophilic regions. The entire set of interactions forms a pharmacophore model, which can be used for rapid VS in external screening platforms, like Catalyst, MOE, and Phase. Furthermore, the pharmacophore models derived from different complexes – whether from different experimental X-ray structures or from docking experiments – can be overlayed either by their pharmacophore features or by the alignment of the α-carbon atoms of the amino acids forming the binding site, thus allowing the creation of both common feature models and merged feature models, which incorporate features obtained from different complex structures.[73] An example of the creation of a structure-based common feature model is shown for two different COX-1 inhibitors (PDB codes 1EQG and 1PGE) in Figure 3.2.

Ortuso *et al.*[95] recently published their GRID-based pharmacophore model (GBPM) approach for pharmacophore model generation. The authors use the GRID force field[63] calculation of interaction energies for the elucidation of hot spots encoding favorable regions for protein–ligand interaction. The approach aims at the automated generation of unbiased pharmacophore models starting from protein–ligand structural data. The large number of GRID probes allows the generation of highly potent and selective pharmacophore models that can successfully identify lead structures during virtual high-throughput screening as well as to optimize lead structures. The approach has been evaluated with pharmacophore models generated for X-linked inhibitors of apoptosis (XIAPs) as well as for the interleukin 8 dimer interface. Another GRID-based approach that uses four-point pharmacophore fingerprints has been described by Baroni *et al.*[96] "Fingerprints for ligands and proteins" (FLAP) is a three-dimensional pharmacophore profiling technique for ligands and proteins based on their common frame of reference. It allows ligand–ligand, ligand–protein and protein–protein comparisons at each atom, once classified in its corresponding GRID-type. Four-point pharmacophores are built considering features for hydrophobicity, hydrogen-bond donor or/and acceptor and positive or negative charge. The exhaustive combination of all of the atoms provides the information about the four-point pharmacophores together with the chirality to be stored in an appropriate file. This method was able to distinguish between the three serin proteases factor Xa, trypsin, and thrombin, a dataset that had been reported previously by one of the authors for a comparison with a similar method.[97] In a second application example, FLAP was able to correctly classify 23 kinase X-ray structures belonging to five different kinase targets (CDK2, GSK3β, P38α, and LCK) by their four-point pharmacophore fingerprints. Similarly, the Pocket module in the LigBuilder program analyzes the binding site with different probes in a grid-based manner, but it also uses information from the co-crystallized ligand for selecting the essential binding-spots among all the detected,[98,99] an approach for which also Ahlström *et al.* reported that it gave better retrieval rates of active compounds in their study of GRID-derived pharmacophores.[100]

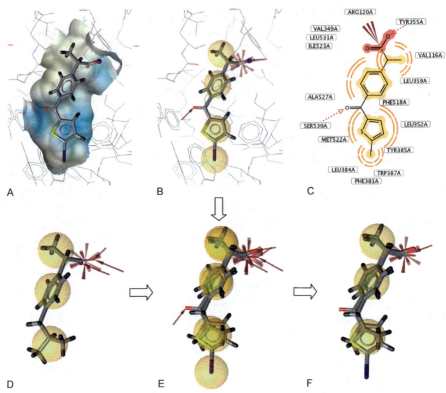

Figure 3.2 Generation of structure-based models for COX-1 with LigandScout: (*A*) binding pocket of PDB entry 1PGE, with the surface of the binding pocket coloured by aggregated lipophilicity (high: yellow; low: blue). (*B*) Automatically generated pharmacophore model in the same pocket. Yellow spheres: hydrophobic, red star: negative ionizable, red arrows: hydrogen-bond acceptor features. (*C*) 2D depiction of the ligand and its interactions with the binding site amino acids. (*D*) Ligand and pharmacophore model extracted from PDB entry 1EQG. (*E*) Models and ligands from 1PGE and 1EQG are aligned. (*F*) A common feature pharmacophore model is generated that has the features present in both binding-site models. For clarity, excluded volume spheres are not shown throughout this figure.

Finally, "dynamic" structure-based pharmacophore models can be derived through a method first described by Carlson *et al.*[101] that uses multiple conformations of the target protein, which are obtained either by molecular dynamics simulation or by the use of multiple experimentally determined conformations. The binding sites of the respective snapshots are flooded with small molecular probes (*e.g.*, methanol for hydrogen-bond interactions and benzene for aromatic hydrophobic interactions) and while the protein structure is held rigid the probe molecules are subjected to a low-temperature Monte Carlo minimization where they undergo multiple, simultaneous gas-phase minimizations and cluster into regions that define their complementary binding regions. The results for the different protein conformations are overlayed and pharmacophore features are placed at the conserved clustering spots, thus identifying the feature locations that are independent of the binding-site conformation.[101–105]

Other new approaches include the application of pharmacophore fingerprints for lead identification, *e.g.*, as implemented in the methods FLIP (fingerprint-based lead identification protocol[106]) and SIFt (structural interaction fingerprint[107]). Such SIFts are 1D binary representations of the interaction patterns derived from a 3D protein–inhibitor complex. The fingerprint representation of the interaction patterns is compact, and allows for rapid clustering and analysis of massive numbers of complexes. For a group of structures binding to the same target protein, the receptor site is defined as the list of residues comprising the union of all residues involved in ligand binding over the entire library of structures. By default, seven different types of interactions occurring at each binding residue are extracted and classified. The resulting binary fingerprint can be extended to so-called p-SIFts (profile-based structural interaction fingerprints)[108] and used for enabling the researcher to describe the conservation of interactions between a set of protein–ligand receptor complexes. The use of profiles provides a sensitive means to compare and contrast multiple inhibitors binding to a drug target. A p-SIFt thus represents the degree to which interactions are conserved across a set of ligand–receptor complexes.

Using a modification of the PharmPrint methodology,[109–111] McGregor has identified pharmacophore patterns for 220 X-ray crystal structures of protein kinases in complex with ligands.[112] The protein structures and their respective pharmacophore models were then overlayed to generate a pharmacophore map, which can be inspected to visualize the interactions made by all ligands with their receptors or to highlight differences between the kinases. Furthermore, a fitting and scoring algorithm was described that can retrieve conformations close to the crystal structure pose for most ligands, starting from 2D ligand structure and no knowledge of the crystal structure from which it was derived. The algorithm also gave a useful enrichment of active compounds in a training set.

To test the flexibility and reliability of the structure-based pharmacophore design approach, Spitzer *et al.*[113] used pharmacophore models to describe interactions between DNA and minor groove binding ligands. The study focused on the implementation of sequence specific properties encoded by the minor groove. The pharmacophore models were created by using exclusively DNA structure information: the bases facing the minor groove floor were decorated with hydrogen bonding features decisive for selective ligand interaction. To enable the resulting DNA pharmacophore model to be compared with the ligand, feature types and directions were inverted and the model was divided into small overlapping parts. Adding excluded volume spheres that described the shape of the groove was found to be useful. By mapping the ligand on each single part of the model a stepwise screen of the whole minor groove could be simulated. The collected Fit values of all mappings showed a clear preference of the ligand for the experimentally confirmed binding site, showing the great potential of the method for future application in VS experiments.

3.4.3   Inclusion of Shape Information

One of the most prominent reasons for the retrieval of false-positives in pharmacophore screens is the lack of spatial restriction of the models: the chances that a compound will have the required features somewhere present in its structure increases with its size and flexibility. Therefore, most 3D pharmacophore methods have the possibility to add either inclusion or shape volumes that must be filled by the ligands, or "forbidden" or excluded volumes that describe the space that is occupied by atoms forming the binding site.

Creation of an inclusion shape feature describing the allowed maximum or required minimum volume is usually straightforward by converting the best-fitting conformation of a highly active ligand matched against the pharmacophore. Depending on the desired restrictivity of the model, the smallest as well as the biggest high affinity ligand may be chosen. The problem with inclusion shapes is that the rigidity of the inclusion volume is the same in every direction, meaning that some large actives may be missed. Still, there are many examples of successful screening enrichments with this method.[61,114,115] Furthermore, VS methods relying primarily on shape similarity, as used in OpenEye's ROCS, have been proven to be quite effective for VS.[116–119]

The advantage of the excluded volume approach is that features can be used to only partially limit the space around the ligands, which describes the actual situation in a binding site in a more accurate way. For ligand-based models, the placement of excluded volume spheres can be induced by the inclusion of compounds that are inactive despite their ability to map all the required features, and for which a good reason exists to assume that this lack of affinity is caused by steric constraints (another possible reason for this could be negative interactions, such as repulsing charges of the same type close to each other in both ligand and binding site). Such methods for the automated placement of excluded volume spheres are, *e.g.*, included in the HypoRefine and HipHopRefine modules of Catalyst and have successfully been applied for the identification of P450 19 (aromatase) and PTP1B inhibitors, respectively.[120,121] A similar method exists in Phase, where the user can also create an excluded volume shell around the aligned ligands, which consists of multiple excluded volumes placed at rectangular grid points within a given distance of the molecules.[40]

In structure-based pharmacophore modeling, excluded volume spheres can be placed at atoms forming the binding site, a feature that has also been included in LigandScout. For faster screening and less restrictive models, LigandScout alternatively allows for the placement of only a few excluded volume spheres at the lipophilic side-chain residues that are in contact with hydrophobic features in the ligand.[119,122] We have found that excluded volume coats of the binding site can drastically increase specificity, but this is usually also accompanied by a certain decrease of sensitivity, which supports the findings of Pandit *et al.*, who have applied a similar approach for HIV protease inhibitors in MOE.[123] Both included and excluded volume shapes have been used to create a pharmacophore model for the fitting of small molecule intercalators of double stranded DNA.[124]

3.4.4 Qualitative *vs.* Quantitative Pharmacophore Models

If a set of diverse molecules with measured affinities spanning multiple orders of magnitude is available for a given target, one can create pharmacophore models that can predict the binding affinity of the investigated compounds. The most widely-used method for this task is the Catalyst/DiscoveryStudio module HypoGen.[125] HypoGen tries to find models that are common among the active compounds of the training set but do not reflect the inactive ones. Pharmacophores that correlate best the three-dimensional arrangement of features in a given set of training compounds with the corresponding pharmacological activities (IC_{50} or K_i) are constructed and ranked. Certain guidelines for 3D QSAR model generation in Catalyst must be respected: to ensure the statistic relevance of the calculated model, the training set should contain at least 16 compounds together with their activity values derived from comparable binding assays, *i.e.*, from an equivalent analytical method, similar species, and similar tissue. Activities should spread equally over at least four orders of magnitude.[125] The most active compound available must be included, and each order of magnitude should be represented by at least three compounds. Each compound should provide new structural information. An uncertainty factor (default value 3) for each compound is defined, representing the ratio range of uncertainty in the activity value based on the expected statistical straggling of biological data collection. The generation of pharmacophore models (called hypotheses in Catalyst) with HypoGen consists of three steps: the hypotheses created in the initial phase, the constructive phase, consider all possible pharmacophore configurations of the most active compounds to imply pharmacophore demands. In the second phase, the subtractive phase, all possible pharmacophore configurations of the constructive phase are analyzed and only those models are kept that are not mapping the least active training set members as well. In the final optimization phase, small perturbations are applied to the remaining pharmacophore models, and scored based on geometric fit, activity, error estimation, and cost calculation. The hypothesis generation process stops when no better score of the hypothesis can be accomplished.[2] Usually, ten hypotheses are output. Reliable models are characterized by high cost difference (70 or higher), low error cost, low root-mean-square divergence values, and high correlation coefficients. Using the module CatScramble, the molecular spreadsheets of the training set is modified by arbitrary scrambling of the affinity data for all compounds. These randomized spreadsheets should yield hypotheses without statistical significance; otherwise, the original model is also random. To achieve, *e.g.*, a statistical significance level of 95%, 19 random spreadsheets have to be generated from each training set. Each feature of the resulting models occupies a certain weight that is proportional to its relative contribution to biological activity. HypoGen therefore constructs pharmacophore models that correlate best with measured activities and that consist of as few features as possible. The final hypotheses can then be evaluated by predicting the affinity of compounds from a test set. Various applications of this method have been reported, as, *e.g.*, our paper on the discovery of high-affinity ligands of sigma-1 receptor, ERG2, and emopamil binding protein (EBP).[61] The HypoRefine module can be used to generate quantitative models with excluded volume spheres by adding compounds that are inactive for sterical reasons to the training set.[93,120]

An explicit specification of inactive compounds as well as the maximum number of excluded volume spheres, and softening of excluded volume spheres during fitting can be adjusted by the user.

Notably, the obtained quantitative models are somewhat different from qualitative ones, as produced by the Catalyst module HipHop:[27] while the qualitative model seeks to find common features present in all models of the training set, the quantitative method looks primarily for those features that can explain the high affinity of the most active compounds, and which are not present in the lesser active ones. Often publications report the use of structurally closely related compounds for the generation of quantitative pharmacophore models. This may be largely because during optimization of initial lead structures, where this method is preferentially applied, medicinal chemists usually work on a distinct chemical class of compounds. A qualitative model based on highly similar compounds identifies all those features that were not changed (for whatever reason) as being equally important for the binding, and therefore often identifies also structurally quite similar compounds in a database screen, as we showed in a recent study for 11β-hydroxysteroid dehydrogenase (11β-HSD) inhibitors.[126] A quantitative model based on highly similar compounds, however, may be well suited for investigating further derivatives of the investigated compounds, but may miss some features that are important for the general affinity. Two recent examples for such models are given for ligands of the kappa opioid receptor[77] and of the SARS-CoV 3C-like protease:[127] in these two cases, only 2 out of 5 and 1 out of 3 features (the latter model is enhanced by two excluded volume spheres), respectively, are placed at the common scaffold – obviously as anchor points for the alignment of those features that are important to explain the structure-affinity relationship. The range of applicability of quantitative pharmacophore models that were created from compounds showing the same scaffolds and have activities spanning less than three orders of magnitude remains questionable to us.[128,129] For pharmacophore models aiming to identify novel scaffolds by VS, we recommend to use a high ligand diversity within the training set.

Even with quantitative models that have been generated with the upmost care to the guidelines given above, affinity prediction can often be off by two or more orders of magnitude for certain compounds. It has thus become a widely-used method to assess the quality of a model in a semi-quantitative way,[130] which may meet the requirements for many applications: three or more activity classes are defined at certain cutoff-values for K_i or IC_{50}, such as high activity ($+++$), medium activity ($++$), low activity ($+$) and inactive ($-$). The models are then evaluated for their ability to place a compound into the correct activity class.[77,114,127,131–133] Vadivelan *et al.*[93] evaluated their HypoRefine model for CDK2-cyclin A inhibitors on a database of 302 molecules. Of these compounds, 213 were classified as highly active ($IC_{50} < 1\,\mu M$), 54 as moderately active ($IC_{50} = 1–10\,\mu M$), and 35 as having low activity ($IC_{50} > 10\,\mu M$). For this classification, 16 false positives and 12 false negatives were found. While it might have been advantageous to shift the cutoff values to get a more balanced distribution of the activity classes for the test set compounds, whose IC_{50} values are well distributed across almost six orders of magnitude, the calculated r^2 value of 0.912 gives another good measure for the overall quality of the model. In another recent paper,[134] the same group evaluated their HypoGen models for histone deacetylase inhibitors based on three different chemotypes. In this case, 378 molecules were again divided into three sets of molecules: 109 highly active ($IC_{50} < 20\,nM$), 108 moderately active ($IC_{50} = 20–200\,nM$), and 161 with low activity ($IC_{50} > 200\,nM$). For these, 36 false positives and 12 false negatives were found, with a correlation factor of 0.897.

While HypoGen calculates the affinity based on Fit values with different weights for the features, the 3D QSAR method included in Phase uses a method similar to classical 3D QSAR methods: a rectangular grid is placed around the aligned molecules, and present pharmacophore features are converted into volume bits that are treated as a pool of independent variables for QSAR model development by applying partial least squares (PLS) regression. Cubes of favorable or unfavorable interaction are highlighted upon alignment of a test molecule.[40] This method was extensively evaluated against HypoGen for eight different targets, and while the grid technique of Phase showed equal or better results than the HypoGen method, the overall quality of test set prediction remained disappointing for both programs, with what the authors classified as "good" models being obtained for only four out of eight data sets by Phase and two out of eight by Catalyst.[42] Comparison with a recent paper[135] led the authors to the assumption that in terms of R^2 predictivity alone, fingerprint-descriptor-based QSAR methods might provide superior results.

3.4.5   Validation of Models for Virtual Screening

A useful pharmacophore model must be able to correctly classify – and in the case of quantitative models, correctly predict the affinity of – a so-called test or validation set, which consists of active compounds that were not used during the generation of the model. Often the ability of the model to retrieve known actives from a larger drug-like database is also assayed. It has been confirmed by several studies that the characteristics of the known inactives or decoys chosen for VS assessments have a significant impact on the enrichment of VS approaches.[136–138]

Both known inactives and decoys are required to meet some essential pre-requisites to achieve meaningful results. The most important need for decoys is the comparability of their physicochemical properties to the actives set. Probably one of the best examples of the hidden impact of decoy characteristics on the enrichment of VS techniques is the direct dependence of docking scores on the molecular weight of the ligands.[139] Verdonk *et al.*[136] have demonstrated that docking campaigns conducted against smaller decoys than actives achieve significantly higher enrichment than with larger decoys. The authors provide evidence that it is not sufficient to just use a random library (*e.g.*, subsets of public databases) for performance assessments, but it is essential to build up a so-called focused library that reflects the physicochemical properties of the actives set. The largest public available database of decoys that considers comparable 1D properties is the Directory of Useful Decoys (DUD),[137] which is available from http://dud.docking.org/. The DUD is a collection of 36 decoys for each of the 2950 collected actives of 40 different targets (95 316 in total, after duplicate removal). The compounds, representing a subset of the ZINC database, are of similar physical properties (*e.g.*, molecular weight, calculated Log *P*).[140] Decoys should be topologically dissimilar with respect to the active compounds. Otherwise, it is likely that several decoys are actually actives, which would lead to a significant amount of false positive hits, an unwanted bias. The DUD also meets this requirement.

Different descriptors have been found to answer the different questions that may be asked to address the quality of the VS run: how has the method improved the percentage of retrieved hits? Does the model find all actives? Does it filter out the inactives? Does a high score mean that the compound is highly active? Or does it describe the probability that the compound is active in a qualitative way? Some of the most common metrics will be described in the following short overview, though a more exhaustive discussion is given elsewhere.[141,142]

The general aim of VS methods is to retrieve from a molecular database a fraction of true positives that is significantly larger than that of a random compound selection. If a VS method selects *n* molecules from a database with *N* entries, the selected hit list consists of active compounds (true positive compounds, *TP*) and decoys (false positive compounds, *FP*). Active molecules that are not retrieved by the VS method are defined false negatives (*FN*), whereas the unselected database decoys represent the true negatives (*TN*) (Figure 3.3).

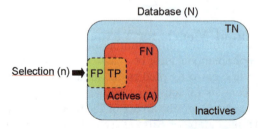

Figure 3.3   Selection of *n* molecules from a database containing *N* entries by a virtual screening protocol.

One of the most popular descriptors for evaluating VS methods is the yield of actives (Equation 3.1). This descriptor quantifies the probability that one of n selected compounds is active. In other words, it represents the hit rate if all compounds selected by the VS protocol are tested for activity.[62,143,144] However, it contains no information about the consistence of the database and the increase of the ratio of active molecules to inactives within a VS compound selection compared to a random compound selection:

$$Ya = \frac{TP}{n} \qquad (3.1)$$

This issue is addressed by the enrichment factor *EF* (Equation 3.2). This descriptor takes into account the improvement of the hit rate by a VS protocol compared to a random selection:[62,144,145]

$$EF = \frac{TP/n}{A/N} \qquad (3.2)$$

One disadvantage of the *EF* is its high dependency on the ratio of active molecules of the screened database.[146,147] This descriptor can be used to decide which VS method possesses the best performance if the same database of actives and decoys is utilized for evaluation. In contrast, comparisons of *EFs* derived from VS workflow evaluations using compound sets with different ratios of active molecules are less reliable.[147] Another disadvantage is that all actives contribute equally to the value. On that account, the *EF* does not distinguish highest-ranked active molecules from actives ranked at the end of a rank-ordered list.[147] Thus, the *EF* belongs to the classic enrichment descriptors that do not consider the "early recognition problem".

Many of the commonly used enrichment descriptors are based on two values. The first value is the sensitivity (*Se*, true positive rate, Equation 3.3), which describes the ratio of the number of active molecules found by the VS method to the number of all active database compounds:[62,146]

$$Se = \frac{N \text{ selected actives}}{N \text{ total actives}} = \frac{TP}{TP + FN} \qquad (3.3)$$

The second value is the specificity (*Sp*, false positive rate, Equation 3.4), which represents the ratio of the number of inactive compounds that were not selected by the VS protocol to the number of all inactive molecules included in the database:[146]

$$Sp = \frac{N \text{ discarded inactives}}{N \text{ total inactives}} = \frac{TN}{TN + FP} \qquad (3.4)$$

The "Goodness of hit list" (*GH*) was designed by Güner and Henry for evaluation of the discriminatory power of pharmacophore models (Equation 3.5).

With respect to the presence of active molecules that bind to another site of the target which cannot be represented by the pharmacophore model, the descriptor favors the *Ya* over *Se*:[148b]

$$GH = \left(\frac{3}{4}Ya + \frac{1}{4}Se\right) \cdot Sp \tag{3.5}$$

The Receiver Operating Characteristic (ROC) curve analysis describes *Se* for any possible change of *n* as a function of $(1-Sp)$.[146] The second term describes the percentage of false positives obtained compared to all inactives. If all molecules scored by a VS protocol with sufficient discriminatory power are ranked according to their score, starting with the best-scored molecule and ending with the molecule that got the lowest score, most of the actives will have a higher score than the decoys. Since some of the actives will be scored lower than decoys, an overlap between the distribution of active molecules and decoys will occur, which will lead to the prediction of false positives and false negatives. The selection of one score value as a threshold strongly influences the ratio of actives to decoys and therefore the validation of a VS method. The ROC curve method avoids the selection of a threshold by considering all *Se* and *Sp* pairs for each score threshold.[146] The ROC curve representing the ideal distribution, where no overlap between the scores of active molecules and decoys exists, proceeds from the origin along the ordinate to the upper-left corner until all the actives are retrieved and *Se* reaches the value of 1. There-after, only decoys can be found using the VS method. In contrast, the ROC curve for a set of actives and decoys with randomly distributed scores tends towards the $Se = 1-Sp$ line asymptotically with increasing number of actives and decoys. Finally, ROC curves between the random graph and the ideal curve are plotted for VS workflows that can score more active molecules higher than decoys and cause overlapping distributions, which represents the usual case in VS (Figure 3.4).[146] If the ROC curves do not cross each other, the curve that is located closer to the upper-left corner represents the VS workflow with the better performance in discriminating actives from decoys. On that account, ROC curves allow an intuitive visual comparison of the discriminatory power of different VS methods over the whole spectrum of *Se* and *Sp* pairs.[147]

The final validation of the usefulness of the model is of course its ability to find new active compounds by VS of a large database, which may contain, *e.g.*, compounds from commercial suppliers or compounds from an in-house collec-tion. Often the hit-lists are too large to screen all compounds reported by the model, and further filtering has to be applied, *e.g.*, by the rank order presented by the pharmacophore, the Lipinski rules,[11] ADME filters, diversity of the hits by 2D fingerprints, high quality docking, and also by visual inspection by a trained medicinal chemist. Pharmacophore models are frequently applied as an early filtering method, when other methods would demand too much effort to be applied to the whole database. The fact that pharmacophore screening methods often leave place for – or demand – certain decisions to be made by the scientists, must be taken into consideration when trying to quantitatively compare VS protocols against each other or against high-throughput screening methods.

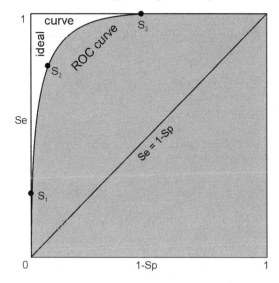

Figure 3.4 ROC curves for ideal and overlapping distributions of actives and decoys. The three ROC curve points S1, S2, S3 represent different cut-off values, depending on the overall screening purpose (few false positives *vs.* no false negatives). A random distribution causes a ROC curve that tends towards the Se = 1 − Sp line asymptotically with increasing number of actives and decoys.

3.5 Application of Pharmacophore Models in Virtual Screening

As shown in the previous section, there have been an impressive number of new approaches and tools appearing in the field of structure- and ligand-based pharmacophore modeling. All of them address slightly different issues and therefore a combination of several methods, as shown in different application examples, can significantly enhance the chances of success.

In our study on the discovery of novel ligands for the sigma-1 receptor and related proteins we were able to find compounds binding to the targets up to the subnanomolar concentration range by using a ligand-based VS approach, combining molecular shape descriptors with pharmacophore constraints for database mining.[61] In addition to several synthetic compounds, interesting natural products like tomatidine and solanidine were also identified and experimentally confirmed to be potent sigma-1 binders. Furthermore, we were able to show the similarities between the sigma-1 receptor and the homologous ERG2 sterol isomerase by comparing the similar pharmacophore models and pointing out the matching of the assumed high-energy intermediate of sterol isomerization to the ERG2 model.

Schuster *et al.* developed ligand-based pharmacophore models for 11β-hydroxysteroid dehydrogenase (11β-HSD) inhibitors.[126] By VS they were able to identify inhibitors of this enzyme active in the nanomolar range.

Some hits also revealed sufficient selectivity of type 1 inhibition *versus* the type 2 isoform, which is advantageous for the side-effect profile of these compounds. Comparison of the model for 11β-HSD1 inhibitors with the X-ray crystal structure (which was published shortly after model generation and VS) showed good correlation of the chemical features responsible for ligand binding. In another study, a combination of common feature-based qualitative and quantitative models was used as 3D pharmacophore search query to successfully detect novel endothelin-A antagonistic lead structures.[148a]

Steindl *et al.*[149] used structure-based pharmacophore modeling together with statistical analysis of molecular descriptors to identify new inhibitors of the human rhino virus coat protein. Barreca *et al.*[150] have described successful computational strategies in discovering novel non-nucleoside HIV-1 reverse transcriptase (RT) inhibitors. They generated a three-dimensional common feature pharmacophore model using the X-ray structure of RT/non-nucleoside inhibitor (NNRTI) complexes. Starting from the pharmacophore model and the structure of the lead compound TBZ, new NNRTIs were designed and synthesized, possessing the benzimidazol-2-one system as a scaffold. Docking experiments showed that these molecules docked in a position and orientation similar to that of known inhibitors. HIV-1 integrase (IN) is an essential enzyme for viral replication and the discovery of β-diketo acids was crucial in the validation of IN as a legitimate target in drug discovery against HIV infection. In their study, Dayam *et al.*[151] discovered a novel class of IN inhibitors using a 3D pharmacophore guided database search procedure. The recently published 3D structure of an isoform of angiotensin converting enzyme (ACE2) led Rella *et al.*[60] to perform a structure-based 3D pharmacophore search and to identify and experimentally confirm several novel scaffolds that will serve as interesting starting points for the development of selective ACE2 inhibitors.

Schlegel *et al.*[152] compared the performance of three different Catalyst pharmacophore models, each based on the assumed bio-active conformation of a single ligand, against GOLD docking into a homology model for the human histamine H_3 receptor. While the docking method obtained good results in scoring known active compounds it was outperformed by the pharmacophore models, leading to the suggestion of a combined approach for VS. Another homology model-derived pharmacophore was used by Edwards *et al.*[153] for the identification of formylpeptide receptor (FPR) antagonists. The initial database of about 480 000 compounds was reduced to 4324 compounds, which were physically screened with the HyperCyt flow cytometry platform in high-throughput, no-wash assays. 52 compounds (1.2% of the selected compounds) were confirmed as hits, which corresponds to an enrichment factor of 12 compared to a previous study screening a random collection (1 active out of 880).[154]

The pharmacophore-based screening approach has recently been shown to be also very successful for the identification of bio-active natural products. Rollinger *et al.*[155,156] derived a structure-based pharmacophore model on the co-crystal structure of acetylcholine esterase (AChE) with its ligand galanthamine, and used it for an *in silico* screening of a multi-conformational database consisting of more than 110 000 natural products. From the obtained hit list, promising candidates were selected, namely scopoletin and its glucoside scopolin. Their AChE inhibitory effect was verified first from the crude extract of *Scopolia carniolica* roots as in the *in vivo* test. The i.c.v. application of both coumarins on rats resulted in a long-lasting, pronounced and – in the case of the glucoside – even in a two-fold higher increase of the neurotransmitter's concentration than the one caused by the positive control galanthamine. In another study, the objective was to determine the cyclooxygenase (COX) inhibitory activity of Morus root bark, applying two different methods for their discovery. Firstly, the computer-aided approach with VS filtering experiments was used to identify the compounds able to interact with the pharmacophore models for COX-1 and COX-2. Secondly, a bioassay-guided fractionation was conducted for the isolation of the COX-inhibiting constituents. This resulted in the isolation of nine compounds belonging to the chemical classes of sanggenons and moracins. In the enzyme assay, the isolates showed moderate to potent inhibitory activities on COX-1 and COX-2. While the five sanggenons were correctly predicted, the four moracins were not detected by the pharmacophore model because they could not map with all the set features.[157]

3.5.1 Pharmacophore Models as Part of a Multi-step Screening Approach

Recent years have seen increasing use of pharmacophore models as one step in a multistep screening approach. In such screening runs, large databases are successively reduced to smaller and smaller compound sets, until in the end only a handful of compounds remains for *in vitro* testing. Accordingly, it seems reasonable that the first steps consist of fast filters that quickly remove the clearly undesired compounds, and that the complexity and accuracy – and with that often the computational efforts needed to virtually screen the remaining molecules – of each step increases. Here we give a few examples of successful VS protocols where pharmacophore screening was one of multiple successive steps that led to the identification of novel lead-structures.

Tintori *et al.*[158] reported the identification of HIV-1 integrase inhibitors by first screening the Asinex Gold database of over 200 000 compounds with the electron–ion interaction potential (EIIP) method,[159] which describes the long-range interaction of biological molecules, then with filters for Lipinki's rule of five[11] and rotatable bonds <10, followed by a ligand-based pharmacophore model. Finally, the remaining 15 000 compounds were docked and scored, leading to the selection of 12 compounds, one of which displayed significant inhibitory potency.

Lu *et al.*[160] screened a prefiltered set of 11 000 compounds from the NCI database with a pharmacophore model based on the crystal structure of murine double minute 2 (MDM2) oncoprotein in complex with the p53 tumor suppressor and several non-peptide small inhibitors. The remaining compounds were docked, scored, and 67 compounds were selected for testing, ten of which were active in a competitive binding assay. Similarly, Yu *et al.*[92] used a Catalyst HipHop model combined with the molecular shape of an experimental X-ray

structure to search the Maybridge database for inhibitors of vascular endothelial growth factor receptor tyrosine kinase. Eleven of the obtained 39 hits
were rejected by the Lipinski filter, and docking into the binding site finally
identified one hit with both high docking score and high Catalyst Fit value,
which could be experimentally shown to inhibit the target. Charlier *et al.*[161]
were able to discover new human 5-lipoxygenase inhibitors acting at concentrations in the nanomolar range by a combined ligand- and target-based
approach. Additionally, by comparing the results from pharmacophore modeling and docking of ligands into the binding site, they obtained structural
insights into the mode of action of such compounds.

Barreca *et al.*[162] screened for non-nucleoside reverse transcriptase inhibitors of
HIV-1 by searching the Derwent World Drug Index (WDI) and the Chemicals
Available for Purchase (CAP) databases with a Catalyst model generated by
LigandScout. A Fit value cutoff of 3.0 and Lipinski filters, followed by docking,
led to the selection of three compounds, two of which were commercially
available. A search for available close analogues of all three compounds finally
led to the purchase of six compounds, of which five were shown to be active.

Evers and Klabunde[163] applied a hierarchical screening approach to search
the Aventis in-house database for antagonists of the alpha1 adrenergic receptor. After filtering for compounds with a maximum number of nine rotatable
bonds and a molecular weight below 600, compounds were screened with two
pharmacophore models. The retrieved hits were docked into a homology model
of the alpha1 receptor, and clustered by their Unity fingerprint similarity, which
led to the selection of 80 diverse compounds, 37 of which revealed K_i values
better than $10\,\mu M$, with the most active compound displaying $1.4\,nM$ affinity.

Desai *et al.*[164] applied a slightly different approach in their attempt to identify
inhibitors for parasitic cysteine proteases falcipain-2 and falcipain-3: as there
were only a few nonpeptide inhibitors of falcipain-2 reported so far, with IC_{50}
values in the micromolar range, they generated homology models of the two
targets and screened a filtered set from the Available Chemical Directory
(ACD) database with the GOLD docking program. From the 100 selected
compounds, 22 were detected to be active against at least one of the investigated targets. Here, pharmacophore models were used in a follow-up investigation to rationalize the common pharmacophore features shared among the
identified hits coming from different structural classes.

3.5.2   Antitarget and ADME(T) Screening
Using Pharmacophores

In the search for an optimal lead compound, not only activity on the desired target
but also activity on other targets should be considered. Sometimes, multi-protein-
targeting is desired, *e.g.*, for multi-kinase inhibitors in the oncology field.[165]
However, also activity on side effect-related targets – so-called antitargets –
is to be investigated. Compounds with no or very low affinity to proteins related
to cardiovascular, cytotoxic, or metabolic effects are more likely to pass

subsequent phases of drug development. In this context, complementary to *in vitro* profiling for potential side effects, large-scale *in silico* activity profiling is now emerging. Compounds of interest are screened against hundreds or thousands of parameters to predict their pharmacokinetic and pharmacodynamic profile. Parameters such as adsorption, distribution (binding to plasma proteins, drug transporters), metabolism (*e.g.*, by the cytochrome P450 system), and toxicity (*e.g.*, affinity to the cardiac potassium channel hERG) can be derived from such a screening. The use of computational tools can guide the chemical optimization of novel lead series lacking antitarget-mediated side effects.

In recent publications, Klabunde *et al.*[166,167] describe their efforts to generate anti-target pharmacophore models to avoid GPCR-mediated side effects. Pharmacophores derived for the rational design and synthesis of alpha receptor ligands as described by Barbaro *et al.*[168] can in principle also be used in such a counter-screening strategy to eliminate compounds that might bind to these targets. Also for modeling of pharmacokinetic properties of drug candidates, the pharmacophore approach has been used successfully. Chang and Ekins summarize the results of numerous studies aimed at building pharmacophores for human ADME/Tox related proteins.[169]

A pharmacophore-based cytochrome P450 (P450) profiler was presented by Schuster *et al.*[170] This parallel screening-based system includes models for P450s 1A2, 2C9, 2C19, 2D6, and 3A4 substrates and inhibitors, respectively. These isoenzymes contribute mainly to xenobiotic metabolism and can possibly lead to drug–drug interactions resulting in severe side effects. However, metabolism-mediated side effects not only occur by direct interaction at the enzyme. P450 expression is mediated by several nuclear receptors, the most prominent of them being the pregnane X receptor (PXR). Pharmacophore models that can successfully identify potential ligands for PXR[171] are therefore also useful anti-target filters in the early drug discovery phase. Notably, though, inhibitors of certain cytochromes, *e.g.*, P450 19 (aromatase), might be of interest in the treatment of breast cancer. In this case, the enzyme may be seen as a target, and pharmacophore modeling can be of use for finding new potential inhibitors in an efficient way.[120] The application scenario of a pharmacophore model, whether for the identification of a few highly active compounds for lead finding or for reliable prevention of critical side effects will of course largely influence the optimization of the model towards either high specificity (no false-positives) or high sensitivity (no false-negatives).

3.5.3 Pharmacophore Models for Activity Profiling and Parallel Virtual Screening

Today, large-scale counter screening can be technically handled using parallel screening systems. However, these methods are quite expensive and so, especially for academic groups and small companies, parallel VS for novel drug candidates can be a cost-effective alternative to identify new lead structures with good selectivity profiles.

Steindl *et al.*[25,26] have recently published a Pipeline Pilot-based parallel screening system using structure-based pharmacophore models: the parallel screening of 100 antiviral compounds against 50 models belonging to five different targets led to a correct activity profile in 89% of the cases. In a second experiment,[172] they determined the selectivity of HIV protease inhibitor models against other protease inhibitors and inactive compounds. The results showed a clear trend toward most extensive retrieval of known actives followed by general protease inhibitors and lowest recovery of inactive compounds. Cleves and Jain[24] presented broad multi-target ligand-based profiling based on molecular similarity and imprinting methods.

In silico activity profiling can also be applied to so-called target fishing. In this context, possible targets (or anti-targets) for a compound are searched for. Markt *et al.*[122] performed a validation study of the target fishing approach using 357 compounds with known activity on peroxisome proliferator activated receptor (PPAR) isoforms. They screened all compounds against all models from the Inte:Ligand pharmacophore database. From 181 targets screened, PPAR targets were ranked first more often than any other target. This approach is also highly relevant especially in the field of natural products,[173,174] where constituents of proven herbal drugs can be subjected to virtual *in silico* profiling to identify the active compound. More generally, virtual natural product databases[175] can be screened using computer-aided methods. Another application example would be to profile the whole compound library in stock of a pharmaceutical company with the goal of finding the "hidden treasures": compounds with excellent activity, maybe even selectivity, for targets that have not been screened yet.

3.6 Pharmacophore Method Extensions and Comparisons to Other Virtual Screening Methods

In the following section, we discuss the differences between pharmacophore modeling approaches and some other VS methods, highlighting the advantages and short-comings of the respective methods.

3.6.1 Topological Fingerprints

Topological fingerprints consider the connection table of a target molecule and ignore the respective atom coordinates. The rapid calculation and efficiency of such fingerprint methods implicate their wide acceptance in industry.[117] An overview of similarity-based 2D fingerprint methods and their performance during VS is provided in ref. 176.

Structurally similar molecules will often bind to the same group of proteins. While this hypothesis may be violated in specific cases – a small change in chemical structure can dramatically change binding affinity – chemical similarity is often a good guide to the biological action of an organic molecule.[177] Keiser *et al.* have recently presented a method to relate receptors to each other quantitatively, based solely on the chemical similarity, as described by Daylight fingerprints, between their respective ligands.[178] In their method, called the Similarity Ensemble Approach (SEA), two sets of ligands are often judged similar even though no single identical ligand is shared between them.

Pharmacophore fingerprints proved to be superior to chemical fingerprints when used by von Korff and Steger[179] as descriptors for the creation of self organizing-maps (SOMs) for distinguishing small molecules that bind to GPCR subtypes from those that bind to other proteins. McGaughey and co-workers[117] found in a comparative evaluation of topological, shape, and docking methods for VS on 11 targets (carbonic anhydrase, cyclin-dependent kinase 2, cyclooxygenase-2, dihydrofolate reductase, estrogen receptor, HIV-1 protease, HIV reverse transcriptase, neuraminidase, protein tyrosine phosphatase 1B, thrombin and thymidylate synthetase) that simple methods based on topological descriptors keep up surprisingly well with 3D VS methods, as they can retrieve the largest number of actives with the least amount of computational power. However, the work also demonstrates that hits obtained from topological screening methods are less diverse than hits retrieved from 3D VS. Since the diversity of lead structure candidates is of extraordinary importance for VS, the higher computational demands of more sophisticated approaches may be justified.

Recently, Nettles *et al.*[180] have reported the development of the Novartis in-house software FEPOPS (FEature POint PharmacophoreS) which uses a "fuzzy" molecular representation: a compound is reduced to (usually four) feature points, onto which the pharmacophoric properties are encoded.[180,181] The authors analyzed subsets of 47 505, 2351 and 109 457 annotated ligands from the WOMBAT (World of Molecular Bioactivity) database,[182] respectively, by both 2D fingerprint methods, using ECFP-6 and MDL fingerprints, and FEPOPS 3D descriptors. While the 2D methods were better in the overall ability to identify a neighbor molecule that shared the same primary protein target as the query structure, the 3D method proved to be superior in capturing compounds of the same activity in those cases where the query structures showed low structural similarity to any other compound in the database (so-called singletons).

CATS[48] (chemically advanced template search) specifically targets the problem of scaffold hopping. It is based on the idea of generating an exhaustive molecular 2D fingerprint based on topological pharmacophore models for pair-wise comparison of molecules. CATS3D[81] is the three-dimensional, computationally more demanding counterpart of CATS. It is based on the correlation vector representation of a 3D conformation – in contrast to the topological representation of CATS – and has been successfully applied to identifying inhibitors of metabotropic glutamate receptors (mGluR) 1 and 5, using both supervised and unsupervised neural networks.[183] As ligand binding to the protein is a 3D problem, the exploitation of this additional information in CATS3D is supposed to increase predictive accuracy. SURFCATS[81] is another extension of CATS3D, considering molecular surfaces.

One of the major strengths of pharmacophore descriptors and searches is that they are well suited for scaffold hopping, which is finding compounds with similar bioactivity to a reference ligand but with different chemotype.[48] A different chemical scaffold can yield a favorable side effect profile compared to the reference ligand. Additionally, it enables patenting of compounds for already established mechanisms of action. Atom-based topological descriptors, on the other hand, are frequently used to assay the structural diversity of compounds used for pharmacophore model generation, and to evaluate the obtained hits from a pharmacophore screening run. However, recent methods like group fusion methods,[184,185] which combine several different reference structures for 2D similarity searching, or clique detection applied to reduced graphs[186] have shown their ability to overcome the short-comings of conventional 2D chemical fingerprints and 2D graphs with respect to scaffold hopping.

3.6.2   Shape-based Virtual Screening

As molecular shape plays a central role during ligand binding, this property is used in several approaches as a metric for molecular similarity (see Haigh *et al.*[187]). Moreover, several docking algorithms geared towards high-throughput screening use shape-guided procedures for rapid ligand placement at the binding site (*e.g.*, Fred[188,189] and LigandFit[190]). ROCS[191–193] is currently the most commonly used shape-based, ligand-centric VS platform available. It uses a smooth Gaussian function to define molecular volumes of small organic molecules. In our recent comparative study on structure-based pharmacophores and ROCS we found comparable performance of both approaches.[119] Sykes *et al.*[194] used ROCS for the prediction of nonspecific binding of drugs to hepatic microsomes. Thereby, they found that the color force field (based on a series of SMARTS patterns) considerably enhances the performance of ROCS.

3.6.3   Docking Methods

Today, protein–ligand docking is the most prominent approach for structure-based VS. The docking process is divided into two major steps: first, the correct placement of the ligand at the protein binding-site and, second, the estimation of the ligand affinity by a scoring function. In contrast to rapid VS methods like pharmacophore modeling, the performance of docking methods is always a trade-off between computational demands and accuracy. This is reflected by the plentitude of very different docking approaches available that aim at different fields of application: incremental construction approaches (*e.g.*, FlexX[195]), shape-based algorithms (*e.g.*, DOCK[196,197]), genetic algorithms (*e.g.*, GOLD[198]), systematic search (*e.g.*, Glide[199,200]), Monte Carlo simulations (*e.g.*, LigandFit[190]), and surface-based molecular similarity methods (*e.g.*, Surflex[201]). Most exhaustive algorithms focus on the accurate prediction of a binding pose; more efficient algorithms on the docking of small ligand databases within reasonable time, and rapid algorithms on the virtual high-throughput screening of millions of compounds. Warren *et al.*[202] have investigated the performance of ten docking programs and 37 scoring functions on eight different targets. They found that docking programs are in general able to generate ligand poses that are similar to the experimentally determined ligand pose bound to the protein. The performance is highly dependent on the target. Moreover, they found no statistically significant correlation between docking scores and ligand affinity. As we have shown in the previous section, pharmacophore screening and docking methods are frequently combined as two different approaches towards the given problem. Often the pharmacophore model is used as a quick first filter. Similarly to docking, the problem of poor correlation between scores and affinity can be seen for pharmacophore models, even for quantitative models.[42] However, similarly to the scores given by a pharmacophore model, docking scores can sometimes be useful to describe the general likeliness of affinity of a compound, rather than trying to predict the affinity of the compound itself, as can be seen by the shape of retrieved ROC or enrichment curves.[146,152,202,203] While pharmacophore models are thus able to provide good enrichment of candidate molecules from the screening of large database with less computational efforts, docking methods may sometimes still be better fit for analyzing smaller sets of compounds, especially in retrospective analysis. Pharmacophore models are sometimes too coarse to pick up subtle differences induced by small structural variations in the ligands. Furthermore, docking studies try to include all the possible interactions – both positive and negative – at the binding site at the same time, thus allowing the detection of unexpected binding modes as well as unfavorable interactions. One major advantage of pharmacophore models is that they provide the user with a large number of options for model refinement: models can be trained with the help of a training set and optimized through automated or manual refinement of feature weights, variation of feature definitions, addition or deletion of features, the placement of excluded volume spheres, and many more. Changing the parameters for the docking algorithms and docking scores, on the other hand, is a much more demanding task. The recent inclusion of protein flexibility, as, *e.g.*, in Schrödinger's Induced Fit Methodology,[204,205] provides an interesting extension of the docking approach, especially for pose-prediction and explanation of received binding-affinities. It remains to be seen, though, whether the increased computational demands and the provision of even more possibilities for compounds to fit the binding site will make this method useful for the successful VS of large databases.

Finally, while pharmacophore and docking methods are still two distinct methods for VS, the distance between them appears to be growing smaller: structure-based pharmacophore methods are trying to include more and more information about the binding site (Sections 3.4.2 and 3.4.3), while some docking programs have successfully incorporated pharmacophore constraints, which we discuss in the next section.

3.6.4 Pharmacophore Constraints Used in Docking

One of the most important applications for docking methods is the correct prediction of a possible binding pose, an ability that has been exhaustively studied for various docking programs, and ways to optimize the assessment of binding pose quality have been discussed.[136,206–208]

Sometimes, the available data on a given binding pocket and a corresponding set of known ligands may provide a very good reasoning for the assumption as to where exactly a certain part of the ligand is located inside the binding pocket. In such a case, pharmacophore constraints describing the well-known or expected interactions may speed up the docking process while at the same time minimizing the number of wrong pose predictions. Simple pharmacophore-like sterical constraints can be applied, *e.g.*, in the docking packages GOLD[198] and Glide.[199,200] Verdonk *et al.*[136] showed that pharmacophore restraints and modified scoring functions that include a pharmacophore mapping term improved the enrichment during screening of libraries against CDK2 and neuraminidase. The pharmacophore approach has been applied to the FlexX docking method, resulting in the development of the FlexX Pharm program.[209,210] Other examples include GEMDOCK,[211] LibDock,[212] and SP-Dock.[213] While we have not found a reported application for this procedure yet, it should be noted that, *e.g.*, DiscoveryStudio[29] allows the quick filtering of obtained docking poses with a structure-based pharmacophore model by taking the absolute coordinates of each docked pose and checking whether the conformation matches the pharmacophore model.

3.7 Further Reading

Pharmacophore modeling methods have been in use for about 20 years now and have found broad application in modern drug discovery. One book chapter can thus only give a small overview of the research carried out so far. Two books have been published that deal exclusively with pharmacophore modeling.[62,143] Two recent book chapters by Martin discuss methods and applications of pharmacophore models.[214,215] Van Drie has given a critical review of pharmacophore methods, including a historical overview.[216] The performance of Catalyst, DISCO and GASP has been investigated by Patel *et al.*[217] Recently, Evans *et al.*[42] have published a comparative study on Phase and Catalyst, and we have investigated the screening performance of Catalyst using LigandScout pharmacophore queries in direct comparison to ROCS screening.[119] Applications of pharmacophore models for drug-transporters[218] and ion channels[219] have been reviewed. A two-part review discussing different methods and applications for VS was published recently by Ekins, Mestres, and Testa.[7,8]

3.8 Summary and Conclusion

The pharmacophore concept is a successful and well-known approach – both ligand- and structure-based – for drug design as well as for VS. Several methods for describing pharmacophores have been established, showing significant differences and capabilities in the way in which they describe chemical features as building blocks for pharmacophores. It is important that the chemical feature representation used reflects the interactions that are relevant for the target being represented, and some chemical feature representations are more universal than others. The computational part of pharmacophore modeling (different alignment techniques used for pharmacophore elucidation and VS) has significantly improved with the availability of new software packages. The use of new algorithms has led to performance optimizations over the past few years, leading to modern pattern recognition approaches that can superposition pharmacophores and molecules in a fraction of the time needed by earlier approaches.

The number of papers published within the last years together with the increasing interest of researchers into the re-emerging field of pharmacophore modeling in drug discovery is obviously a consequence of other approaches like structure-based docking not fully meeting the expectations people had of them. While considerable progress has been made in docking with respect to speed and accuracy of binding pose prediction, the biggest issue still remains the correct prediction of the free binding energy. The scoring functions used for this task may work well in each special application cases for which they were tuned to. In other target families, they will probably fail. Since docking and scoring is computationally expensive and since ranking of hits is still not possible with the desired accuracy, the simple concept of 3D pharmacophores has become of interest again. The pharmacophore concept is used while always keeping in mind the need to understand, explain and predict molecular interactions with the targets as well as structure–activity relationships. Its practical applicability for medicinal chemists makes it an excellent communication tool between modelers and synthetic chemists. Pharmacophores are of unambiguous simplicity and usefulness for searching structural databases.[143,220] In our opinion, due to their computational efficiency in database mining, their importance will largely increase when parallel screening software based on pharmacophores will become available together with publicly or commercially available collections of pharmacophore models covering important target as well as anti-targets. This, in fact, will allow for rapid bio-activity profiling of compounds even before they are synthesized and also will drastically enhance the library design process. However, there is still a lot of room for research aimed at improving methods or the design of novel algorithms. Some examples that are not yet (or at least not broadly) included are the possibility to include multi-point features, the automated detection and inclusion of tautomeric forms during both model generation and screening, description of hydrophobic regions by non-spherical features, the use of projection points instead of fixed locations for features other than hydrogen-bonds (thus allowing for a more protein-centered pharmacophore model), inclusion of negative (forbidden) features, and a broad description of different ligand–metal interactions.

Both in the pharmaceutical industry and software companies specialized in computer-aided molecular design the demand for experts in the field interfacing medicinal chemistry and computer sciences will increase within the next decade. There is no doubt that we will experience an exciting period of substantial progress in pharmacophore-based VS technologies in the near future.

References

1. H. Kubinyi, *EFMC-Yearbook 2003*, 2003, 14–28.
2. Y. Kurogi and O. F. Güner, *Curr. Med. Chem.*, 2001, **8**, 1035–1055.
3. T. Langer and E. M. Krovat, *Curr. Opin. Drug. Discov. Dev.*, 2003, **6**, 370–376.
4. O. Güner, O. Clement and Y. Kurogi, *Curr. Med. Chem.*, 2004, **11**, 2991–3005.
5. T. Langer and G. Wolber, *Drug Discov. Today*, 2004, **1**, 203–207.
6. O. F. Güner, *IDrugs*, 2005, **8**, 567–572.
7. S. Ekins, J. Mestres and B. Testa, *Br. J. Pharmacol.*, 2007, **152**, 9–20.
8. S. Ekins, J. Mestres and B. Testa, *Br. J. Pharmacol.*, 2007, **152**, 21–37.
9. W. P. Walters, M. T. Stahl and M. A. Murcko, *Drug Discov. Today*, 1998, **3**, 160–178.
10. D. P. Hristozov, T. I. Oprea and J. Gasteiger, *J. Comput. Aided Mol. Des.*, 2007, **21**, 617–640.
11. C. A. Lipinski, F. Lombardo, B. W. Dominy and P. J. Feeney, *Adv. Drug Delivery Rev.*, 2001, **46**, 3–26.
12. J. Huuskonen, J. Rantanen and D. Livingstone, *Eur. J. Med. Chem.*, 2000, **35**, 1081–1088.
13. J. Zuegge, G. Schneider, P. Coassolo and T. Lave, *Clin. Pharmacokinet.*, 2001, **40**, 553–563.
14. R. Abagyan and M. Totrov, *Curr. Opin. Chem. Biol.*, 2001, **5**, 375–382.
15. D. J. Diller and K. M. Merz Jr., *Proteins: Struct., Funct., Genet.*, 2001, **43**, 113–124.
16. G. Schneider and H.-J. Bohm, *Drug Discov. Today*, 2002, **7**, 64–70.
17. M. Krier, J. X. De Araujo-Junior, M. Schmitt, J. Duranton, H. Justiano-Basaran, C. Lugnier, J.-J. Bourguignon and D. Rognan, *J. Med. Chem.*, 2005, **48**, 3816–3822.
18. M. Stahl and M. Rarey, *J. Med. Chem.*, 2001, **44**, 1035–1042.
19. E. M. Krovat, T. Steindl and T. Langer, *Curr. Comput. Aided Drug Des.*, 2005, **1**, 93–102.
20. D. Wilton, P. Willett, K. Lawson and G. Mullier, *J. Chem. Inf. Comput. Sci.*, 2003, **43**, 469–474.
21. P. Willett, *Drug Discov. Today*, 2006, **11**, 1046–1053.
22. B. Chen, R. F. Harrison, G. Papadatos, P. Willett, D. J. Wood, X. Q. Lewell, P. Greenidge and N. Stiefl, *J. Comput. Aided Mol. Des.*, 2007, **21**, 53–62.
23. H. Li, C. W. Yap, C. Y. Ung, Y. Xue, Z. R. Li, L. Y. Han, H. H. Lin and Y. Z. Chen, *J. Pharm. Sci.*, 2007, **96**, 2838–2860.
24. A. E. Cleves and A. N. Jain, *J. Med. Chem.*, 2006, **49**, 2921–2938.
25. T. M. Steindl, D. Schuster, C. Laggner and T. Langer, *J. Chem. Inf. Model.*, 2006, **46**, 2146–2157.
26. T. M. Steindl, D. Schuster, G. Wolber, C. Laggner and T. Langer, *J. Comput. Aided Mol. Des.*, 2006, **20**, 703–715.

27. D. Barnum, J. Greene, A. Smellie and P. Sprague, *J. Chem. Inf. Comput. Sci.*, 1996, **36**, 563–571.
28. E. A. Hecker, C. Duraiswami, T. A. Andrea and D. J. Diller, *J. Chem. Inf. Comput Sci.*, 2002, **42**, 1204–1211.
29. *DiscoveryStudio*, available from Accelrys Inc., San Diego, CA, USA, www.accelrys.com.
30. Y. C. Martin, M. G. Bures, E. A. Danaher, J. DeLazzer, I. Lico and P. A. Pavlik, *J. Comput. Aided Mol. Des.*, 1993, **7**, 83–102.
31. N. J. Richmond, C. A. Abrams, P. R. Wolohan, E. Abrahamian, P. Willett and R. D. Clark, *J. Comput. Aided Mol. Des.*, 2006, **20**, 567–587.
32. S. J. Cottrell, V. J. Gillet, R. Taylor and D. J. Wilton, *J. Comput. Aided Mol. Des.*, 2004, **18**, 665–682.
33. G. Jones, P. Willett and R. C. Glen, *J. Comput. Aided Mol. Des.*, 1995, **9**, 532–549.
34. A. W. R. Payne and R. C. Glen, *J. Mol. Graph.*, 1993, **11**, 74–91, 121–123.
35. N. J. Richmond, P. Willett and R. D. Clark, *J. Mol. Graph. Model.*, 2004, **23**, 199–209.
36. Y. C. Martin, *J. Med. Chem.*, 1992, **35**, 2145–2154.
37. D. P. Marriott, I. G. Dougall, P. Meghani, Y.-J. Liu and D. R. Flower, *J. Med. Chem.*, 1999, **42**, 3210–3216.
38. T. N. Lokhande, C. L. Viswanathan, A. Joshi and A. Juvekar, *Bioorg. Med. Chem.*, 2006, **14**, 6022–6026.
39. S. L. Dixon, A. M. Smondyrev and S. N. Rao, *Chem. Biol. Drug. Des.*, 2006, **67**, 370–372.
40. S. L. Dixon, A. M. Smondyrev, E. H. Knoll, S. N. Rao, D. E. Shaw and R. A. Friesner, *J. Comput. Aided Mol. Des.*, 2006, **20**, 647–671.
41. S. M. Dixon, P. Li, R. Liu, H. Wolosker, K. S. Lam, M. J. Kurth and M. D. Toney, *J. Med. Chem.*, 2006, **49**, 2388–2397.
42. D. A. Evans, T. N. Doman, D. A. Thorner and M. J. Bodkin, *J. Chem. Inf. Model.*, 2007, **47**, 1248–1257.
43. C.-G. Wermuth, C. R. Ganellin, P. Lindberg and L. A. Mitscher, *Annu. Rep. Med. Chem.*, 1998, **33**, 385–395.
44. C. G. Wermuth, in *Methods and Principles in Medicinal Chemistry, Vol. 32: Pharmacophores and Pharmacophore Searches*, T. Langer and R. D. Hoffmann eds., Wiley-VCH, Weinheim, 2006, pp. 3–13.
45. E. C. Dodds and W. Lawson, *Proc. R. Soc. London, Ser. B*, 1938, **125**, 122–132.
46. F. W. Schueler, *Science*, 1946, **103**, 221–223.
47. C. G. Wermuth and T. Langer, in *3D QSAR in Drug Design. Theory, Methods and Applications*, ed. H. Kubinyi, ESCOM, Leiden, 1993, pp. 117–136.
48. G. Schneider, W. Neidhart, T. Giller and G. Schmid, *Angew. Chem. Int. Ed. Engl.*, 1999, **38**, 2894–2896.

49. G. R. Marshall, C. D. Barry, H. E. Bosshard, R. A. Dammkoehler and D. A. Dunn, in *Computer-Assisted Drug Design*, American Chemical Society, Washington DC, 1979, vol. **112**, pp. 205–225.
50. J. Greene, S. Kahn, H. Savoj, P. Sprague and S. Teig, *J. Chem. Inf. Comput. Sci.*, 1994, **34**, 1297–1308.
51. E. M. Krovat and T. Langer, *J. Med. Chem.*, 2003, **46**, 716–726.
52. G. Wolber and T. Langer, *J. Chem. Inf. Comput. Sci.*, 2005, **45**, 160–169.
53. *Catalyst*, available from Accelrys Inc., San Diego, CA, USA, www. accelrys.com.
54. *Phase*, available from Schrödinger, LLC, New York, NY, USA, www. schrodinger.com.
55. *MOE*, available from Chemical Computing Group, Montreal, QC, CA, www.chemcomp.com.
56. G. Wolber and R. Kosara, in *Methods and Principles in Medicinal Chemistry, Vol. 32: Pharmacophores and Pharmacophore Searches*, T. Langer and R. D. Hoffmann eds., Wiley-VCH, Weinheim, 2006, pp. 131–150.
57. *LigandScout*, available from Inte:Ligand GmbH, Maria Enzersdorf, Austria, www.inteligand.com.
58. SMARTS - A Language for Describing Molecular Patterns, www. daylight.com/dayhtml/doc/theory/theory.smarts.html.
59. T. Steindl, C. Laggner and T. Langer, *J. Chem. Inf. Model.*, 2005, **45**, 716–724.
60. M. Rella, C. A. Rushworth, J. L. Guy, A. J. Turner, T. Langer and R. M. Jackson, *J. Chem. Inf. Model.*, 2006, **46**, 708–716.
61. C. Laggner, C. Schieferer, B. Fiechtner, G. Poles, R. D. Hoffmann, H. Glossmann, T. Langer and F. F. Moebius, *J. Med. Chem.*, 2005, **48**, 4754–4764.
62. T. Langer and R. D. Hoffmann, *Methods and Principles in Medicinal Chemistry, Vol 32: Pharmacophores and Pharmacophore Searches*, Wiley-VCH, Weinheim, 2006.
63. P. J. Goodford, *J. Med. Chem.*, 1985, **28**, 849–857.
64. O. Nicolotti, M. Pellegrini-Calace, A. Carrieri, C. Altomare, N. B. Centeno, F. Sanz and A. Carotti, *J. Comput. Aided Mol. Des.*, 2001, **15**, 859–872.
65. O. Dror, A. Shulman-Peleg, R. Nussinov and H. J. Wolfson, *Front. Med. Chem.*, 2006, **3**, 551–581.
66. Y. C. Martin, in *Pharmacophore Perception, Development and Use in Drug Design*, ed. O. F. Güner, International University Line, La Jolla, CA, 2000, pp. 49–68.
67. G. Jones and P. Willet, in *Pharmacophore Perception, Development and Use in Drug Design*, ed. O. F. Güner, International University Line, La Jolla, CA, 2000, pp. 85–106.
68. *GALAHAD*, available from Tripos, St. Louis, MO, USA, *www.tripos. com*.
69. C. Lemmen and T. Lengauer, *J. Comput. Aided Mol. Des.*, 2000, **14**, 215–232.

70. J. Kirchmair, C. Laggner, G. Wolber and T. Langer, *J. Chem. Inf. Model.*, 2005, **45**, 422–430.
71. J. M. Blaney, G. M. Crippen, A. Dearing and J. S. Dixon, in *QPCE Number 590, Quantum Chemistry Program Exchange*, Indiana University, Bloomington, IN, 1990.
72. C. Bron and J. Kerbosch, *Commun. ACM*, 1973, **16**, 575–577.
73. G. Wolber, A. A. Dornhofer and T. Langer, *J. Comput. Aided Mol. Des.*, 2006, **20**, 773–788.
74. A. Evers, G. Hessler, H. Matter and T. Klabunde, *J. Med. Chem.*, 2005, **48**, 5448–5465.
75. Q. Zhang and I. Muegge, *J. Med. Chem.*, 2006, **49**, 1536–1548.
76. S. Kortagere and W. J. Welsh, *J. Comput. Aided Mol. Des.*, 2006, **20**, 789–802.
77. N. Singh, G. Cheve, D. M. Ferguson and C. R. McCurdy, *J. Comput. Aided Mol. Des.*, 2006, **20**, 471–493.
78. D. D. Beusen and G. R. Marshall, in *Pharmacophore Perception, Development and Use in Drug Design*, ed. O. F. Güner, International University Line, La Jolla, CA, USA, 2000, pp. 21–46.
79. *Unity*, available from Tripos, St. Louis, MO, US, www.tripos.com.
80. J. Feng, A. Sanil and S. S. Young, *J. Chem. Inf. Model.*, 2006, **46**, 1352–1359.
81. S. Renner, U. Fechner and G. Schneider, in *Methods and Principles in Medicinal Chemistry, Vol 32: Pharmacophores and pharmacophore searches*, T. Langer and R. D. Hoffmann eds., Wiley-VCH, Weinheim, 2006, pp. 49–79.
82. M. Rarey and J. S. Dixon, *J. Comput. Aided Mol. Des.*, 1998, **12**, 471–490.
83. M. Rarey, S. Hindle, P. Maaß, G. Metz, C. Rummey and M. Zimmermann, in *Methods and Principles in Medicinal Chemistry, Vol. 32: Pharmacophores and Pharmacophore Searches*, T. Langer and R. D. Hoffmann (eds.), Wiley-VCH, Weinheim, 2006, pp. 81–116.
84. *FTrees*, available from BioSolveIT GmbH, Sankt Augustin, Germany, www.biosolveit.de.
85. L. Jia, J. Zou, S. S. So and H. Sun, *J. Chem. Inf. Model.*, 2007, **47**, 1545–1552.
86. H. S. Kim, Y. Kim, M. R. Doddareddy, S. H. Seo, H. Rhim, J. Tae, A. N. Pae, H. Choo and Y. S. Cho, *Bioorg. Med. Chem. Lett.*, 2007, **17**, 476–481.
87. M. L. Barreca, S. Ferro, A. Rao, L. De Luca, M. Zappala, A. M. Monforte, Z. Debyser, M. Witvrouw and A. Chimirri, *J. Med. Chem.*, 2005, **48**, 7084–7088.
88. S. Tasler, J. Kraus, S. Pegoraro, A. Aschenbrenner, E. Poggesi, R. Testa, G. Motta and A. Leonardi, *Bioorg. Med. Chem. Lett.*, 2005, **15**, 2876–2880.
89. S. Collina, G. Loddo, M. Urbano, L. Linati, A. Callegari, F. Ortuso, S. Alcaro, C. Laggner, T. Langer, O. Prezzavento, G. Ronsisvalle and O. Azzolina, *Bioorg. Med. Chem.*, 2007, **15**, 771–783.
90. G. Klebe, in *Drug Discovery Series*, CRC Press LLC, Boca Raton, 2005, vol. **1**, pp. 3–24.

91. H. M. Berman, J. Westbrook, Z. Feng, G. Gilliland, T. N. Bhat, H. Weissig, I. N. Shindyalov and P. E. Bourne, *Nucleic Acids Res.*, 2000, **28**, 235–242.
92. H. Yu, Z. Wang, L. Zhang, J. Zhang and Q. Huang, *Chem. Biol. Drug. Des.*, 2007, **69**, 204–211.
93. S. Vadivelan, B. N. Sinha, S. J. Irudayam and S. A. Jagarlapudi, *J. Chem. Inf. Model.*, 2007, **47**, 1526–1535.
94. D. S. Patel and P. V. Bharatam, *J. Comput. Aided Mol. Des.*, 2006, **20**, 55–66.
95. F. Ortuso, T. Langer and S. Alcaro, *Bioinformatics*, 2006, **22**, 1449–1455.
96. M. Baroni, G. Cruciani, S. Sciabola, F. Perruccio and J. S. Mason, *J. Chem. Inf. Model.*, 2007, **47**, 279–294.
97. J. S. Mason and D. L. Cheney, *Pac. Symp. Biocomput.*, 1999, 456–467.
98. J. Chen and L. Lai, *J. Chem. Inf. Model.*, 2006, **46**, 2684–2691.
99. X. W. Zhang, Y. L. Yap and R. M. Altmeyer, *Eur. J. Med. Chem.*, 2005, **40**, 57–62.
100. M. M. Ahlström, M. Ridderstrom, K. Luthman and I. Zamora, *J. Chem. Inf. Model.*, 2005, **45**, 1313–1323.
101. H. A. Carlson, K. M. Masukawa, K. Rubins, F. D. Bushman, W. L. Jorgensen, R. D. Lins, J. M. Briggs and J. A. McCammon, *J. Med. Chem.*, 2000, **43**, 2100–2114.
102. A. L. Bowman, Z. Nikolovska-Coleska, H. Zhong, S. Wang and H. A. Carlson, *J. Am. Chem. Soc.*, 2007, **129**, 12809–12814.
103. M. G. Lerner, A. L. Bowman and H. A. Carlson, *J. Chem. Inf. Model.*, 2007, **47**, 2358–2365.
104. A. L. Bowman, M. G. Lerner and H. A. Carlson, *J. Am. Chem. Soc.*, 2007, **129**, 3634–3640.
105. J. Deng, K. W. Lee, T. Sanchez, M. Cui, N. Neamati and J. M. Briggs, *J. Med. Chem.*, 2005, **48**, 1496–1505.
106. P. Karnachi and A. Kulkarni, in *Methods and Principles in Medicinal Chemistry, Vol. 32: Pharmacophores and Pharmacophore Searches*, T. Langer and R. D. Hoffmann eds., Wiley-VCH, Weinheim, 2006, pp. 193–206.
107. Z. Deng, C. Chuaqui and J. Singh, *J. Med. Chem.*, 2004, **47**, 337–344.
108. C. Chuaqui, Z. Deng and J. Singh, *J. Med. Chem.*, 2005, **48**, 121–133.
109. M. J. McGregor and S. M. Muskal, *J. Chem. Inf. Comput. Sci.*, 1999, **39**, 569–574.
110. M. J. McGregor and S. M. Muskal, *J. Chem. Inf. Comput. Sci.*, 2000, **40**, 117–125.
111. F. Deanda and E. L. Stewart, *J. Chem. Inf. Comput. Sci.*, 2004, **44**, 1803–1809.
112. M. J. McGregor, *J. Chem. Inf. Model.*, 2007, **47**, 2374–2382.
113. G. M. Spitzer, B. Wellenzohn, C. Laggner, T. Langer and K. R. Liedl, *J. Chem. Inf. Model.*, 2007, **47**, 1580–1589.
114. E. M. Krovat, K. H. Fruehwirth and T. Langer, *J. Chem. Inf. Comput. Sci.*, 2005, **45**, 146–159.

115. J. Singh, C. E. Chuaqui, P. A. Boriack-Sjodin, W. C. Lee, T. Pontz, M. J. Corbley, H. K. Cheung, R. M. Arduini, J. N. Mead, M. N. Newman, J. L. Papadatos, S. Bowes, S. Josiah and L. E. Ling, *Bioorg. Med. Chem. Lett.*, 2003, **13**, 4355–4359.

116. J. J. Sutherland, R. K. Nandigam, J. A. Erickson and M. Vieth, *J. Chem. Inf. Model.*, 2007, **47**, 2293–2302.

117. G. B. McGaughey, R. P. Sheridan, C. I. Bayly, J. C. Culberson, C. Kreatsoulas, S. Lindsley, V. Maiorov, J. F. Truchon and W. D. Cornell, *J. Chem. Inf. Model.*, 2007, **47**, 1504–1519.

118. P. C. Hawkins, A. G. Skillman and A. Nicholls, *J. Med. Chem.*, 2007, **50**, 74–82.

119. J. Kirchmair, S. Ristic, K. Eder, P. Markt, G. Wolber, C. Laggner and T. Langer, *J. Chem. Inf. Model.*, 2007, **47**, 2182–2196.

120. D. Schuster, C. Laggner, T. M. Steindl, A. Palusczak, R. W. Hartmann and T. Langer, *J. Chem. Inf. Model.*, 2006, **46**, 1301–1311.

121. M. O. Taha, Y. Bustanji, A. G. Al-Bakri, A. M. Yousef, W. A. Zalloum, I. M. Al-Masri and N. Atallah, *J. Mol. Graph. Model.*, 2007, **25**, 870–884.

122. P. Markt, D. Schuster, J. Kirchmair, C. Laggner and T. Langer, *J. Comput. Aided Mol. Des.*, 2007, **21**, 575–590.

123. D. Pandit, S. S. So and H. Sun, *J. Chem. Inf. Model.*, 2006, **46**, 1236–1244.

124. L. B. Hendry, V. B. Mahesh, E. D. Bransome Jr and D. E. Ewing, *Mutat. Res.*, 2007, **623**, 53–71.

125. H. Li, J. Sutter and R. Hoffmann, in *Pharmacophore Perception, Development, and Use in Drug Design*, ed. O. F. Güner, International University Line, La Jolla, CA, 2000, pp. 171–189.

126. D. Schuster, E. M. Maurer, C. Laggner, L. G. Nashev, T. Wilckens, T. Langer and A. Odermatt, *J. Med. Chem.*, 2006, **49**, 3454–3466.

127. K. C. Tsai, S. Y. Chen, P. H. Liang, I. L. Lu, N. Mahindroo, H. P. Hsieh, Y. S. Chao, L. Liu, D. Liu, W. Lien, T. H. Lin and S. Y. Wu, *J. Med. Chem.*, 2006, **49**, 3485–3495.

128. B. Gopalakrishnan, V. Aparna, J. Jeevan, M. Ravi and G. R. Desiraju, *J. Chem. Inf. Model.*, 2005, **45**, 1101–1108.

129. A. K. Bhattacharjee, *Lett. Drug Des. Discov.*, 2006, **3**, 219–235.

130. A. K. Debnath, *J. Med. Chem.*, 2002, **45**, 41–53.

131. M. Y. Li, K. C. Tsai and L. Xia, *Bioorg. Med. Chem. Lett.*, 2005, **15**, 657–664.

132. M. Chopra and A. K. Mishra, *J. Chem. Inf. Model.*, 2005, **45**, 1934–1942.

133. K. Bharatham, N. Bharatham and K. W. Lee, *Arch. Pharm. Res.*, 2007, **30**, 533–542.

134. S. Vadivelan, B. N. Sinha, G. Rambabu, K. Boppana and S. A. Jagarlapudi, *J. Mol. Graph. Model.*, 2008, **26**, 935–946.

135. P. Gedeck, B. Rohde and C. Bartels, *J. Chem. Inf. Model.*, 2006, **46**, 1924–1936.

136. M. L. Verdonk, V. Berdini, M. J. Hartshorn, W. T. M. Mooij, C. W. Murray, R. D. Taylor and P. Watson, *J. Chem. Inf. Comput. Sci.*, 2004, **44**, 793–806.

137. N. Huang, B. K. Shoichet and J. J. Irwin, *J. Med. Chem.*, 2006, **49**, 6789–6801.

138. Y. Pan, N. Huang, S. Cho and A. D. MacKerell Jr., *J. Chem. Inf. Comput. Sci.*, 2003, **43**, 267–272.

139. H. Chen, P. D. Lyne, F. Giordanetto, T. Lovell and J. Li, *J. Chem. Inf. Model.*, 2006, **46**, 401–415.

140. J. J. Irwin and B. K. Shoichet, *J. Chem. Inf. Model.*, 2005, **45**, 177–182.

141. N. Triballeau, H.-O. Bertrand and F. Acher, in *Methods and Principles in Medicinal Chemistry, Vol. 32: Pharmacophores and Pharmacophore Searches*, T. Langer and R. D. Hoffmann eds., Wiley-VCH, Weinheim, 2006, pp. 325–364.

142. J. Kirchmair, P. Markt, S. Distinto, G. Wolber and T. Langer, *J. Comput. Aided Mol. Des.*, 2008, **22**, 213–228.

143. O. F. Güner, *Pharmacophore Perception, Development and use in Drug Design*, International University Line, La Jolla, CA, 2000.

144. M. Jacobsson, P. Liden, E. Stjernschantz, H. Bostroem and U. Norinder, *J. Med. Chem.*, 2003, **46**, 5781–5789.

145. D. J. Diller and R. Li, *J. Med. Chem.*, 2003, **46**, 4638–4647.

146. N. Triballeau, F. Acher, I. Brabet, J.-P. Pin and H.-O. Bertrand, *J. Med. Chem.*, 2005, **48**, 2534–2547.

147. J.-F. Truchon and C. I. Bayly, *J. Chem. Inf. Model.*, 2007, **47**, 488–508.

148. (a) O. F. Güner and D. R. Henry, in *Pharmacophore Perception, Development and Use in Drug Design*, ed. O. F. Güner, International University Line, La Jolla, CA, USA, 2000, pp. 193–212; (b) O. Funk, V. Kettmann, J. Drimal and T. Langer, *J. Med. Chem.*, 2004, **47**, 2750–2760.

149. T. M. Steindl, C. E. Crump, F. G. Hayden and T. Langer, *J. Med. Chem.*, 2005, **48**, 6250–6260.

150. M. L. Barreca, A. Rao, L. De Luca, M. Zappala, A. M. Monforte, G. Maga, C. Pannecouque, J. Balzarini, E. De Clercq, A. Chimirri and P. Monforte, *J. Med. Chem.*, 2005, **48**, 3433–3437.

151. R. Dayam, T. Sanchez, O. Clement, R. Shoemaker, S. Sei and N. Neamati, *J. Med. Chem.*, 2005, **48**, 111–120.

152. B. Schlegel, C. Laggner, R. Meier, T. Langer, D. Schnell, R. Seifert, H. Stark, H.-D. Hoeltje and W. Sippl, *J. Comput. Aided Mol. Des.*, 2007, **21**, 437–453.

153. B. S. Edwards, C. Bologa, S. M. Young, K. V. Balakin, E. R. Prossnitz, N. P. Savchuck, L. A. Sklar and T. I. Oprea, *Mol. Pharmacol.*, 2005, **68**, 1301–1310.

154. S. M. Young, C. Bologa, E. R. Prossnitz, T. I. Oprea, L. A. Sklar and B. S. Edwards, *J. Biomol. Screen.*, 2005, **10**, 374–382.

155. J. M. Rollinger, A. Hornick, T. Langer, H. Stuppner and H. Prast, *J. Med. Chem.*, 2004, **47**, 6248–6254.

156. J. M. Rollinger, P. Mocka, C. Zidorn, E. P. Ellmerer, T. Langer and H. Stuppner, *Curr. Drug Discov. Technol.*, 2005, **2**, 185–193.
157. J. M. Rollinger, A. Bodensieck, C. Seger, E. P. Ellmerer, R. Bauer, T. Langer and H. Stuppner, *Planta Med.*, 2005, **71**, 399–405.
158. C. Tintori, F. Manetti, N. Veljkovic, V. Perovic, J. Vercammen, S. Hayes, S. Massa, M. Witvrouw, Z. Debyser, V. Veljkovic and M. Botta, *J. Chem. Inf. Model.*, 2007, **47**, 1536–1544.
159. V. Veljkovic, *A Theoretical Approach to Preselection of Carcinogens and Chemical Carcinogenesis*, Gordon & Breach, New York, 1980.
160. Y. Lu, Z. Nikolovska-Coleska, X. Fang, W. Gao, S. Shangary, S. Qiu, D. Qin and S. Wang, *J. Med. Chem.*, 2006, **49**, 3759–3762.
161. C. Charlier, J. P. Henichart, F. Durant and J. Wouters, *J. Med. Chem.*, 2006, **49**, 186–195.
162. M. L. Barreca, L. De Luca, N. Iraci, A. Rao, S. Ferro, G. Maga and A. Chimirri, *J. Chem. Inf. Model.*, 2007, **47**, 557–562.
163. A. Evers and T. Klabunde, *J. Med. Chem.*, 2005, **48**, 1088–1097.
164. P. V. Desai, A. Patny, J. Gut, P. J. Rosenthal, B. Tekwani, A. Srivastava and M. Avery, *J. Med. Chem.*, 2006, **49**, 1576–1584.
165. N. W. Choong and E. E. W. Cohen, *Exp. Opin. Ther. Targets*, 2006, **10**, 793–797.
166. T. Klabunde, in *Methods and Principles in Medicinal Chemistry, Vol 32: Pharmacophores and pharmacophore searches*, T. Langer and R. D. Hoffmann eds., Wiley-VCH, Weinheim, 2006, pp. 283–297.
167. T. Klabunde and A. Evers, *ChemBioChem*, 2005, **6**, 876–889.
168. R. Barbaro, L. Betti, M. Botta, F. Corelli, G. Giannaccini, L. Maccari, F. Manetti, G. Strappaghetti and S. Corsano, *J. Med. Chem.*, 2001, **44**, 2118–2132.
169. C. Chang and S. Ekins, in *Methods and Principles in Medicinal Chemistry, Vol 32: Pharmacophores and pharmacophore searches*, T. Langer and R. D. Hoffmann eds., Wiley-VCH, Weinheim, 2006, pp. 299–324.
170. D. Schuster, C. Laggner, T. M. Steindl and T. Langer, *Curr. Drug Discov. Technol.*, 2006, **3**, 1–48.
171. D. Schuster and T. Langer, *J. Chem. Inf. Model.*, 2005, **45**, 431–439.
172. T. M. Steindl, D. Schuster, C. Laggner, K. Chuang, R. D. Hoffmann and T. Langer, *J. Chem. Inf. Comput. Sci.*, 2007, **47**, 563–571.
173. J. M. Rollinger, T. Langer and H. Stuppner, *Curr. Med. Chem.*, 2006, **13**, 1491–1507.
174. J. M. Rollinger, T. Langer and H. Stuppner, *Planta Med.*, 2006, **72**, 671–678.
175. J. M. Rollinger, S. Haupt, H. Stuppner and T. Langer, *J. Chem. Inf. Comput. Sci.*, 2004, **44**, 480–488.
176. J. Hert, P. Willett, D. J. Wilton, P. Acklin, K. Azzaoui, E. Jacoby and A. Schuffenhauer, *Org. Biomol. Chem.*, 2004, **2**, 3256–3266.
177. H. Matter, *J. Med. Chem.*, 1997, **40**, 1219–1229.
178. M. J. Keiser, B. L. Roth, B. N. Armbruster, P. Ernsberger, J. J. Irwin and B. K. Shoichet, *Nat. Biotechnol.*, 2007, **25**, 197–206.

179. M. von Korff and M. Steger, *J. Chem. Inf. Comput. Sci.*, 2004, **44**, 1137–1147.
180. J. H. Nettles, J. L. Jenkins, A. Bender, Z. Deng, J. W. Davies and M. Glick, *J. Med. Chem.*, 2006, **49**, 6802–6810.
181. J. H. Nettles, J. L. Jenkins, C. Williams, A. M. Clark, A. Bender, Z. Deng, J. W. Davies and M. Glick, *J. Mol. Graph. Model.*, 2007, **26**, 622–633.
182. *WOMBAT*, available from Sunset Molecular Discovery LLC, Santa Fe, NM, sunsetmolecular.com.
183. S. Renner, M. Hechenberger, T. Noeske, A. Bocker, C. Jatzke, M. Schmuker, C. G. Parsons, T. Weil and G. Schneider, *Angew. Chem. Int. Ed. Engl.*, 2007, **46**, 5336–5339.
184. J. Hert, P. Willett, D. J. Wilton, P. Acklin, K. Azzaoui, E. Jacoby and A. Schuffenhauer, *J. Chem. Inf. Comput Sci.*, 2004, **44**, 1177–1185.
185. C. Williams, *Mol. Divers*, 2006, **10**, 311–332.
186. E. J. Barker, D. Buttar, D. A. Cosgrove, E. J. Gardiner, P. Kitts, P. Willett and V. J. Gillet, *J. Chem. Inf. Model.*, 2006, **46**, 503–511.
187. J. A. Haigh, B. T. Pickup, J. A. Grant and A. Nicholls, *J. Chem. Inf. Model.*, 2005, **45**, 673–684.
188. T. Schulz-Gasch and M. Stahl, *J. Mol. Mod.*, 2003, **9**, 47–57.
189. M. R. McGann, H. R. Almond, A. Nicholls, J. A. Grant and F. K. Brown, *Biopolymers*, 2003, **68**, 76–90.
190. C. M. Venkatachalam, X. Jiang, T. Oldfield and M. Waldman, *J. Mol. Graph. Model.*, 2003, **21**, 289–307.
191. J. A. Grant, M. A. Gallard and B. T. Pickup, *J. Comput. Chem.*, 1996, **17**, 1653–1666.
192. T. S. Rush III, J. A. Grant, L. Mosyak and A. Nicholls, *J. Med. Chem.*, 2005, **48**, 1489–1495.
193. A. Nicholls and J. A. Grant, *J. Comput. Aided Mol. Des.*, 2005, **19**, 661–686.
194. M. J. Sykes, M. J. Sorich and J. O. Miners, *J. Chem. Inf. Model.*, 2006, **46**, 2661–2673.
195. M. Rarey, B. Kramer, T. Lengauer and G. Klebe, *J. Mol. Biol.*, 1996, **261**, 470–489.
196. T. J. A. Ewing, S. Makino, A. G. Skillman and I. D. Kuntz, *J. Comput. Aided Mol. Des.*, 2001, **15**, 411–428.
197. T. J. A. Ewing and I. D. Kuntz, *J. Comput. Chem.*, 1997, **18**, 1175–1189.
198. G. Jones, P. Willett, R. C. Glen, A. R. Leach and R. Taylor, *J. Mol. Biol.*, 1997, **267**, 727–748.
199. T. A. Halgren, R. B. Murphy, R. A. Friesner, H. S. Beard, L. L. Frye, W. T. Pollard and J. L. Banks, *J. Med. Chem.*, 2004, **47**, 1750–1759.
200. R. A. Friesner, J. L. Banks, R. B. Murphy, T. A. Halgren, J. J. Klicic, D. T. Mainz, M. P. Repasky, E. H. Knoll, M. Shelley, J. K. Perry, D. E. Shaw, P. Francis and P. S. Shenkin, *J. Med. Chem.*, 2004, **47**, 1739–1749.
201. A. N. Jain, *J. Med. Chem.*, 2003, **46**, 499–511.
202. G. L. Warren, C. W. Andrews, A.-M. Capelli, B. Clarke, J. LaLonde, M. H. Lambert, M. Lindvall, N. Nevins, S. F. Semus, S. Senger,

G. Tedesco, I. D. Wall, J. M. Woolven, C. E. Peishoff and M. S. Head, *J. Med. Chem.*, 2006, **49**, 5912–5931.

203. E. M. Krovat and T. Langer, *J. Chem. Inf. Comput Sci.*, 2004, **44**, 1123–1129.
204. W. Sherman, H. S. Beard and R. Farid, *Chem. Biol. Drug. Des.*, 2006, **67**, 83–84.
205. W. Sherman, T. Day, M. P. Jacobson, R. A. Friesner and R. Farid, *J. Med. Chem.*, 2006, **49**, 534–553.
206. M. Kontoyianni, L. M. McClellan and G. S. Sokol, *J. Med. Chem.*, 2004, **47**, 558–565.
207. E. Kellenberger, J. Rodrigo, P. Muller and D. Rognan, Proteins: Struct., *Funct., Bioinf.*, 2004, **57**, 225–242.
208. R. T. Kroemer, A. Vulpetti, J. J. McDonald, D. C. Rohrer, J.-Y. Trosset, F. Giordanetto, S. Cotesta, C. McMartin, M. Kihlen and P. F. W. Stouten, *J. Chem. Inf. Comput Sci.*, 2004, **44**, 871–881.
209. S. A. Hindle, M. Rarey, C. Buning and T. Lengaue, *J. Comput. Aided Mol. Des.*, 2002, **16**, 129–149.
210. H. Claussen, M. Gastreich, V. Apelt, J. Greene, S. A. Hindle and C. Lemmen, *Curr. Drug Discov. Technol.*, 2004, **1**, 49–60.
211. J. M. Yang and T. W. Shen, *Proteins*, 2005, **59**, 205–220.
212. S. N. Rao, M. S. Head, A. Kulkarni and J. M. Lalonde, *J. Chem. Inf. Model.*, 2007, **47**, 2159–2171.
213. X. Fradera, R. M. Knegtel and J. Mestres, *Proteins*, 2000, **40**, 623–636.
214. Y. C. Martin, in *Comprehensive Medicinal Chemistry II, Vol 4: Computer-Assisted Drug Design*, ed. J. S. Mason, Elsevier, Oxford, 2007, pp. 119–147
215. Y. C. Martin, in *Comprehensive Medicinal Chemistry II, Vol 4: Computer-Assisted Drug Design*, J. S. Mason ed., Elsevier, Oxford, 2007, pp. 515–536.
216. J. H. Van Drie, in *Computational Medicinal Chemistry for Drug Discovery*, P. Bultinck, H. de Winter, W. Langenaecker and J. P. Tollenaere eds., Marcel Dekker, New York, 2004, pp. 437–460.
217. Y. Patel, V. J. Gillet, G. Bravi and A. R. Leach, *J. Comput. Aided Mol. Des.*, 2002, **16**, 653–681.
218. C. Chang, S. Ekins, P. Bahadduri and P. W. Swaan, *Adv. Drug Delivery Rev.*, 2006, **58**, 1431–1450.
219. Y. Li and W. E. Harte, *Curr. Pharm. Des.*, 2002, **8**, 99–110.
220. R. D. Hoffmann, S. Meddeb and T. Langer, in *Computational Medicinal Chemistry for Drug Discovery*, P. Bultinck, H. de Winter, W. Langenaecker and J. P. Tollenaere eds., Marcel Dekker, New York, 2004, pp. 461–482.

CHAPTER 4

Molecular Similarity Analysis in Virtual Screening

LISA PELTASON AND JÜRGEN BAJORATH

Department of Life Science Informatics, B-IT, LIMES Institute, Program Unit Chemical Biology and Medicinal Chemistry, Rheinische Friedrich-Wilhelms-Universität, Dahlmannstr. 2, D-53113 Bonn, Germany

4.1 Introduction

Molecular similarity analysis as we narrowly define and understand it today employs a holistic molecular view and attempts to establish structure–activity relationships between molecules beyond what one can "see" with a chemist's eye. Similarity analysis has become an integral part of the chemoinformatics spectrum.[1] A plethora of conceptually different similarity methods have been, and continue to be, developed and are applied in the hunt for novel active compounds in pharmaceutical and other research settings. Table 4.1 lists exemplary methods that reflect current development trends in molecular similarity research. These and other approaches are often applied to search large virtually formatted compound databases for active molecules using information of known ligands as input, a process commonly referred to as "virtual screening". A characteristic feature of currently available similarity methods is that the underlying algorithms and their complexity often differ in rather significant ways. Nevertheless, every newly introduced methodology is shown to have predictive value, at least in benchmark calculations, and approaches ranging from the most simplistic to complicated ones often produce comparable results when applied to the same test cases. In fact, the impression one typically gets when reviewing this field is that essentially "anything goes", at least to some extent. At the same time, the

Chemoinformatics Approaches to Virtual Screening
Edited by Alexandre Varnek and Alex Tropsha
© Royal Society of Chemistry, 2008
Published by the Royal Society of Chemistry, www.rsc.org

Table 4.1 Exemplary similarity-based methodologies applicable to virtual screening.[a]

	Method	Descriptors	Approach	Reference
compound classification and mapping	clustering	property descriptors, fingerprints	distance-based grouping of compounds	41,42
	decision trees	property descriptors	partitioning of compounds along tree structures	43
	binary kernel discrimination	fingerprints	estimation of class label probabilities	44
	support vector machine (SVM)	property descriptors, fingerprints	prediction of class labels using a maximum-margin hyperplane	45,46
	cell-based partitioning	property descriptors	mapping of compounds to subsections of chemical reference space	47
	DMC	binary transformed property descriptors	mapping of compounds to consensus positions in chemical space	48,49
	MAD, DynaMAD	property descriptors	mapping of compounds to activity-selective descriptor value ranges with iterative dimension extension (DynaMAD)	50,51
similarity searching	fingerprint comparison	BCI: predefined structural fragments	quantitative comparison of bit string representations using a similarity coefficient	52

Table 4.1 (Continued).

Method	Descriptors	Approach	Reference
	Daylight: hashed connectivity pathways		53
	Molprint 2D: layered atom environments		54
	shape fingerprints: set of reference shapes		55
	3D pharmacophore fingerprints: set of pharmacophore patterns		9,56
	PDR-FP: equifrequently binned activity-sensitive 2D descriptors		57
reduced graphs	simplified 2D graphs	determination of graph similarity	20
ROCS	gaussian shape models	determination of volume overlap	58
MolBlaster	random fragment populations	comparison of histograms	22
BDACCS	property descriptors	Bayesian distance function	59

[a]Clustering is the original approach to similarity-based compound classification, for which many different algorithms have been developed. For simplicity, clustering is only referred to as a general approach. Similar considerations apply to cell-based and statistical partitioning for which BCUTs and decision trees are referenced as a prototype, respectively. The table format is adapted from.[5]

performance of similarity methods typically shows a strong compound class dependence that is not predictable. Moreover, in practice, similarity-based methods generally perform less well in virtual screening than suggested by results obtained in conventional benchmark calculations. Despite the intense efforts that go into the development and evaluation of novel methodologies and search tools, the reasons for these more general observations are not well understood and many open questions remain. Therefore, we feel it is timely to look at similarity-based methods from a more principal point of view and evaluate potential reasons for the above-mentioned trends. Accordingly, this chapter focuses on the scientific foundations of molecular similarity analysis that generally affect similarity approaches and their performance in virtual screening, regardless of methodological details and complexity.

4.2 Ligand-based Virtual Screening

Virtual screening (VS) techniques have become an integral part of modern computer-assisted drug discovery.[2] VS consists of a spectrum of methods designed to efficiently search large compound databases *in silico* for molecules likely to have a desired biological activity. Hits from a VS campaign are typically selected for testing in a biological assay and, if novel actives are identified, further development towards lead structures. Thus, VS attempts to rationalize drug candidate testing strategies, in contrast to random screening. VS methods are principally divided into approaches that take the structure of the target protein into account (structure-based virtual screening, SBVS)[3,4] and approaches that rely on structures of known active small molecules (ligand-based virtual screening, LBVS).[5,6] While SBVS aims to identify active small molecules by "docking" them into the target binding site, LBVS uses information of known active molecules to find novel structures with the desired activity. LBVS greatly benefits from its independence of the target structure, because, in many cases, active compounds present the principal source of information. Of course, SBVS and LBVS strategies are not mutually exclusive and can be used in concert.

As stated above, LBVS techniques derive knowledge from given active molecules to estimate the activity of candidate molecules in large databases. Various methods have been developed that are distinguished by the chemical information they use and the methodological paradigm that is followed. Figure 4.1 gives an overview of the distinct approaches. Methods like pharmacophore searching[7–9] or QSAR predictions[10] focus on local features of chemical structure, whereas molecular similarity analysis employs a global view of molecular structure. Chemical similarity searching has one of its origins in substructure searching, where database molecules are screened for the presence of a predefined structural fragment.[11] A substructure search retrieves all molecules that contain the query fragment, irrespective of the structural environment in which it occurs, which corresponds to a purely local method of compound comparison. A pharmacophore search is distantly related to substructure searching in that an ensemble of chemical groups is defined as a query for database screening.

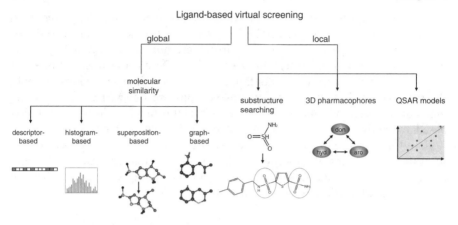

Figure 4.1 Ligand-based virtual screening methods. The figure shows different computational methods for screening compound databases that take either a local or a global view on molecular structure. Molecular similarity methods that operate on molecular descriptors, histogram representations, superposition or (reduced) molecular graphs evaluate molecular structure globally. By contrast, local structural features are explored by substructure and pharmacophore searching or QSAR modeling.

In some ways, pharmacophore models represent an activity-centric refinement of substructure queries guided by three-dimensional (3D) information and knowledge of or hypotheses about target–ligand interactions. In pharmacophore modeling, a set of 3D conformations of active reference molecules is used to derive hypotheses concerning the parts of a molecule that render it active through interactions with its target. This spatial arrangement of interaction points and elements represents the pharmacophore model that is used as a query for database searching. Molecules that exhibit geometric features similar to the pharmacophore are selected as candidate molecules likely to bind to the target of interest.

Another method that derives information about bioactivity from local features of a set of reference molecules is Quantitative Structure–Activity Relationship analysis (QSAR). Conceptually similar to pharmacophore methods, QSAR initially concentrates on parts of a molecule where chemical changes alter biological activity and that are believed or known to form interactions with its target. The effects of local chemical changes on biological activity of molecules are then studied. A mathematical model is derived that relates structural features and molecular properties with bioactivity in a quantitative manner. Properties are typically expressed through the use of chemical descriptors. The model is applied to predict activities of candidate molecules. QSAR requires sufficient knowledge about the parts of a molecule that are relevant for bioactivity and usually depends on the availability of series of closely related molecules or analogs, which also reflects the intrinsically "local" nature of the approach, irrespective of methodological details.

In contrast to pharmacophore or QSAR modeling, chemical similarity searching and other similarity-based methods require "global" molecular representations that take the complete molecular structure into account and do not rely on pre-conceived notions of activity-determining structural features. The global similarity between the structure of each database compound and one or several active reference compound(s) is evaluated. Database compounds are ranked according to their similarity to the reference molecules. Molecules that are overall similar to active reference molecules are thought to have a high probability of displaying similar activity. This requires an appropriate specification of intermolecular similarity that captures molecular characteristics relevant for activity. Concepts and applications of molecular similarity are discussed in the following.

4.3   Foundations of Molecular Similarity Analysis

A theoretical foundation of molecular similarity analysis is the assumption of "neighborhood behavior",[12] which refers to the tendency of molecules with globally similar structures to exhibit similar biological activity. The well-known "similarity-property principle" (SPP) of Johnson and Maggiora[13] expresses this paradigm and promotes a holistic view of molecular structure and properties. Molecular similarity applications assume that chemical similarity can be related to biological activity in a meaningful manner. However, the success of this approach ultimately depends on the way molecular similarity is defined.

4.3.1   Molecular Similarity and Chemical Spaces

For the definition and analysis of molecular similarity, a reference frame is needed that relates molecular structures to each other and facilitates computational comparisons. Hence, an important aspect of similarity analysis is the definition of theoretical chemical spaces into which compounds are projected.[14] A chemical reference space is spanned by a set of molecular descriptors, where each descriptor adds a dimension to the reference space. Molecules are located in reference spaces according to their descriptor values; molecular "coordinates" in the reference space correspond to the values that descriptors adopt for individual compounds. Molecular similarity or dissimilarity is then defined through the intermolecular distance in the reference space. Meaningful chemical space representations map similar compounds to contiguous regions. The major challenge for chemical space design is the choice of molecular representations such that similar biological activity and properties are reflected by small intermolecular distances. For this purpose, various molecular descriptors are currently used.[15,16] Generally, a molecular descriptor is a mathematical representation of a molecule that contains information about structure or physicochemical properties. This information can be conveyed by scalar values, vectors, or bit strings. Descriptors are often classified according to the dimensionality of the molecular representation from which they are calculated. 1D descriptors utilize the chemical formula and represent mostly bulk properties such as molecular weight or atom counts. Descriptors based on the molecular graph representation are classified as 2D and contain, for example, information about connectivity or defined structural fragments, but also estimations of physicochemical properties. 3D descriptors such as molecular volume, surface or pharmacophore patterns require an experimental or predicted 3D conformation of a molecule. A "molecular fingerprint" is a type of descriptor that is widely used in similarity searching.[14] Fingerprints are string representations of chemical structures consisting of bins, each of which contains a scalar descriptor. Most often, fingerprints are bit strings where the molecular information is encoded in a binary format. For example, the popular structural keys (*e.g.* MACCS[17]) constitute a class of binary fingerprints where each bit denotes the presence or absence of a predefined structural fragment. In contrast to this simple type of 2D fingerprint, several fingerprints are available that use 3D geometrical information. For example, pharmacophore fingerprints[18] monitor possible pharmacophore patterns in conformers of a molecule and usually contain very large numbers of bit positions.

4.3.2 Similarity Measures

Whatever molecular representation is used, it is important to note that neighborhood relationships between molecules are not invariant to the choice of chemical reference space. Hence, molecular similarity can be assessed only with respect to a given molecular representation. As stated above, molecular similarity or dissimilarity is measured by intermolecular distance in the chosen reference space. Conventional distance metrics such as Euclidean or Hamming distance measure the distance between molecules in chemical space, whereas similarity coefficients (*e.g.*, Tanimoto, Dice or Cosine coefficient) directly assess intermolecular similarity.[19] Most similarity coefficients yield values that range from 0 (denoting maximum dissimilarity) to 1 (maximum similarity), or can be normalized accordingly, and are referred to as association coefficients. When binary fingerprints are used, bit string overlap serves as a measure of molecular similarity. The association coefficient most widely used in chemical applications is the Tanimoto coefficient (Tc), which counts the number of bits common to two binary fingerprints with respect to the total number of bits that are set in each fingerprint. The Tc for two binary fingerprint representations A and B is calculated as follows:

$$Tc(A, B) = \frac{N_{AB}}{N_A + N_B - N_{AB}} \tag{4.1}$$

where N_{AB} is the number of bits set on in both fingerprints and N_A and N_B refer to the number of bits set on in A and B, respectively.

However, these descriptor-based similarity definitions present only one class of available similarity and distance measures. Approaches to molecular similarity assessment independent of conventional descriptors include, for example, superposition, molecular graph representations,[20] histogram comparisons,[21,22] and Brownian processing of molecules.[23]

4.3.3 Activity Landscapes

A given chemical space representation provides a projection of molecular data into a multi-dimensional reference space. For a set of molecules active against the same target, the magnitude of their individual activities adds another dimension to the given reference space. The projection of molecules into chemical space together with their activities can thus be envisioned as a topographical map, where the magnitude of the activities (potencies) forms the surface of an "activity landscape".[24] For instance, if a 2D projection of chemical space is employed, the activity surface forms 3D shapes comparable to geographic landscapes (Figure 4.2). The notion of activity landscapes illustrates basic relationships between molecular structure and biological activity and helps to characterize them.

4.3.4 Analyzing the Nature of Structure–Activity Relationships

Similarity search calculations are known to show highly varying performance on different compound activity classes. Methods of different design and complexity

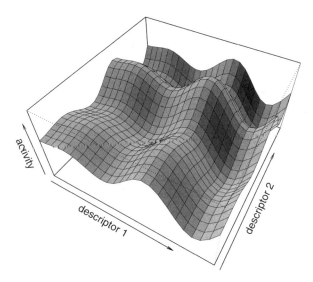

Figure 4.2 Schematic illustration of an activity landscape. A two-dimensional chemical reference space is defined by two molecular descriptors. Biological compound activity adds a dimension to the reference space, forming the surface of an activity landscape.

often succeed comparably well, whereas in other cases considerable differences in compound recall are observed and there is very little overlap between sets of selected compounds.[5] Currently available results suggest that there is no generally preferred methodology, which implies that there should be principal limitations to the success of similarity assessment beyond specific features of computational methods. The strong activity class dependence of similarity search calculations indicates that the nature of underlying structure–activity relationships (SARs) is a major determinant of success or failure of similarity methods.

The conceptual basis for similarity analysis is provided by the similarity-property principle that states that similar molecules have similar biological activity.[13] This rather intuitive principle has been widely accepted and substantiated by a wealth of observations. The success of many similarity-based virtual screening calculations can only be rationalized on the basis of this principle. However, minor modifications in molecular structure can dramatically alter the biological activity of a small molecule.[25] This situation is exploited in lead optimization efforts, but limits the potential of similarity methods. These considerations also suggest that there must be fundamental differences between the structure–activity relationships (SARs). Thus, different types of SARs are expected to critically determine the success of similarity methods and systematic SAR analysis helps to better understand on a case-by-case basis why similarity methods might succeed or fail.

SARs are largely distinguished by the responsiveness of molecules to structural modifications. SAR characteristics are mirrored by the topology of the activity landscape under consideration and such topologies might substantially differ. For example, what we regard as "continuous" SARs are characterized by gradual biological responses to chemical changes, which results in smooth activity landscapes reminiscent of gently rolling hills.[24] The spectrum of active compounds, which we call the "activity radius", covers a range of increasingly diverse structures, including closely related molecules or analogs as well as different chemotypes having comparable biological activity. In certain regions within continuous SAR landscapes molecules with different core structures can be found that have similar potency. This situation is consistent with the similarity-property principle and provides a sound basis for molecular similarity analysis. The holistic nature of similarity methods builds on the presence of different structural motifs displaying the same activity. In particular, continuous SARs enable the departure from given core structures and the identification of novel active compounds, which is generally considered the major challenge for virtual screening.

In contrast to continuous SARs, "discontinuous" SARs are characterized by substantial changes in biological activity in response to small or even minute chemical changes. In the presence of discontinuous SARs, modification of only a single functional group can decrease the biological activity of a molecule dramatically or even abolish it altogether. Conversely, a small structural modification might transform an active molecule into a highly potent lead (Figure 4.3). This situation corresponds to the presence of "activity cliffs" in the

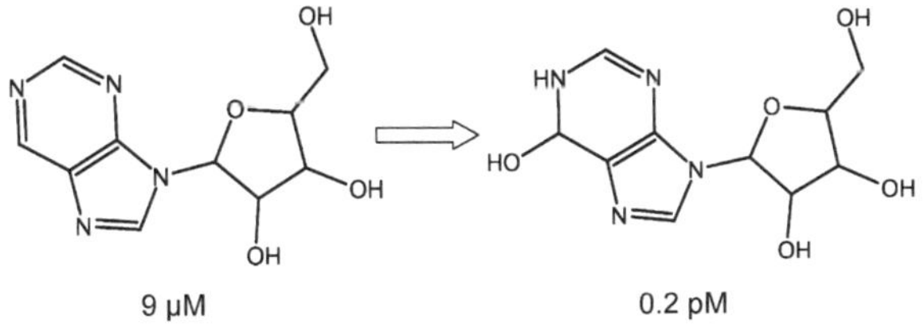

Figure 4.3 Example of an "activity cliff" illustrated by closely related adenosine deaminase inhibitors having dramatic potency differences. The introduction of a hydroxyl group that coordinates a zinc cation in the active site of the enzyme adds several orders of magnitude to the potency of an inhibitor.

activity landscape that produce an area of a rugged canyon-like surface.[24] In these cases, the activity radius is limited by the need for bioactive compounds to contain substructures that are responsible for the formation of crucial binding interactions with a receptor. It follows that discontinuous SARs do not conform to the holistic molecular view of similarity and the similarity-property principle. When discontinuous SARs are prevalent, molecular similarity analysis becomes a futile exercise.

However, notably, the concept of an activity radius does not generally assume a homogeneous distribution of compound properties throughout a given radius. The presence of continuous SARs within a given similarity range does not imply that the corresponding activity radius consists of exclusively continuous SARs. Within the same radius, discontinuous regions can also be encountered, depending on the direction into which we move in chemical space. Consequently, compounds within a given activity radius will in most cases exhibit varying SAR characteristics.

In systematic SAR analysis, molecular structure and similarity need to be represented and related to each other in a measurable form. Just like any molecular similarity approach, SAR analysis critically depends on molecular representations and the way similarity is measured. The nature of the chemical space representation determines the positions of the molecules in space and thus ultimately the shape of the activity landscape. Hence, SARs may differ considerably when changing chemical space and molecular representations. In this context, it becomes clear that one must discriminate between SAR features that reflect the fundamental nature of the underlying molecular structures as opposed to SAR features that are merely an artifact of the chosen chemical space representation. Consequently, activity cliffs can be viewed as either fundamental or descriptor- and metrics-dependent. The latter occur as a consequence of an inappropriate molecular representation or similarity metrics and can be smoothed out by choosing a more suitable representation, *e.g.*, by considering activity-relevant physicochemical properties.[26] By contrast, activity cliffs fundamental to the underlying SARs cannot be circumvented by changing the reference space. In this situation, molecules that should be recognized as similar by any meaningful metrics display different biological activity. Hence, holistic similarity methods cannot solve such discontinuity, irrespective of the molecular representations that are used.

4.3.4.1   *Relationships between Different SARs*

The different prototypic SAR characteristics described above have been observed for various compound classes. It is puzzling at first glance that the ligands of many target proteins display apparently inconsistent SAR features. Often, series of target-specific inhibitors are related by continuous SARs; they gradually depart in structure from a potent molecule but show only moderate losses in potency. However, for the same target, other series of similar molecules are found that display remarkably different potency or selectivity. This apparent inconsistency can be rationalized by systematic SAR analysis. Correlation of the structural similarity of biologically active molecules and their potency reveals that activity landscapes are more variable than often thought and complex relationships exist between different SAR characteristics.[27] There is substantial evidence that activity landscapes frequently contain both smooth and rugged regions that are populated by different series of inhibitors related to each other by either continuous or discontinuous SARs. Thus, different types of SARs might coexist within an active site. Such insights have suggested that many SARs should be "heterogeneous" in nature[5] and this possibility has been thoroughly explored by systematic comparisons of experimental binding conformations, 2D similarity, and potency of series of ligands binding to different enzyme targets.[27] The following sections describe in more detail the investigations that have provided a conceptual framework for qualitative and quantitative characterization of SARs.

4.3.4.2   *SARs and Target–Ligand Interactions*

The biological activity of small molecules ultimately has to be considered in the context of receptor–ligand interactions. In other words, for the study of SARs, it is important to take into account that the biological activity of small molecules always results from their interaction with a macromolecular target, most often an enzyme or receptor protein. For a small molecule, specific binding to a receptor requires the formation of well-defined chemical and physical interactions. Crucial for binding is the chemical and geometrical complementarity of ligand and receptor, and these requirements are not independent of each other. Chemical complementarity is reflected by the formation of hydrogen bonds, electrostatic or ionic interactions, van der Waals interactions and, in addition, by hydrophobic or other entropic effects. Geometrical complementarity requires that the ligand matches the shape of the binding site, as originally postulated using the lock-and-key analogy[28] or the induced-fit model of ligand binding,[29] which is often more appropriate.

This target-centric view of biological activity helps to rationalize the SARs of many different activity classes and to distinguish them from each other. On the one hand, activity cliffs can be envisioned as resulting from critically important target–ligand interactions that are indispensable for binding, irrespective of ligand structure and properties. Such rigid constraints posed by a binding site introduce cliffs into the activity landscape and define the boundaries of an activity radius. On the other hand, continuous SARs are often indicative of permissive binding events and binding sites having plasticity that tolerate ligand variability to a certain extent. Such binding sites might often require a high degree of shape complementarity with a ligand but provide different potential interaction points, or sub-sites, to achieve the required complementarity. However, consistent with the variable nature of activity landscapes,[27] SAR characteristics are not exclusively dictated by binding site elements. The structural and chemical features of ligands also play an important role and co-determine the "language" that proteins and ligands use to communicate. Moreover, continuity or discontinuity of SARs is not simply determined by the region of the molecule where structural modifications are made. Often, affinity changes are achieved by chemical modification of the parts of a molecule that interact with its target. But there are also cases where such modifications do not produce significant affinity variation or trigger the formation of different binding modes, which provides evidence that SARs are not necessarily local molecular characteristics. The global nature of SARs is also reflected by the presence of compounds having different molecular scaffolds but sharing the same biological activity: the "holy grail" of similarity analysis. Furthermore, there are also potency determining effects that do not depend on receptor–ligand interactions, for example, desolvation free energy.

4.3.4.3 Qualitative SAR Characterization

Characterization of SARs was facilitated through systematic comparison of 2D and 3D similarity of ligands with their potency.[27] For this purpose, X-ray structures were collected of different target enzymes for which complexes with multiple ligands had been determined. Table 4.2 summarizes the crystallographic structures that were analyzed. For each enzyme, binding conformations of ligands were compared in a pair-wise manner using alignments based on overlap of atomic property density functions that took positional and conformational differences into account. 2D similarity between ligand pairs was calculated using the Tanimoto coefficient on MACCS structural keys. Moreover, for each ligand pair, the relative potency difference was calculated. This provided the basis for the qualitative assessment of SARs that offered fundamental insights into the nature of activity landscapes. Representative results are discussed below.

A set of inhibitors of the coagulation factor Xa is found to present prototypic continuous SARs. There is detectable correlation between 2D and 3D molecular similarity and most similar 2D structures bind very similarly and with

Table 4.2 Summary of crystallographic data used for qualitative SAR analysis.

Data[a]	Ribonuclease A	Carbonic anhydrase	Factor Xa	Elastase
Number of complex structures	9	27	16	14
PDB codes (hetero ID)	1afk (PAP)	1a42 (BZO)	1ezq (RPR)	1bma (4-mer)
	1afl (ATR)	1avn (HSM)	1f0r (815)	1eas (TFK)
	1jn4 (139)	1bcd (FMS)	1f0s (PR2)	1eat (TFI)
	1o0f (A3P)	1bn1 (AL5)	1fjs (Z34)	1ela (4-mer)
	1o0h (ADP)	1bn3 (AL6)	1g2l (T87)	1elb (4-mer)
	1o0m (U2P)	1bn4 (AL9)	1ksn (FXV)	1elc (4-mer)
	1o0n (U3P)	1bnn (AL1)	1lpg (IMA)	1eld (4-mer)
	1o0o (A2P)	1bnq (AL4)	1lpk (CBB)	1ele (4-mer)
	1qhc (PUA)	1bnt (AL2)	1lpz (CMB)	1gvk (4-mer)
		1bnu (AL3)	1mq5 (XLC)	1h9l (4-mer)
		1bnv (AL7)	1mq6 (XLD)	1inc (ICL)
		1bnw (TPS)	1nfu (RRP)	1qr3 (8-mer)
		1cil (ETS)	1nfw (RRR)	4est (5-mer)
		1cim (PTS)	1nfx (RDR)	5est (3-mer)
		1cin (MTS)	1nfy (RTR)	
		1cnw (EG1)	1xka (4PP)	
		1cnx (EG2)		
		1cny (EG3)		
		1g1d (FSB)		
		1g52 (F2B)		
		1g53 (F6B)		
		1g54 (FFB)		
		1if7 (SBR)		
		1if8 (SBS)		
		1okl (MNS)		
		1okn (STB)		
		1ttm (667)		
2D similarity				
minimum	0.76	0.07	0.24	0.34
maximum	0.99	1	1	0.92
average	0.87	0.59	0.50	0.52
3D similarity				
minimum	0.13	0	0.28	0.09
maximum	0.87	0.99	0.96	0.96
average	0.44	0.60	0.58	0.37
correlation				
2D/3D similarity	0.58	0.79	0.47	0.31
potency (K_i values in nM)				
minimum	27	0.03	0.007	0.46
maximum	82000	125000	131	890000
average	12820	4669	24	120512

[a]Shown are PDB accession codes and similarity and potency data for complex crystal structures subjected to qualitative SAR analysis. The table is adapted from the supplementary material of.[27]

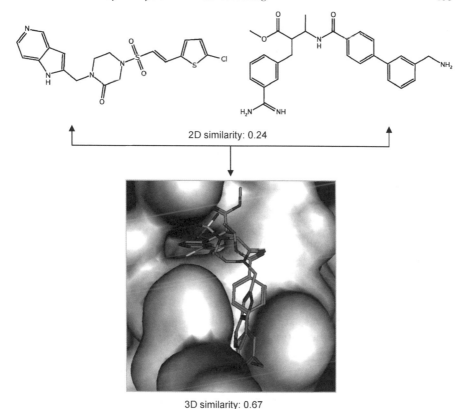

2D similarity: 0.24

3D similarity: 0.67

Figure 4.4 A pair of factor Xa inhibitors with limited structural similarity that adopt very similar binding conformations. 3D similarity is calculated from the overlap of atomic property density functions (for details, see ref. 27).

comparable potency. In addition, there is a significant degree of structural diversity. A perhaps unexpected characteristic of diverse factor Xa inhibitors with distinct chemical scaffolds is their tendency to bind in similar conformations. Diverse structures adopt comparable binding modes that match the shape of the binding pocket (Figure 4.4). This indicates that, in this case, binding to the receptor is largely governed by shape complementarity, which provides the basis for the structural diversity of factor Xa ligands. The active site of factor Xa tolerates structural variations as long as a high degree of spatial complementarity is maintained and a few key interactions are formed.

A different example is provided by a set of elastase inhibitors. These ligands are also related by continuous SARs. This is reflected by the presence of highly potent inhibitors with diverse structures and, in addition, structurally similar ligands that display only minor potency differences. However, in contrast to factor Xa, 3D analysis reveals a more complex picture. Specifically, elastase accepts multiple binding modes, each of which is adopted by structurally

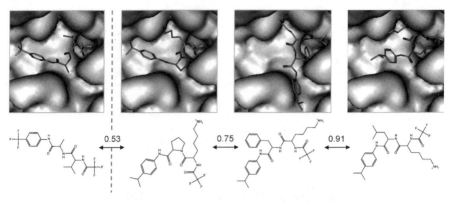

Figure 4.5 Different binding modes of elastase inhibitors. The three inhibitors on the right-hand side have similar 2D structures but adopt distinct binding modes. By contrast, the two structures on the left (separated by the dashed vertical line) have only limited 2D similarity but share the same binding mode. MACCS Tc values are reported for pair-wise comparisons.

diverse ligands. Moreover, structurally similar ligands can bind in distinct conformations and orientations, whereas structurally diverse inhibitors are found to adopt the same binding mode and have very similar 3D conformations (Figure 4.5). Subsets of elastase inhibitors can be identified that show either direct or inverse correlation between 2D and 3D similarity. However, the 3D binding characteristics appear to have no significant influence on the potency of the structures. Hence, substantially different SAR characteristics are observed for different series of compounds that are active against the same target.

By contrast, discontinuous SAR characteristics are present among, for example, ribonuclease A inhibitors. All inhibitors studied here are closely related nucleotide analogs containing one or more phosphate groups and either an adenine or a uracil base. Despite these shared moieties, there are potency differences of several orders of magnitude among analogous structures. In addition, binding conformations exhibit a remarkable degree of variability. Two distinct binding modes are observed that distinguish between inhibitors containing different nucleobases (Figure 4.6a). Hence, structurally closely related inhibitors bind in a completely different manner, dependent on which base they contain. Moreover, even analogs of the same nucleotide can adopt different binding conformations. Inspection of the architecture of the receptor binding site helps to better understand this spatial permissiveness. The active site of ribonuclease A contains a positively charged phosphate binding pocket. Potent inhibitors fill this pocket with a phosphate group. Apart from this severe structural constraint, the enzyme accommodates different binding conformations and allows a remarkable degree of structural variability. However, these findings cannot explain the high potency differences among very similar ligands, because the highest potency differences occur between ligands that

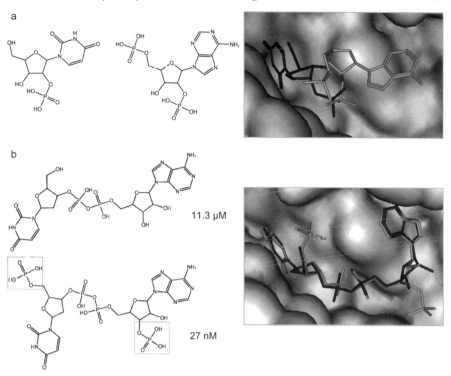

Figure 4.6   Ribonuclease inhibitors. (*a*) Analogs of different nucleotides adopt distinct binding modes. (*b*) Closely related analogs that differ only in the position of two phosphate groups and bind in very similar conformations show significant potency differences.

share the same binding mode (Figure 4.6b). Ribonuclease A inhibitors present an example of discontinuous SARs that manifest themselves on both the 2D and 3D level.

Taken together, the qualitative characterization of SARs on the basis of available structural and potency data reveals that most target sites have a certain degree of permissiveness to structural variation of ligands. Few targets are completely restrictive or permissive in their binding characteristics. There is often more than one way that ligands interact with their target. Accordingly, the corresponding activity landscapes frequently contain multiple regions of strong local activity, each of which is limited by the presence of an activity cliff and a resulting individual activity radius. Ligands adopt specific 3D binding conformations to specifically match chemical and geometric features of a binding site. Analysis of multiple binding conformations often helps to rationalize underlying SARs. Owing to the intrinsic variability of many target–ligand interactions, diverse structures can be accommodated in binding sites through distinct binding modes. Hence, the analysis of target–ligand interactions and binding conformations enables SAR analysis on the basis of experimental structural and binding data. The findings discussed above suggest that many SARs are not exclusively continuous or discontinuous, but rather are heterogeneous in nature. This has important implications for molecular similarity analysis and virtual screening, as further discussed below.

4.3.4.4 Quantitative SAR Characterization

The qualitative analysis of SARs described above has shed light on the highly complex nature of SARs. In medicinal chemistry, SARs are typically analyzed on a case-by-case basis. Thus far, few if any approaches have been introduced to systematically and quantitatively describe SAR characteristics of different compound classes. In the following, two related approaches are presented that provide a quantitative measure of SAR characteristics only based on 2D structural similarity and binding data. Limiting similarity assessment to 2D molecular representations departs from the 3D similarity-oriented correlation studies described above, but makes it possible to extend quantitative SAR analysis to targets for which no, or only few, relevant X-ray structures are available.

SAS Maps. Structure–activity similarity (SAS) maps were originally introduced by Maggiora and colleagues[30] and provide a graphical representation of relationships between structural similarity and "activity similarity" of bioactive compounds. Structural similarity of ligand pairs is evaluated based on the Tanimoto coefficient for MACCS structural keys and plotted against an index for activity similarity derived from IC_{50} values. In the resulting graphs, regions of varying information content can be identified. For instance, regions of high 2D structural similarity and low similarity in activity correspond to rugged activity landscapes, whereas regions of high structural and activity similarity form smooth activity landscapes. According to these features, SAS maps for idealized rough and smooth activity landscapes can be devised to serve as a reference for SAR comparisons. The comparison of different SAS maps is based on an information-theoretic metric that compares the frequency distributions of the structural and activity similarity in theoretical and observed SAS maps. The Kullback–Leibler divergence[31] applied for this purpose provides a quantitative measure of the divergence between these distributions. Using idealized SAS maps as a reference, the smoothness or roughness of activity landscapes of sets of active molecules can be estimated.

SAR Index. The SAR Index (SARI) has recently been introduced to quantitatively capture the continuous, discontinuous, or heterogeneous nature of activity landscapes and SARs.[32] Similar to SAS maps, it exclusively relies on the 2D structural similarity and potency distribution within a set of active compounds. However, SARI aims to categorize the SARs of a population of compound sets without employing idealized reference states. It generates a numerical index between 0 and 1 that reflects the (dis-)continuity of the SARs under consideration. SARI distinguishes between three major categories of

SARs: continuous, discontinuous and heterogeneous. In the following, the conceptual framework of SARI will be presented and compound classes representative for the distinct SAR types are discussed.

SARI is constituted by two individual score components that evaluate the similarity spectrum within a compound class and potency differences between related ligands as the major determinants of SAR characteristics. Two-dimensional structural similarity of compounds is calculated using the Tanimoto coefficient for MACCS structural keys and potency is represented by either pK_i or pIC_{50} values. Both individual scores of the SARI are first calculated in a "raw" numerical form and then transformed into final normalized scores.

The "continuity score" estimates the continuous character of an activity landscape. The essence of continuous SARs is the presence of increasingly diverse structures with similar potency, corresponding to the absence of activity cliffs that are captured by the "discontinuity score", as explained below. Therefore, the continuity score essentially measures the structural diversity within a class of active compounds. It is derived from the potency-weighted mean of pair-wise compound similarity. For each ligand pair, the weights combine the magnitude of their potency and also the potency difference. Accordingly, pairs of ligands with high potency but low potency differences contribute more to the continuity score than ligand pairs with overall low potency and high potency differences. This weighting scheme takes into account that continuity of SARs is often reflected by the presence of comparably potent inhibitors of increasing structural diversity. The potency-weighted mean of ligand similarity is transformed into a diversity measure by subtraction from 1. The "raw" continuity score is defined as:

$$\mathrm{raw_{cont}} = 1 - \frac{\sum\limits_{\mathrm{Ligands}\ i>j} w_{ij}\mathrm{sim}(i,j)}{\sum\limits_{\mathrm{Ligands}\ i>j} w_{ij}} \qquad (4.2)$$

where the weight for each ligand pair (i,j) is set to:

$$w_{ij} = \frac{\mathrm{pot}(i)\cdot\mathrm{pot}(j)}{1+|\mathrm{pot}(i)-\mathrm{pot}(j)|} \qquad (4.3)$$

In this formula, $\mathrm{pot}(i)$ gives the potency value of compound i and $\mathrm{sim}(i,j)$ refers to the MACCS Tanimoto similarity between compounds i and j.

The "discontinuity score" assesses the discontinuous features of an activity landscape.

The most prominent characteristic of discontinuous SARs is the presence of activity cliffs. Accordingly, the discontinuity score considers average potency differences for pairs of similar ligands. Here only ligand pairs are considered that reach a predefined similarity threshold value because the discontinuity of SARs is largely reflected by the presence of similar compounds having significant differences in potency. To emphasize potency differences among closely related compounds, the potency difference of each ligand pair is multiplied by their pair-wise similarity value. The "raw" discontinuity score is defined as:

$$\text{raw}_{\text{disc}} = \frac{\displaystyle\sum_{\{i>j|\,\text{sim}(i,j)>0.6\}} |\text{pot}(i) - \text{pot}(j)| \cdot \text{sim}(i,j)}{|\{i>j|\,\text{sim}(i,j)>0.6\}|} \qquad (4.4)$$

The similarity threshold for ligand pairs that are considered in calculating the discontinuity score is set to 0.6. This relatively "soft" threshold value ensures that also potency differences between remotely similar compounds are taken into account and thus enables a thorough assessment of putative activity cliffs, which is further emphasized by multiplication by pair-wise ligand similarity.

To obtain a common reference frame for SARI analysis, raw scores are converted into z-scores using the sample mean and standard deviation of the scores of a reference set of different activity classes. For initial SARI calculations discussed herein, a set of 16 representative compound classes was used as a reference set[32] (Table 4.3). Using larger reference sets had no significant influence on the results of the analysis. Thus, the classes studied here were sufficient to calculate statistically sound z-scores. Therefore, the mean and

Table 4.3 Summary of quantitative SAR characteristics of 16 reference classes.[a]

Target	SAR characteristics		
	Continuity score	Discontinuity score	SAR index
continuous			
poly(ADP-ribose) polymerase	0.82	0.03	0.89
coagulation factor Xa	0.71	0.12	0.80
cyclin-dependent kinase 2	0.74	0.36	0.69
protein-tyrosine phosphatase 1b	0.75	0.44	0.66
elastase	0.64	0.38	0.63
heterogeneous			
carbonic anhydrase II	0.30	0.08	0.61
cyclooxygenase 2	0.79	0.69	0.55
trypsin	0.37	0.42	0.47
dihydrofolate reductase	0.59	0.67	0.46
thromboxane synthase	0.82	0.89	0.46
acetylcholine esterase	0.82	0.93	0.45
peptidylprolyl isomerase (FKBP-12)	0.17	0.26	0.45
thymidylate synthase	0.16	0.33	0.41
thrombin	0.71	0.92	0.40
discontinuous			
ribonuclease A	0.004	0.68	0.16
adenosine deaminase	0.15	0.85	0.15

[a]SARI scores for 16 sets of active compounds used as reference classes for SARI calculations. The table is adapted from.[32] The SARI scoring range of approximately 0.4–0.6 is considered intermediate and indicates the presence of heterogeneous SARs.

standard deviation derived from this set can also be used as reference for SARI calculations on other compound classes.

The z-scores are then transformed into scores between 0 and 1 by calculating the cumulative probability distribution for each score under the assumption of a normal distribution, which yields the final continuity and discontinuity scores:

$$\text{score}_{\text{cont}} = \Phi(z\text{-score}_{\text{cont}}) = \frac{1}{\sqrt{2\pi}} \int_{-\infty}^{z\text{-score}_{\text{cont}}} \exp\left(-\frac{1}{2}x^2\right) dx \qquad (4.5)$$

$$\text{score}_{\text{disc}} = \Phi(z\text{-score}_{\text{disc}}) = \frac{1}{\sqrt{2\pi}} \int_{-\infty}^{z\text{-score}_{\text{disc}}} \exp\left(-\frac{1}{2}x^2\right) dx \qquad (4.6)$$

The SAR Index is ultimately calculated as the mean of the final scores. High continuity and discontinuity scores are indicative of opposite SAR characteristics and, therefore, the discontinuity score is subtracted from 1 to obtain a complementary value:

$$\text{SARI} = \frac{1}{2}[\text{score}_{\text{cont}} + (1 - \text{score}_{\text{disc}})] \qquad (4.7)$$

Hence, the SAR Index yields values between 0 and 1. Low SARI values indicate discontinuous SARs, high SARI values continuous ones, and intermediate values heterogeneous SARs that combine continuous and discontinuous elements.

Exemplary SAR Analysis. In addition to classifying SARs into the three major categories described above, SARI analysis has made it possible to distinguish between two previously unobserved sub-types of heterogeneous SARs that differ in the way continuous and discontinuous elements are related to each other. Examples are presented in the following section. For each compound class the relationship between structural similarity and potency is represented in a diagram that correlates pair-wise 2D molecular similarity and potency differences, reminiscent of SAS maps. Representative diagrams corresponding to the compound classes discussed below are shown in Figure 4.7. Table 4.3 summarizes SARI and its component scores for the 16 activity classes.

Continuous SARs. The set of factor Xa inhibitors discussed above presents a prime example of continuous SARs. As illustrated in Figure 4.7(a), many compounds show a high degree of structural diversity but have rather similar potency values and, in addition, similar compounds also have similar potency. These characteristics are reflected by a high SARI score of 0.80 produced by high continuity and low discontinuity scores, which is indicative of a continuous SAR. The continuity score of 0.71 reflects high intra-class structural diversity, whereas the discontinuity score of 0.12 indicates that similar ligands

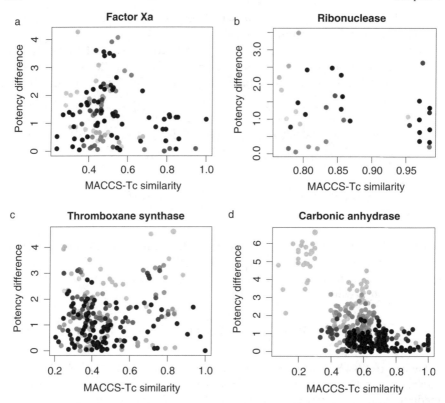

Figure 4.7   Potency difference *versus* 2D similarity of enzyme inhibitors. Each data
point represents a pair-wise comparison of inhibitors within an activity
class. Data points are grayscale-coded according to potency represented as
the sum of their pK_i values using a continuous spectrum from light grey
(lowest combined potency) to black (highest combined potency). Distributions are shown for four sets of enzyme inhibitors that represent
different types of SARs, as discussed in the text: (*a*) factor Xa, (*b*) ribonuclease A, (*c*) thromboxane synthase and (*d*) carbonic anhydrase.

generally have very low potency differences. Figure 4.8 shows a diverse spectrum of inhibitors with potency in the nanomolar range.

Discontinuous SARs.   Consistent with the results of our qualitative SAR
analysis presented above, SARI scoring indicates that ribonuclease A inhibitors are a prototypic example of discontinuous SARs. The potency among
these highly similar nucleotide analogs differs by up to three orders of magnitude (Figure 4.7b). As already mentioned above, these features clearly indicate the presence of discontinuous SARs. The low SARI score of 0.16 mirrors
these findings. It results from the combination of a very low continuity score of
nearly zero (0.004) that reflects the lack of structural diversity within the

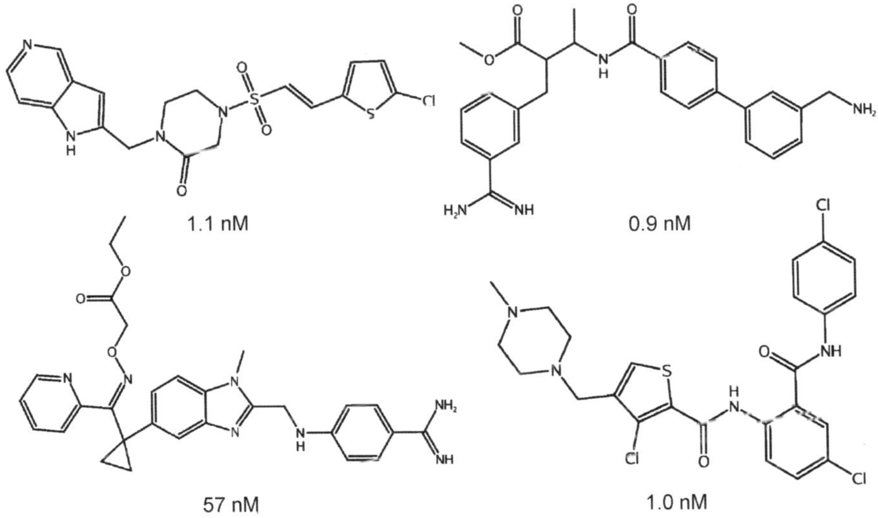

Figure 4.8 A spectrum of structurally diverse factor Xa inhibitors with potencies in the nanomolar range.

compound set and a high discontinuity score of 0.68 that accounts for large potency differences among very similar ligands.

Heterogeneous SARs. Heterogeneous SARs combine continuous and discontinuous features that can be associated in different ways, depending on the nature of the activity class. Systematic SARI calculations revealed two distinct sub-types of heterogeneous SARs. Both sub-types yield intermediate SARI scores but are distinguished by the magnitude of continuity and discontinuity scores. One sub-type is characterized by mutually coexisting continuous and discontinuous SARs, yielding both high continuity and discontinuity scores. This combination is indicative of permissive or adaptable binding sites. Accordingly, the corresponding SAR category is termed "heterogeneous-relaxed". The other sub-type of heterogeneous SARs consists of inhibitors that are related by continuous SARs that exist within the boundaries presented by a binding constraint and the ensuing activity cliff. In this case, the binding constraint is best rationalized as a structural feature within an active site imposing interactions on ligands that are essential for binding. For example, this could be an ion coordination sphere or a charged residue presented in an otherwise hydrophobic binding site environment. This situation produces both low continuity and discontinuity scores and the corresponding SAR sub-type is referred to as "heterogeneous-constrained".

Inhibitors of thromboxane synthase are an example of heterogeneous-relaxed SARs. The ligand spectrum includes diverse and highly potent inhibitors, similar to factor Xa, but also pairs of structurally similar inhibitors having very different potency (Figure 4.7c). In this case, the coexistence of continuous

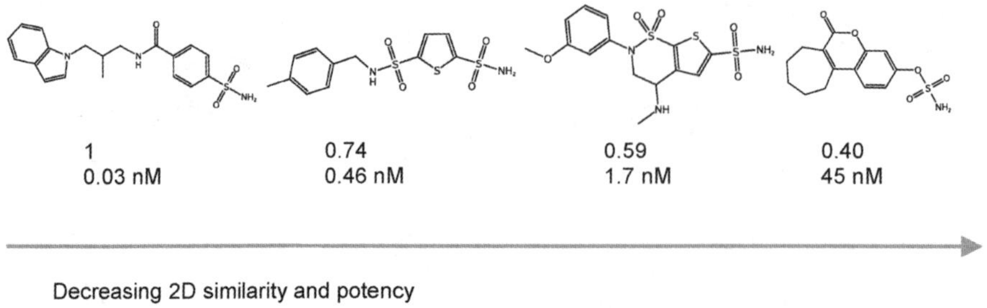

1	0.74	0.59	0.40
0.03 nM	0.46 nM	1.7 nM	45 nM

Decreasing 2D similarity and potency

Figure 4.9   Carbonic anhydrase inhibitors following a continuous SAR. Within a set of sulfonamides, increasing structural diversity is accompanied by gradual changes in potency. 2D similarity is reported on the basis of MACCS Tc values for pair-wise comparisons using the molecule on the left-hand side as the reference structure.

and discontinuous SARs is reflected by high continuity and discontinuity scores of 0.82 and 0.89, respectively.

Carbonic anhydrase inhibitors represent a prototype of a heterogeneous-constrained SAR type, characterized by a low continuity (0.30) and very low discontinuity (0.08) score. The low continuity score indicates that the diversity of inhibitors is limited, especially among highly potent ones (Figure 4.7d). All inhibitors share a sulfonamide group, which is required to complex a catalytically important zinc cation within the active site of carbonic anhydrase. The need to complex this zinc ion presents a binding constraint for the inhibitors. However, this activity cliff does not render the activity landscape completely discontinuous. It is only one important element in a landscape of heterogeneous topology that limits the activity radius of effective inhibitors. Compounds that fall within this activity radius meet the constraint and are related by a restricted continuous SAR. Accordingly, moderate structural variations are tolerated that result in gradual potency changes, consistent with the features of continuous SARs (Figure 4.9).

4.3.4.5   Implications for Molecular Similarity Analysis and Virtual Screening

The systematic qualitative and quantitative analysis of SAR characteristics confirms the assumption that many SARs should be heterogeneous in nature and provides a stringent formalism for classification of SAR categories. Clearly, the results of systematic similarity and potency correlation analysis show that SARs are more complex than often assumed. Multiple SARs can coexist within the same active site, providing evidence that the architecture of the binding site does not always "dictate" SAR features, but rather tolerates different types of SARs depending on the features of ligands. Moreover, even severe constraints on binding permit significant variability of compound binding modes and at least some degree of structural diversity among ligands. These findings partly revise current views that similar ligands display a general tendency to bind in very similar ways to the same target.[33] The confirmed variability of activity landscapes indicates that, for a multitude of target proteins, heterogeneous SARs prevail. Therefore, for many target proteins, the similarity-property principle, one of the cornerstones of similarity analysis, is applicable and active compounds are expected to cover a diverse structural spectrum.

What can SAR analysis teach us about the general nature of compounds that are active against a specific target? As with every knowledge-based approach, SARI calculations are limited to the information that can be derived from available data. SAR characteristics of small populated regions within theoretical activity landscapes do not comprehensively capture their global topology. Specifically, systematic SAR analyses and the fundamental considerations presented above lead to the conclusion that purely continuous or discontinuous SARs are rare. This implies that compound sets showing strong continuous or discontinuous SAR features might only represent a limited repertoire of the overall spectrum of structures that are active against a given target. However, extrapolating from the SAR characteristics of known bioactive compounds is often complicated by the existence of multiple SARs for which available compound data might be, at least in part, limited or biased. Nonetheless, for activity classes in which continuous or heterogeneous-relaxed SARs are observed, the opportunity exists to discover structurally diverse active compounds. In contrast, in compound classes that exhibit discontinuous or heterogeneous-constrained features, activity cliffs severely restrict activity radii, which limits the success of similarity search calculations.

In a typical LBVS scenario, ligands with known biological activity serve as reference molecules. These reference compounds often provide the only information available for SAR analysis. However, as described above, knowledge about the nature of the SARs can be obtained. This knowledge has implications for the potential success or failure of molecular similarity methods. Again, the topology of an activity landscape presents the major determinant for the success or failure of global similarity analysis. Hence, the existence of continuous SAR elements is a prerequisite for the successful application of similarity methods and the identification of structurally diverse molecules having a desired bioactivity. In contrast, discontinuous SARs are inconsistent with the similarity-property principle and do not provide a basis for global similarity evaluation and correlation with biological activity. In this case, due to the presence of activity cliffs in discontinuous activity landscapes and small activity radii, similarity methods are prone to produce many false-positives and their results become essentially meaningless. However, Table 4.3 shows that most SARs studied thus far fall into the intermediate SARI range that is characteristic for heterogeneous SARs. Ultimately, the success of holistic similarity methods in virtual screening critically depends on the presence of continuous or heterogeneous SARs. Only for compound classes where underlying SARs have such features can we expect to identify diverse structures having similar activity. Going beyond hit identification through molecular similarity analysis, heterogeneous SARs are of high interest for lead optimization: in the presence of heterogeneous SARs, there is not only a high probability of finding novel active molecules; moreover, it is also likely that these compounds can be optimized, provided analog generation can be focused on a region with an activity cliff that disrupts the rolling hill topology of the activity landscape. By contrast, although purely or mostly continuous SAR landscapes are prime targets for LBVS, they can present severe problems for medicinal chemistry, because it might be rather difficult to substantially improve the potency of hits by generating series of analogs and converting them into leads. This gives rise to a phenotype often called "flat SAR", which is much disliked by medicinal chemists.

4.4   Strengths and Limitations of Similarity Methods

The heterogeneous nature of many SARs helps to rationalize why rather different similarity methods succeed in many cases and are indeed capable of identifying structurally diverse active compounds. Moreover, it also provides an explanation for the strong compound class dependence of similarity methods, albeit only in part. The existence of distinct SAR categories does not readily explain why various methods frequently perform rather differently on a given compound class. However, we also need to consider that one of the fundamental challenges for similarity analysis is to find local SAR regions of continuous nature by selecting reference molecules that account for these features. At this stage, we can expect alternative methods to either succeed or fail, dependent on the molecular representations and chemical reference spaces they utilize and the way SARs are established within these reference frames. The choice of molecular representations and reference spaces significantly influences the topology of activity landscapes and ultimately determines whether or not continuous SAR features can be abstracted from reference compounds. The rather intricate relationships between coexisting individual continuous and discontinuous SARs and their dependence on chosen representations and the way similarity relationships are "measured" complicate similarity analysis in many ways. The fact that diverse structures can have similar activity whereas closely related analogs may differ substantially in their activity presents a "similarity paradox", at least at first glance, and illustrates principal caveats for similarity methods at the molecular level of detail (Figure 4.10).

One should also consider that the potential success of similarity methods is not exclusively determined by representation- and reference frame-dependent SAR characteristics. Importantly, the success of similarity methods is also influenced by more technical factors. In other words, SAR categories principally determine if similarity search calculations can succeed; whether they actually do is another question. For example, the intrinsic structural diversity of active compounds related by many continuous SARs makes similarity calculations particularly difficult because remote similarity relationships must be

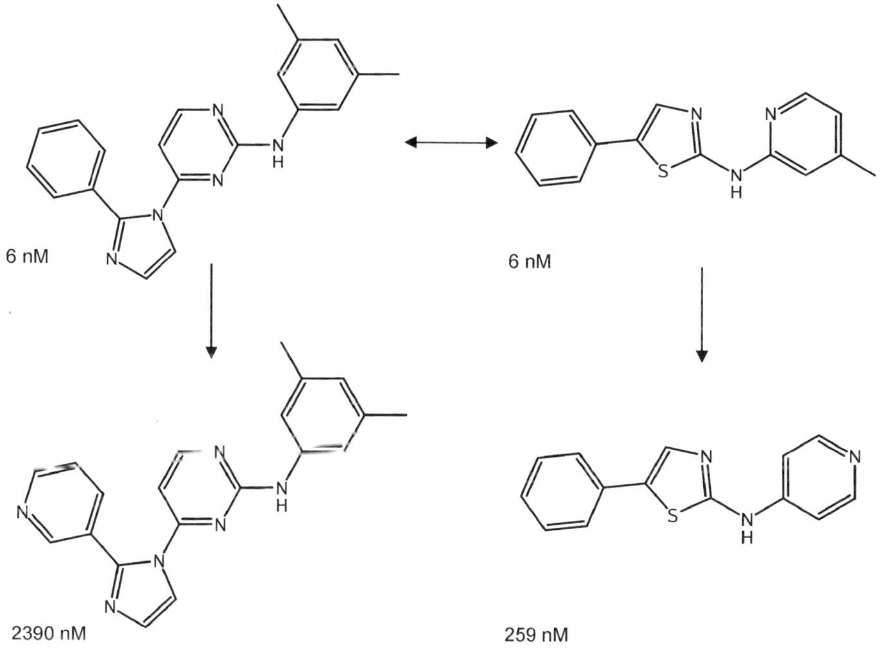

Figure 4.10 Example of a heterogeneous SAR. Four inhibitors of vascular endothelial growth factor receptor (VEGFR-2) tyrosine kinase are shown. The two inhibitors at the top have different core structures but equally high potency and are thus part of a continuous SAR. By contrast, closely related analogs of each inhibitor shown at the bottom have several orders of magnitude lower potency, which is a characteristic feature of discontinuous SARs.

detected, which is much more difficult than searching for analogs or closely related actives. The detection of remote similarity relationships is also highly sensitive to chosen molecular representations, pre-selected structural features or descriptors, and the composition and dimensionality of chemical reference spaces, just like the categorization of SARs.

Molecular similarity methods are set apart from other LBVS approaches by the holistic view they take on molecular structure and similarity. This global approach contributes to the attraction but, at the same time, presents a major drawback of such approaches. In fact, a common criticism is that the global similarity approach takes molecular features into account that do not contribute to biological activity. However, similarity searching is particularly attractive if no information about target–ligand interactions or mechanisms of bioactivity is available. Because similarity analysis considers the entire molecular structure, no knowledge or hypotheses about an activity-dependent pharmacophore is required and no information is needed concerning details of the underlying receptor–ligand interactions. Moreover, although similarity searching benefits from the use of multiple active reference structures,[34–36] it can also be applied when only single reference molecules are available (in contrast to similarity-based compound classification or machine learning methods[2]). Thus, similarity searching and related methods provide valuable tools during the early stages of a drug discovery campaign, when little additional information is available. The global assessment of molecular similarity adds a degree of "fuzziness" to the analysis, which facilitates the departure from given reference structures and enables "scaffold hops",[37] *i.e.*, the transition from one series of active compounds to another. However, the comparably low resolution of similarity analysis causes a general tendency to produce false-positives, which also explains why benchmark calculations on hand-selected activity classes consisting of optimized molecules added to random decoys are typically much more successful than "real life" virtual screening trials where the targets are hits, not optimized leads. Nevertheless, many practical applications also show the value of similarity analysis in virtual screening and computer-aided drug discovery. If we take into account that millions of test compounds are often screened on the computer to ultimately select only tens or hundreds of candidates for testing, akin to a "needles in haystacks" scenario, then the identification of novel active compounds for further study and optimization presents a considerable success, even if only a few are found.

4.5   Conclusion and Future Perspectives

In this chapter we have discussed the foundations of molecular similarity analysis that are largely responsible for success or failure of similarity-based methods in the context of virtual compound screening. Furthermore, we have described how crucial underlying SAR characteristics are for molecular similarity analysis and presented a comprehensive methodological framework for the qualitative and quantitative analysis of SARs and the classification of different SAR types.

The classical application scenario for similarity methods is the retrieval of candidate molecules with a desired bioactivity from a compound database. While similarity searching continues to be of great use in this field, new application areas are beginning to emerge. For example, the development of chemogenomics[38] as a relatively young interdisciplinary area of research brings up new challenges for similarity analysis in target and lead discovery. Chemogenomics aims at the exploration of therapeutically relevant target families using small molecules and mapping of "pharmacological space",[39] *i.e.*, a systematic analysis of the universe of specific target–ligand interactions. Thus, the major aspect of such studies is to relate small molecule chemical space to target space, *e.g.*, by profiling small molecules against arrays of target proteins. Here, similarity analysis can be of particular use for identifying active compounds that are similar to reference molecules and also show activity against other targets within the same family. Going beyond compounds that have differential activity against members of target families or sub-families, "selectivity searching" can be applied to detect compounds that are selective against individual targets among related ones.[40] Clearly, despite its principal limitations and caveats, the opportunities of molecular similarity analysis reach beyond its classical application scenarios and we can expect that there will be much more to come.

References

1. J. Bajorath, *Drug Discov. Today*, 2004, **9**, 13.
2. J. Bajorath, *Nature Drug Discov. Rev.*, 2002, **1**, 337.
3. D. B. Kitchen, H. Decornez, J. R. Furr and J. Bajorath, *Nature Rev. Drug Discov.*, 2004, **3**, 935.
4. P. D. Lyne, *Drug Discov. Today*, 2002, **7**, 1047.
5. H. Eckert and J. Bajorath, *Drug Discov. Today*, 2007, **12**, 225.
6. F. L. Stahura and J. Bajorath, New methodologies for ligand-based virtual screening, *Curr. Pharm. Des.*, 2005, **11**, 1189.
7. P. Gund, in *Progress in Molecular and Subcellular Biology*, ed. F. E. Hahn, Springer-Verlag, Berlin, 1977, p. 117.
8. R. P. Sheridan, A. Rusinko III, R. Nilakantan and R. Venkataraghavan, *Proc. Natl. Acad. Sci. U.S.A.*, 1989, **86**, 8165.
9. J. S. Mason, A. C. Good and E. J. Martin, *Curr. Pharm. Des.*, 2001, **7**, 567.
10. E. X. Esposito, A. J. Hopfinger and J. D. Madura, *Methods Mol. Biol.*, 2004, **275**, 131.
11. J. M. Barnard, *J. Chem. Inf. Comput. Sci.*, 1993, **33**, 532.
12. D. E. Patterson, R. D. Cramer, A. M. Ferguson, R. D. Clark and L. E. Weinberger, *J. Med. Chem.*, 1996, **39**, 3049.
13. In *Concepts and Applications of Molecular Similarity*, ed. M. A. Johnson, G. M. Maggiora, Wiley, New York, 1990.
14. J. Bajorath, *J. Chem. Inf. Comput. Sci.*, 2001, **41**, 233.
15. D. J. Livingstone, *J. Chem. Inf. Comput. Sci.*, 2000, **40**, 195.
16. L. Xue and J. Bajorath, *Comb. Chem. High Throughput Screening*, 2000, **3**, 363.
17. MACCS structural keys, Symyx Software, San Ramon, CA.
18. J. S. Mason, I. Morize, P. R. Menard, D. L. Cheney, C. Hulme and R. F. Labaudiniere, *J. Med. Chem.*, 1999, **42**, 3251.
19. P. Willett, J. M. Barnard and G. M. Downs, *J. Chem. Inf. Comput. Sci.*, 1998, **38**, 983.
20. V. J. Gillet, P. Willett and J. Bradshaw, *J. Chem. Inf. Comput. Sci.*, 2003, **43**, 338.
21. R. P. Sheridan and S. K. Kearsley, *Drug Discov. Today*, 2002, **7**, 903.
22. J. Batista, J. W. Godden and J. Bajorath, *J. Chem. Inf. Model.*, 2006, **46**, 1937.
23. D. J. Graham, C. Malarkey and M. V. Schulmerich, *J. Chem. Inf. Comput. Sci.*, 2004, **44**, 1601.
24. G. M. Maggiora, *J. Chem. Inf. Model.*, 2006, **46**, 1535.

25. H. Kubinyi, *Perspect. Drug Discovery Des.*, 1998, **9–11**, 225.
26. F. Bonachera, B. Parent, F. Barbosa, N. Froloff and D. Horvath, *J. Chem. Inf. Model*, 2006, **46**, 2457.
27. L. Peltason and J. Bajorath, *Chem. Biol.*, 2007, **14**, 489.
28. E. Fischer, *Ber. Dtsch. Chem. Ges.*, 1894, **27**, 2985.
29. D. E. Koshland Jr., *Proc. Natl. Acad. Sci. U.S.A.*, 1958, **44**, 98.
30. V. Shanmugasundaram, and G. M. Maggiora, 222nd American Chemical Society National Meeting, Division of Chemical Information, American Chemical Society, Washington, DC, 2001; Abstract no. 77.
31. S. Kullback and R. A. Leibler, *Ann. Math. Statist.*, 1951, **55**, 79.
32. L. Peltason and J. Bajorath, *J. Med. Chem.*, 2007, **50**, 5571.
33. J. Boström, A. Hogner and S. Schmitt, *J. Med. Chem.*, 2006, **49**, 6716.
34. J. W. Godden, L. Xue, F. L. Stahura and J. Bajorath, *Pac. Symp. Biocomput.*, 2000, **8**, 566.
35. L. Xue, J. W. Godden, F. L. Stahura and J. Bajorath, *J. Chem. Inf. Comput. Sci.*, 2003, **43**, 1218.
36. A. Schuffenhauer, P. Floersheim, P. Acklin and E. Jacoby, *J. Chem. Inf. Comput. Sci.*, 2003, **43**, 391.
37. G. Schneider, P. Schneider and S. Renner, *QSAR Comb. Sci.*, 2006, **25**, 1162.
38. M. Bredel and E. Jacoby, *Nat. Rev. Genet.*, 2004, **5**, 262.
39. G. V. Paolini, R. H. Shapland, W. P. van Hoorn, J. S. Mason and A. L. Hopkins, *Nat. Biotechnol.*, 2006, **24**, 805.
40. I. Vogt, D. Stumpfe, H. E. A. Ahmed and J. Bajorath, *Chem. Biol. Drug Des.*, 2007, **70**, 195.
41. P. Willett, *J. Med. Chem.*, 2004, **48**, 4183.
42. F. L. Stahura and J. Bajorath, *Combin. Chem. High Throughput Screen.*, 2004, **7**, 259.
43. X. Chen, A. Rusinko III and S. S. Young, *J. Chem. Inf. Comput. Sci.*, 1998, **38**, 1054.
44. G. Harper, J. Bradshaw, J. C. Gittins, D. V. S. Green and A. R. Leach, *J. Chem. Inf. Comput. Sci.*, 2001, **41**, 1295.
45. M. K. Warmuth, J. Liao, G. Ratsch, M. Mathieson, S. Putta and C. Lemmen, *J. Chem. Inf. Comput. Sci.*, 2003, **43**, 667.
46. R. N. Jorissen and M. K. Gilson, *J. Chem. Inf. Model.*, 2005, **45**, 549.
47. R. S. Pearlman and K. M. Smith, *Perspect. Drug Discov. Design*, 1998, **9**, 339.
48. J. W. Godden, J. W. Furr, L. Xue, F. L. Stahura and J. Bajorath, *J. Chem. Inf. Comput. Sci.*, 2004, **44**, 21.
49. J. W. Godden, F. L. Stahura and J. Bajorath, *J. Med. Chem.*, 2004, **47**, 4286.
50. H. Eckert and J. Bajorath, *J. Med. Chem.*, 2006, **49**, 2284.
51. H. Eckert, I. Vogt and J. Bajorath, *J. Chem. Inf. Model.*, 2006, **46**, 1623.
52. J. M. Barnard and G. M. Downs, *J. Chem. Inf. Comput. Sci.*, 1997, **37**, 141.
53. C. A. James and D. Weininger, *Daylight Theory Manual*, Daylight Chemical Information Systems, ch. 6, 2006.

54. A. Bender, H. Y. Mussa, R. C. Glen and S. Reiling, *J. Chem. Inf. Comput. Sci.*, 2004, **44**, 1708.
55. J. A. Haigh, B. T. Pickup, J. A. Grant and A. Nicholls, *J. Chem. Inf. Model.*, 2005, **45**, 673.
56. J. C. Saeh, P. D. Lyne, B. K. Takasaki and D. A. Cosgrove, *J. Chem. Inf. Model.*, 2005, **45**, 1122.
57. H. Eckert and J. Bajorath, *J. Chem. Inf. Model.*, 2006, **46**, 2515.
58. T. S. Rush III, J. A. Grant, L. Mosyak and A. Nicholls, *J. Med. Chem.*, 2005, **48**, 1489.
59. M. Vogt, J. W. Godden and J. Bajorath, *J. Chem. Inf. Model.*, 2007, **47**, 39.

CHAPTER 5

Molecular Field Topology Analysis in Drug Design and Virtual Screening

EUGENE V. RADCHENKO, VLADIMIR A. PALYULIN
AND NIKOLAY S. ZEFIROV

Department of Chemistry, Moscow State University,
Moscow 119991, Russia

5.1 Introduction: Local Molecular Parameters in QSAR, Drug Design and Virtual Screening

During the search for effective drugs and other bioactive compounds, attention is usually focused on specific activity stemming from the "receptor-like" interactions of small organic ligand molecules with a well-defined biological target (enzyme, receptor, etc.).[1] The nature and strength of such interactions are obviously controlled by the local physicochemical features of a molecule, related in structural terms to the properties of its atoms and bonds. Thus, structure–activity relationships can be analyzed by correct comparison of activity and local molecular properties, both within a single structure and between various congeneric structures. Once such a comparison is achieved, one can build a predictive statistical model linking these properties to the bioactivity parameters. This model can then be used as a virtual screening filter to select a manageable subset of promising structures among a large body of conceivable or accessible structures of the same chemical class. In addition, such a model can lead to

Chemoinformatics Approaches to Virtual Screening
Edited by Alexandre Varnek and Alex Tropsha
© Royal Society of Chemistry, 2008
Published by the Royal Society of Chemistry, www.rsc.org

important conclusions with respect to the mechanism of drug action as well as the directions for further optimization of activity profile.

Generally speaking, existing approaches to the QSAR analysis based on local molecular properties can be classified into 3D-based and topology-based, depending on the underlying structure representation. The first group of approaches starts with a 3D model of atom positions. From it, some uniform representation of structural features is derived to make different molecules comparable. Instead of individual atoms, this representation is linked to the molecular axes of inertia or to an abstract spatial grid, thus avoiding the problem of matching the atoms of different structures. However, the path from molecular structure to uniform representation is not well defined and often requires some heuristics and manual intervention, which complicates not only the construction of a model but also its application to virtual screening.

Another, topological, group of approaches is directly based on the structural formulae of the compounds, representing types of atoms and bonds (in addition, atoms and bonds may be labeled with the physicochemical and stereochemical data). At first glance, 3D-based approaches seem more precise and better reflect the actual 3D nature of biotarget and ligand. However, practical experience shows that consideration of a 3D model is not always beneficial. In part, topological approaches are easier to synchronize to the mentality of organic and medicinal chemists, thus facilitating the design of novel promising structures and synthesis planning. In addition, 3D techniques involve a lot of data on the particular details of molecular structures, conformational behavior and physico-chemical parameters of the compounds. Unfortunately, in many cases they fail to create a holistic picture of the ligand–target interaction, serving instead as a kind of "info-noise" that complicates the structure–activity analysis.

The most commonly used and possibly even "classical" method of 3D QSAR analysis is the Comparative Molecular Field Analysis (CoMFA) technique intro-duced by R. Cramer *et al.* in 1988.[2] In almost 20 years since, it has seen substantial development and enhancement, as well as the creation of several related approa-ches. In general terms, it aims to identify the spatial regions around the molecule where certain local properties have a positive or negative effect on activity.

The foundation of the CoMFA approach lies in the fact that the interaction between the biotarget and organic ligand is usually non-covalent and sub-stantially controlled by the shape of molecules. In addition, van der Waals and Coulomb forces in most cases provide an adequate description of non-covalent interactions within a molecular mechanics framework. Thus, the authors assumed that the biological action of compounds can be explained by the shape and electrostatic field of their molecules.

A key feature of this approach is the comparison of the quantitative measures of these fields following the spatial alignment of 3D structures of similar com-pounds. It involves populating the descriptor matrix with calculated energies of the van der Waals (steric) and Coulomb (electrostatic) interaction of a molecule with a probe species in each node of a rectangular 3D grid. Depending on the spe-cific problem, various probe species can be used, such as proton, sp^3 carbon atom with a unit positive charge, *etc.* The QSAR models are usually derived from these megavariate data matrices by means of the cross-validated Partial Least Squares Regression (PLSR).[3] In contrast to the traditional multiple linear regression, it allows the predictive statistical relationship to be detected even if the number of descriptors is much greater than the number of the experimental data points.

Despite some problems,[4] this approach rapidly gained widespread acceptance in the QSAR field. First, this can be attributed to the appeal of 3D activity maps derived from the PLSR model and representing regions of favorable and unfavorable interactions. In addition, after patenting the CoMFA technique,[5] it was implemented in a popular Tripos SYBYL molecular modeling package.[6] Many hundreds of works have been published concerning the application and further development of the CoMFA approach to the prediction of bioactivity of organic compounds.[7] The GRID/GOLPE technique,[8,9] an approach conceptually similar to CoMFA, is also commonly used.

In addition to the steric and electrostatic descriptors, it was proposed to use other 3D molecular fields characterized by the sampling over the rectangular grid – in particular, the hydrophobic field/molecular lipophilic potential (MLP),[10–12] hydrogen bonding[13] and quantum-chemical parameters, *e.g.*, orbital densities.[14,15] Descriptor selection techniques are often recommended to enhance the stability, predictivity and interpretability of the CoMFA models.[16–18]

Nevertheless, the practical application of CoMFA is frequently hampered by the large number of descriptors and by the problem of alignment of the 3D molecular structures,[19] especially for flexible molecules where many accessible low-energy conformations exist and the induced ligand–biotarget fit may give rise to ligand conformations substantially different from optimal conformations of the isolated molecules. In some cases, simple topological models (*e.g.*, fragment-based) can provide a better prediction of activity.[20,21]

Several approaches were proposed to alleviate these problems. Modifications of the original CoMFA procedure[2] involve flexible instead of rigid molecular alignment[22,23] as well as the alternative statistical analysis techniques.[24] In addition, some approaches aim to minimize model sensitivity to alignment or eliminate altogether this step from the analysis. For instance, steric molecular fields can be characterized by the intersection volume of the van der Waals spaces of the ligand molecule and probe species. The distance dependence of such volumes is smoother than the standard Lennard-Jones potential function.[25] The spatial auto- and cross-correlation parameters[4,26,27] as well as the mass and charge distribution moments[28] and vibrational modes[29] provide 3D molecular descriptors invariant to the translation and rotation of molecules and less sensitive to their conformations. Unfortunately, they are also less intuitive and interpretable.

Within a series of congeneric compounds, some canonical (*i.e.*, formal) rules for conformer selection and alignment often are required and are sufficient to obtain useful CoMFA results.[19,30,31] Further development of this concept has led to the creation of the Topomer CoMFA approach[32–34] that can be regarded as an automated "2.5D" alignment followed by the 3D QSAR analysis. While it is undoubtedly a very promising technique, addressing many common obstacles and particularly well adapted to virtual screening, one potential issue lies in the very fact that alignment is built without taking into account any specific features of the receptor.[32] This might create a risk of overlooking important differences in binding modes, as well as weaken the (already indirect) correspondence between 3D QSAR model and the actual target structure. Now that the Topomer CoMFA is publicly available, it will be very interesting to see it applied to a wide range of QSAR problems.

Generally speaking, 3D QSAR approaches provide useful tools for drug design and virtual screening. However, in many cases they require one to "go back" to topology-based (2D or 2.5D) structure representation rather than analyze the 3D molecular models directly.

5.2 Supergraph-based QSAR Models

5.2.1 Rationale and History

As we have shown, topological (2D-based) approaches to QSAR modeling are free from several complications typical of 3D analysis. If the compounds under study have sufficiently similar structures, the comparison of the local molecular properties should allow one to reveal structural features critical for activity. In a sense, this concept can be traced back to Hansch analysis[35–37] (parameters of the substituents in specific positions) and Free–Wilson approach[38,39] (indicator variables representing particular substituents present in specific positions). However, these substituent-based approaches lose their applicability if we go beyond a limited number of simple substituents to more practically useful compound series. Thus, most topological QSAR approaches rely on some characteristics of the molecule as a whole, *e.g.*, presence or occurrence number of certain fragments (substructures).[40] In this case, the information on the arrangement of various molecular features is almost lost, limiting the application of these approaches to the modeling of the receptor-type activity.

Nevertheless, several proposed approaches are based on the concept of a superstructure spanning the variability of molecules in a dataset. It can be thought of as a topological network that allows superimposing of every dataset structure. This brings their atoms and bonds into the same frame of reference, yielding the uniform representation of the local molecular properties.

Notably, the concept of super- or hyperstructure is sometimes used in chemo-informatics as a way to build the most compact representation of a series of structures for easy database storage and retrieval.[41,42] Some attempts were made to take qualitative activity data into account during the construction of a hyperstructure.[43] However, the primary goal of compactness dictates certain hyperstructure features and algorithms that complicate its chemical interpretation and use in QSAR modeling.

On the other hand, several superstructural approaches were designed specifically for the QSAR analysis and lead optimization for organic compounds. Let us consider them in more detail.

The DARC/PELCO method (Méthode de perturbation d'environnements limités concentriques ordonnés)[44,45] for predicting the properties of organic compounds was developed by J.-E. Dubois and C. Mercier. In this approach, the structures are considered as a combination of the core (common central substructure of a series) and the environment (all other atoms and bonds partitioned into a sequence of concentric levels based on their distance from the core). A trace of the population unites all the sites present at least in a single compound of a series and basically represents the superstructure of a series. In the simplest form of this method the indicator variables encoding the presence of the particular atom and bond types in each site of the environment are used as the structural descriptors. The activity of a structure as a whole is represented as a sum of some basic activity of the core and the "perturbations" caused by all the occupied environment sites. These contributions are determined by means of the regression analysis and can be visualized by the "activity maps". Thus, the DARC/PELCO method may be viewed as a detailed and generalized version of the Free–Wilson approach.[38,39]

If necessary for the analysis, secondary descriptors may be included that are constructed as logical combinations of the primary descriptors.[46] In further development of the method, it was proposed to include the so-called "external" physicochemical properties of a whole molecule, e.g., lipophilicity and quantum-chemical parameters important for the mechanism of action.[45,47] Unfortunately, in recent years, the progress of this approach seems to have ceased.

In the approach proposed by G. Menon and A. Cammarata[48,49] a series of congeneric structures is classified by the activity type using the principal components of a set of local molecular descriptors based on a superstructure constructed by the simplest "chemically consistent" superposition of the structures.

The Positional Analysis approach proposed by P. Magee[50,51] aims to detect the structural features responsible for the ligand–biotarget interaction by constructing the simplest common hyperstructure for a series of compounds and analyzing the presence and interaction parameters for each hyperstructure position by means of the multiple linear regression.

The Minimum Topological Difference (MTD) approach[52,53] developed by Z. Simon and T. Oprea, especially in its newer MTD-PLS variant employing the PLS regression analysis, successfully relates the activity to the presence and physicochemical parameters of the atoms and fragments defined over a quasi-3D hypermolecule[54] as well as global physicochemical descriptors.[55] Despite some methodological problems and limitations, this approach is promising and in active current development.[56]

5.2.2 Molecular Field Topology Analysis (MFTA)

5.2.2.1 General Principles

The Molecular Field Topology Analysis (MFTA) technique was proposed by us[57,58] as a generalization and extension of several superstructural approaches to the QSAR analysis of organic compounds. It is in line with the modern trend of QSAR studies that involves application of the topological and quasi-topological (2D and 2.5D) methods to the modeling of drug–biotarget interactions.

In a sense, MFTA may be viewed as a topological analogue of the CoMFA approach.

The method is based on the assumption that, in many cases, the alignment of the topological (2D) rather than the spatial (3D) structures of the compounds could alleviate the problems inherent to the 3D QSAR approaches and provide rather general methodology for the prediction of bioactivity of organic compounds based on the specific (receptor-like) action mechanism. As molecular descriptors, this approach uses the local physicochemical parameters – atom and bond properties that can be quickly estimated from the structure of the compound. The uniform frame of reference for their analysis is provided by a so-called molecular supergraph (MSG) – simple graph automatically constructed in such a way that any training set structure can be represented as its subgraph.

The process of the QSAR analysis based on the MFTA approach[59] and implemented in the convenient MFTAWin software involves two related procedures: the construction of a structure–activity model and the prediction of bioactivity for the new, as yet untested compounds. Figure 5.1 presents the general flow-chart of the analysis.

In the model construction phase, the molecular supergraph is first constructed for a training set of compounds with known experimental activity data, and the local molecular descriptors are calculated. By superimposing each structure onto the supergraph, the uniform descriptor matrix is obtained that can be analyzed by means of the PLS regression or other statistical learning techniques, yielding the predictive QSAR model. During the prediction phase, the new structures are also superimposed onto the supergraph, and the resulting uniform descriptor vectors are used to estimate the predicted activity values from the statistical model. These values can be used to select the promising structures in the virtual screening study. In addition, the MFTA model itself can provide valuable insights into the critical features of the active

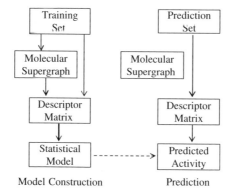

Figure 5.1 The general flow-chart of the MFTA QSAR analysis involves the construction of a model and the prediction of activity.

molecules, helping to identify the potential drug–biotarget interactions as well as the directions of lead optimization.

Each of these steps is discussed in more detail in the following sections.

5.2.2.2   Local Molecular Descriptors: Facets of Ligand–Biotarget Interaction

One of the advantages of the MFTA approach is the support of the open descriptor set that can be modified or extended to take into account the specific features of a problem. The currently available descriptors can provide an adequate description of the major types of ligand–biotarget interactions:[59]

- *Electrostatic descriptors*, in particular, effective atomic charge Q estimated using the electronegativity equalization approach[60,61] and absolute charge Qa.
- *Steric descriptors*, in particular, Bondi's van der Waals radius of an atom R,[62] effective van der Waals radius of a group Rg (taking into account the steric requirements of a central non-hydrogen atom and connected hydrogen atoms), effective van der Waals radius of atom's first environment Re (taking into account the steric requirements of a central non-hydrogen atom and all the connected atoms, both hydrogen and non-hydrogen), atomic contribution to the molecular van der Waals surface S (the surface of the atom's van der Waals sphere excluding the areas intersected by other atoms, neglecting the possible ternary intersections) and relative steric accessibility $A = S/S_{free}$ (where S_{free} is the van der Waals surface of the isolated atom of the same type).
- *Lipophilicity descriptors*, in particular, the atomic lipophilicity contribution La in Ghose and Crippen's system, taking into account the environment of an atom,[63] and group lipophilicity Lg defined as a sum of contributions for a non-hydrogen atom and attached hydrogens.
- *Hydrogen bonding descriptors*, in particular, the ability of an atom in a given environment to be a donor (Hd) and acceptor (Ha) of a hydrogen bond characterized by the Abraham's constants.[64]
- *Indicator variables* taking into account, for instance, position occupancy, bond presence and/or local stereochemistry.

In each particular problem, only a subset of the available descriptors should be used that reflects the factors important for the activity in question.

5.2.2.3   Construction of a Molecular Supergraph

Molecular supergraph for the MFTA analysis is a graph allowing the superimposition of the training set structures. The supergraph is not required to be minimal (although, for the sake of model reliability, it is desirable to use the most compact supergraph consistent with the chemical reason and a specific

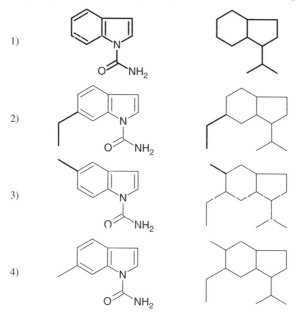

Figure 5.2 Construction of a molecular supergraph (MSG) for a simple series of structures. In each step the bold lines mark the fragments missing from the current structure-MSG intersection that are to be added to the supergraph.

problem) or unique. The basic procedure for the supergraph construction involving the stepwise processing of the training set structures is illustrated in Figure 5.2. In each step, the intersection is found between the currently constructed (initially empty) supergraph and the molecular graph of a structure. Then the supergraph is augmented by the vertices and edges corresponding to the atoms and bonds missing from the intersection (such as the ethyl substituent at step 2 in Figure 5.2).

The algorithm for the detection of intersections[59] combines the approaches based on the vertex-by-vertex expansion and on the search for the maximum cliques (complete subgraphs) in the modular graph product.[65] It provides efficient and flexible detection of the maximum connected graph intersections. The set of possible mappings between the MSG and the structures is determined by the atom and bond matching rules for a specific problem. If, for a given atom or vertex, several possible mappings are available, the preferred mapping is selected so as to achieve the maximum similarity of the local property distributions in their immediate environment.[59] This is an important feature of our algorithm that, in most cases, facilitates quick identification of the most suitable mapping of the entire structure to the supergraph, taking into account the atomic parameters critical for the interaction with the biotarget. However, one should keep in mind that the exact topology of a supergraph and the correspondence between the structure atoms also depends to some extent on the local descriptors used. Thus, it can be controlled explicitly for each particular problem instead of relying on some formal protocols.

5.2.2.4 Formation of Descriptor Matrix

The molecular supergraph enables the formation of uniform descriptor vectors for all the structures of a training set. Once the structure is superimposed onto the MSG, the descriptor vector is filled by assigning the local descriptor values (*e.g.*, atomic charge Q and its van der Waals radius R) for the structure atoms and bonds to the corresponding MSG vertices and edges. This procedure is illustrated in Figure 5.3 for the fourth structure of Figure 5.2. As is usual in the topological analysis of molecular structures, the properties of the hydrogen atoms are taken into account as additional descriptors for the corresponding non-hydrogen atom rather than handled explicitly. If some vertex or edge is not occupied in a particular structure, the "neutral" descriptor values are assigned (*e.g.*, Q0, R0 in Figure 5.3). In contrast to some approaches using zero values for empty positions, the neutral values provide a model of properties in the un-occupied regions of space around the molecule. Thus, they should not be considered as "missing values" in the statistical sense. If necessary, these parameters may be optimized for a particular problem; nevertheless, our tests show that the model is not affected by small variations in their values if a qualitative picture is correct.

5.2.2.5 Statistical Analysis

The descriptor matrix formed by the uniform descriptor vectors for all the training set structures, as well as the experimental activity values, serves as the source data for the statistical analysis and construction of the QSAR model.

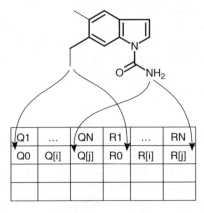

Figure 5.3 Formation of a uniform descriptor vector for a structure by superimposing it onto the MSG.

This model can subsequently be used to predict the activity estimates for new structures. In addition, it can provide information concerning the influence on activity for each local descriptor in different positions of the structure, which can be useful for the lead optimization, the analysis of action mechanism, and the detection of relevant 3D alignment anchors.

In most cases, the MFTA models are built using the Partial Least Squares Regression (PLSR)[3] technique that is suitable for the stable modeling based on the excessive and/or correlated descriptors (under-defined data sets). However, the MFTA approach is not limited to the PLSR models and can successfully employ other statistical learning techniques such as the Artificial Neural Networks (ANN) supporting the detection of the nonlinear structure–activity relationships.[66]

In addition, we have developed several extensions of the basic PLSR procedure[67] to enhance the quality and reliability of the models. The first extension is a so-called Stable Cross-Validation (Stable-CV) procedure designed to provide a more objective and reliable estimate of the model predictivity, *i.e.*, the statistical expectation of prediction error within the model's applicability domain.[68] In practice, one has to assume (for lack of better data) that the available set of compounds represents a reasonable sampling of this region of the chemical universe. This set serves as a basis for the construction as well as the validation of a model. Taking into account the unavoidable irregularities of the sampling and the absence of the *a priori* knowledge of factors affecting the activity, the traditional training/test set approach (external validation) is arguably not the best strategy for identifying the most predictive QSARs. This problem is resolved to some extent by the traditional cross-validation approaches using several complementary splits of a dataset into the training and test subsets. However, experience shows that the results are still significantly affected by the actual grouping of the compounds. This is especially important if the predictivity measures are to be used as guidance for model selection and/or optimization (*e.g.*, by means of a descriptor selection), increasing the risk of a chance correlation.

From the above general principle, the best model should provide the lowest average error of prediction for all reasonable splits of a dataset. We propose to use the Q^2 values averaged over several random reshufflings of the compounds between the fixed-size cross-validation groups. Such a reshuffling and accumulation is repeated iteratively until the resulting average Q^2 is stabilized within a specified precision. The tests show that a few dozen iterations are sufficient to bring the variation below the 0.001 threshold.

Figure 5.4 illustrates typical behavior of the traditional and stable Q^2 values. The values for the individual reshufflings indeed vary substantially, sometimes even changing the preferred number of factors. However, the average values are quickly stabilized.

The computational experiments with the artificially constructed data sets containing varied amounts of both Y-noise (random offset from the true function) and X-noise (additional descriptors unrelated to the true function) show that the Stable-CV procedure (especially with higher precision requirements)

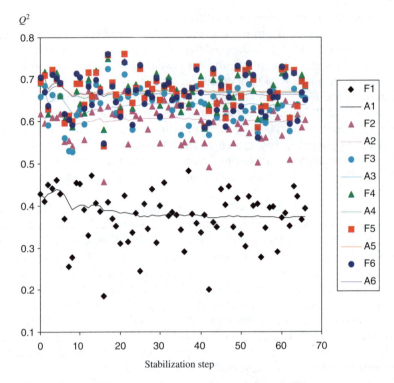

Figure 5.4   Sample Stable-CV run for a dataset of 60 compounds and 159 descriptors. Point series (F1–F6) represent the Q^2 values for the individual reshufflings and the lines (A1–A6) show the dynamics of the averaged Q^2 values for one- to six-factor models, respectively.

allows one to minimize the risk of chance correlations for normal modeling as well as for the model/descriptor selection.

Notably, the Stable-CV Q^2 is, strictly speaking, a different quantity than a conventional Q^2, and the received rules-of-thumb concerning the acceptable value ranges require some refinement. Additional research on this topic is still needed, but the models with Stable-CV Q^2 above 0.4–0.5 seem to provide useful and interpretable models consistent with independently constructed models as well as the biotarget structure.

Another extension is the *descriptor selection procedure* designed to enhance the stability and predictivity of the PLSR models. Its aim is to minimize the info-noise that can dilute and distort the true structure–activity relationship. The procedure[69] involves two phases. The first phase consists of the elimination of the low-variable descriptors that have the same value for all but a few (2–3) compounds in the training set. Such descriptors cannot provide useful statistical information and instead simply help to fit these particular compounds into a model, thus decreasing its predictivity. This filtering is performed entirely in the X-space, without regard for the activity values. In the optional second phase, the descriptor subset included in a model is optimized for better predictivity using the Q^2-guided descriptor selection by means of a genetic algorithm. Measures should be taken (*e.g.*, by using the Stable-CV procedure) to avoid obtaining correlations of insufficient reliability with artificially inflated Q^2 values. Without Stable-CV, Q^2-guided descriptor selection often finds the "quasi-optimal" descriptor subset for a particular split of a data set. The model based on a full descriptor set, despite a lower Q^2 value, may prove more reliable and more useful in the screening of prospective structures thanks to taking into account the influence and inter-correlations of all descriptors. On the other hand, a correct subset-based model provides more focused picture of factors critical for activity and may be more interpretable. Thus, the Q^2-guided descriptor selection is a powerful tool that should be used with care. In most cases some experimenting may be required.

5.2.2.6 Applicability Control

In the virtual screening context, the detection of compounds falling outside the applicability range of a model is of utmost importance, since a misleading prediction is worse than a failed one. For the MFTA models, such a detection is performed on two levels. First, the structure in question must allow super-imposition onto the molecular supergraph (or at least the mismatches should be minimal and located in the parts of a structure not critical for activity), veri-fying that the compound belongs to the same broad chemical class (scaffold) as the training set structures.

Second, the descriptors for a compound should fit the pattern of inter-descriptor correlations implicit in the PLSR model. The compounds with a substantially different descriptor values are detected using the outlier rating[67] that is based on a relative residual descriptor variance for a prediction object i compared to the residual training variance:

$$OR2X_i = \frac{S2PX_i}{S2CX} \tag{5.1}$$

where the residual X-block (descriptor) variance for a prediction object i is:

$$S2PX_i = \frac{1}{DF} \sum_{j}^{M} e_{ij}^2 \tag{5.2}$$

where the number of degrees of freedom $DF = M - F$. The total residual X-block (descriptor) variance for the training objects is:

$$S2CX = \frac{1}{DF} \sum_{i}^{N} \sum_{j}^{M} e_{ij}^2 \tag{5.3}$$

where the number of degrees of freedom $DF = M*(N-1)$. The number of training objects is N, the number of descriptors is M and the number of factors is F.

This parameter is similar and closely related to the $DModX_{norm}$ parameter used in the SIMCA software[70] [in fact, $DModX_{norm} = \sqrt{(OR2X)}$].

Depending on the primary goal of a study, the applicability constraints may be more conservative (to obtain a small number of promising candidates) or more liberal (to extrapolate, explore wider chemical space and iteratively refine the models).

5.3 From MFTA Model to Drug Design and Virtual Screening

5.3.1 MFTA Models in Biotarget and Drug Action Analysis

In a sense, the MFTA model is just the beginning of a road to better active compounds and a deeper understanding of the activity. Let us consider some ways in which it can be helpful.

First of all, the coefficients of the back-rotated PLSR model[67] may be transformed into the normalized descriptor impacts according to Equation (5.4):[59]

$$I_j = \frac{b_j \, \mathrm{range}(x_j)}{\mathrm{range}(y)} \tag{5.4}$$

where b_j is the coefficient at the descriptor x_j, range(x_j) is the range of x_j, and range(y) is the range of activity in the training set.

The signs and values of the descriptor impacts may be graphically encoded and overlaid on the corresponding molecular supergraph vertices, creating the visual activity map that summarizes the most important factors affecting the activity. This data allows one to draw conclusions concerning the possible ligand binding site as well as to identify the directions for beneficial structure modifications.

As an example, let us consider the MFTA model[71,72] of the HIV-1 reverse transcriptase inhibition by the tetrahydroimidazobenzodiazepinone (TIBO) derivatives.[73–75] The model is based on the atomic charge Q, atomic van der Waals radius R and group lipophilicity Lg as the local descriptors (N=73, N_F= 5, R^2 = 0.887, Q^2 = 0.686). Figure 5.5 shows the molecular supergraph with the superimposed structure of one training set compound.

Figure 5.6 gives the major local descriptor contributions to the activity. For comparison, the 3D molecular structure of the ligand-binding site of the HIV-1 reverse transcriptase in complex with a TIBO derivative (Protein Data Bank,[76] structure 1TVR) is shown. The molecular surface of the protein is colored according to the values of electrostatic and lipophilic potential using the SYBYL software.[6] Figure 5.6(a) shows the most important contributions of local charge Q on the ligand atoms matching the electrostatic potential (EP) on the molecular surface of the protein. The positions where the activity tends to increase with increasing charge in the ligand molecule (red circles) correspond

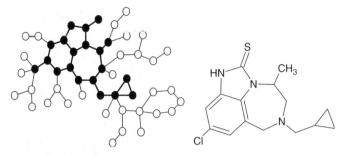

Figure 5.5 Molecular supergraph for a series of tetrahydroimidazobenzodiazepinone (TIBO) derivatives with superimposed structure of a representative training set compound.

to relatively negative local electrostatic potential in the protein (preference of interaction with positively charged groups). In contrast, the positions where the activity decreases with increasing charge (blue circles) correspond to more positive EP values (preference of interaction with more negatively charged groups).

Figure 5.6(b) compares the effect of occupancy and steric bulkiness in the particular positions of a ligand (characterized by the van der Waals radii R) to the steric requirements in the ligand-binding site of a protein (represented by its molecular surface). In particular, areas where steric bulk leads to decrease in activity (blue circles) correspond to a tight binding pocket, while a positive effect of steric bulk on activity (red circles) is found in the open pocket area.

Finally, Figure 5.6(c) demonstrates the correspondence between effect of local lipophilicity Lg on activity (identified by the MFTA model) and the molecular lipophilic potential (MLP) of the protein. An increase of activity with increasing lipophilicity of ligand atoms (red circles) is found in the hydrophobic pocket area. In contrast, positions in the supergraph where activity is decreased by an increase in lipophilicity (blue circles) interact with relatively hydrophilic areas of the protein.

In a study of non-peptide inhibitors of the measles virus entry[77] the findings from the MFTA model were also consistent with the 3D molecular model of the binding site of virus fusion protein. In addition, they suggest the structural modifications to improve the activity. Similar results were obtained in a study of structure–activity and structure-selectivity relationships for anticholinesterase O-phosphorylated oximes.[78]

5.3.2 MFTA Models in Virtual Screening

Virtual screening of the novel promising compounds is among the most important applications of the MFTA technique. Basically, the MFTA model allows one to filter the large set of structures to identify the structures having high estimates of activity predicted with sufficient reliability. Generally

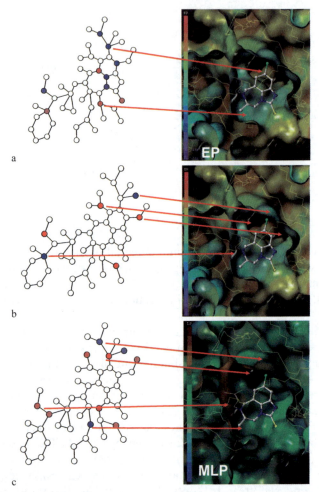

Figure 5.6 Complementarity between the major MFTA descriptor contributions to activity of the TIBO inhibitors of HIV-1 reverse transcriptase and the molecular properties of the biotarget protein: (*a*) atomic charge (*Q*) and electrostatic potential (EP); (*b*) atomic van der Waals radius (*R*) and molecular surface; (*c*) local lipophilicity (*Lg*) and molecular lipophilic potential (MLP) – see text for details.

speaking, three possible sources of structures can be used for virtual screening: (i) limited sets of structures proposed intuitively by a medicinal chemist taking into account the SAR and accessibility considerations; (ii) databases of the compounds available in-house or from commercial suppliers; (iii) virtual structure libraries built by means of structure generators. Each of these sources has some specific features that affect the screening procedure. The limited, manually constructed series simplify the prediction of activity and the analysis of results; in addition, the chemical stability and accessibility of the compounds should be handled implicitly. However, it is very easy to overlook some interesting regions of the available chemical universe, producing a sub-optimal set of candidate structures (especially when the structure–activity relationships are complex). Thus, more automated approaches consistent with the synthetic and medicinal chemistry are desirable. Let us consider their principles and illustrations in more detail.

5.3.2.1 MFTA-based Virtual Screening of Compound Databases

Numerous organic compounds, from simple to rather complex, are now available from many suppliers. Databases of such compounds in various formats can, usually, be obtained directly from the suppliers and/or from the publicly available ZINC project[79] site. In addition, most organizations have their own in-house databases of compounds. The important advantage of this source of screening structures is that the compounds in question are already synthesized and can be purchased and tested rather quickly. The major disadvantage is that the selection of compounds for them is more or less accidental. Thus, it may be difficult or impossible to find the compounds having a particular scaffold and/or substitution pattern; moreover, most of the structures fall outside the applicability domain of a QSAR model and should be filtered out at the preprocessing stage (if possible).

In general, the virtual screening procedure for the database of available compounds involves the following steps:

- Normalization of the structure data set (*e.g.*, removal of explicit hydrogen atoms, standard representation of nitro, azido and similar groups, "deionization", and removal of duplicate structures).
- Preliminary filtering by structural class/scaffold (if desired for the extrapolation or the targeted exploration purposes, wider chemical domain can be covered compared to the training set of a model).
- Prediction of bioactivity endpoints by means of the MFTA models.
- Post-filtering of structures that fall outside the model applicability domain and have unreliable predicted values.
- Selection of candidate compounds having the desired combination of activity endpoints.
- Evaluation of candidate compounds with respect to availability, stability and other relevant factors.

The candidate compounds should also be evaluated for the possibilities of further refinement of target activity as well as extension of the training set.

As an example, let us consider the virtual screening of the potential adenosine receptor agonists selective to the A_3 receptor subtype *vs.* the A_{2A} subtype. Using the structure and activity data for a training set of 29 compounds,[80] the MFTA models with the following parameters were constructed for the affinity to A_3 subtype [represented as $\log(1/K_{i,A3})$; K_i, nM] and the selectivity to A_3 over A_{2A} [represented as $S(A_3/A_{2A}) = \log(1/K_{i,A3}) - \log(1/K_{i,A2A})$; K_i, nM] based on the values of the atomic charge Q, the effective van der Waals radius of the atom's first environment Re and the group lipophilicity Lg as the descriptors[81] (the Q^2 values were estimated using the Stable-CV procedure):

$\log(1/K_{i,A3})$: $N=29$, $N_F=6$, $R^2=0.915$, RMSE=0.290, $Q^2=0.612$, RMSEcv=0.631;
$S(A_3/A_{2A})$: $N=29$, $N_F=4$, $R^2=0.809$, RMSE=0.428, $Q^2=0.537$, RMSEcv=0.678.

Figure 5.7 shows the major contributions of the local descriptors to activity and selectivity. As can be seen, the activity and selectivity maps are quite similar but also differ. Such fine distinctions are difficult to take into account to design new promising structures intuitively (especially if selectivity with respect to other receptor subtypes is also required). Thus, the automated virtual screening approach is preferred.

For the virtual screening, the Asinex, IBScreen and Maybridge vendor subsets were obtained from the ZINC site.[82] In total, these databases contain 876 930 structures. After the normalization and structural pre-filtering steps, a focused screening subset of only 58 structures was obtained. We used somewhat relaxed structural filtering criteria in the hope of identifying possible distantly

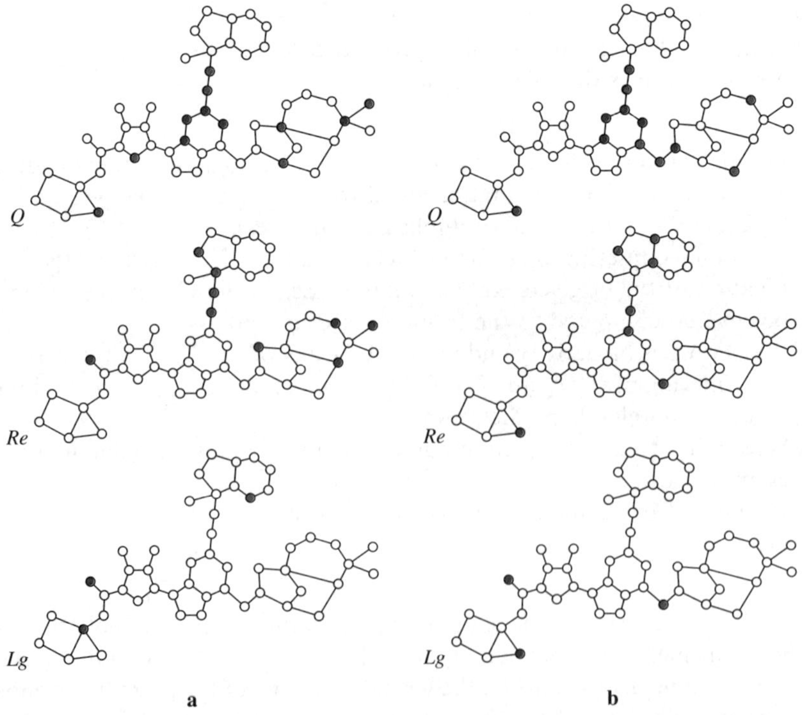

Figure 5.7   Major local descriptor contributions to the MFTA models of activity (*a*) and selectivity (*b*) for the A_3 adenosine receptor agonists.

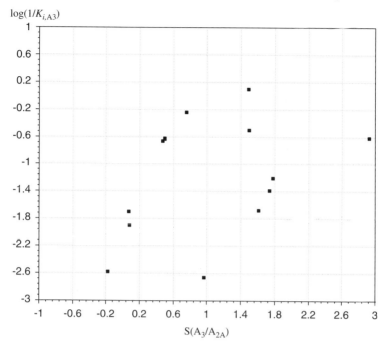

Figure 5.8 Distribution of the predicted values of selectivity and activity to the A_3 adenosine receptor for the filtered virtual screening dataset.

similar ligands. Nonetheless, all the structures in the focused subset were quite close derivatives of the adenosine.

After predicting the activity and selectivity values by means of the MFTA models, the compounds with unreliable predictions were filtered from the virtual screening set using the outlier rating values (Section 5.2.2.6) for two predictions (the threshold value was 30). As a result, a set of 50 structures was obtained. Figure 5.8 shows the distribution of the predicted values. Only one compound is predicted to provide better selectivity than the training set structures as well as reasonable activity. The local descriptor values for this adenosine derivative indeed match the descriptor impacts identified by the MFTA-based selectivity model.

Another example deals with the virtual screening of the potential selective indole ligands of the melatonin (MLT) receptors. Using the structure and activity data for a training set of 80 compounds,[83–85] MFTA models with the following parameters were constructed for the affinity and intrinsic activity with respect to human MT_1 and MT_2 receptor subtypes (MT_1 relative affinity $pRA_1 = pK_{i,1} - pK_{i,1}[MLT]$; MT_1 intrinsic activity relative to melatonin IAr_1; MT_2 relative affinity $pRA_2 = pK_{i,2} - pK_{i,2}[MLT]$; MT_2 intrinsic activity relative to melatonin IAr_2; K_i, nM) based on the values of the atomic charge Q and the effective van der Waals radius of the atom's first environment Re (the Q^2 values were estimated using the Stable-CV procedure):

pRA_1: N=80, N_F=6, R^2=0.882, RMSE=0.495, Q^2=0.718, RMSEcv=0.769;
pRA_2: N=80, N_F=6, R^2=0.903, RMSE=0.476, Q^2=0.767, RMSEcv=0.742;
IAr_1: N=40, N_F=6, R^2=0.934, RMSE=0.103, Q^2=0.693, RMSEcv=0.225;
IAr_2: N=40, N_F=6, R^2=0.919, RMSE=0.113, Q^2=0.537, RMSEcv=0.273.

The structure of the molecular supergraph is shown in Figure 5.9 with several examples of superimposition of the training set structures. It reflects substantial diversity of the training set compounds and, despite apparent complexity, supports reasonable comparison of different structures. Simpler MSGs could

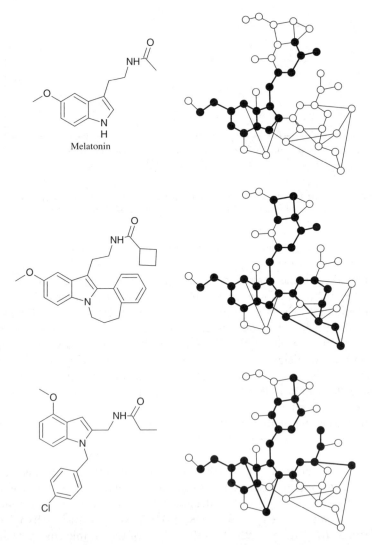

Figure 5.9 Molecular supergraph for a series of melatonin receptor ligands with superimposed structures of several representative training set compounds.

be obtained by splitting the dataset into more narrow structural subsets; however, in this case the statistical stability and applicability of the resulting models would also be diminished. Moreover, the superposition of the similar side groups at the expense of partial mismatch in the indole core seems consistent with relative indifference of the melatoninergic system to the exact nature of central bicyclic aromatic moiety.[86] The affinity and intrinsic activity maps are rather complex and difficult to take into account to design new promising structures intuitively. Thus, the automated virtual screening approach is preferred.

For the virtual screening, the Asinex vendor subset was obtained from the ZINC site.[82] The database contains 322 992 structures. After the normalization and structural pre-filtering steps, the focused screening subset of 4972 structures containing the indole fragment was derived.

At the MFTA prediction step, we used quite strict superimposition requirements: the structures with incomplete supergraph fit were filtered out. This brings the number of compounds down to 387. After predicting the affinity and intrinsic activity endpoints by means of the MFTA models, the compounds with unreliable predictions were filtered from the virtual screening set using the outlier rating values (Section 5.2.2.6) for the four predictions (the threshold value was 20). As a result, a set of 151 structures was obtained. Figure 5.10 shows their distribution with respect to the predicted values of affinity and activity to both target receptors as well as the cross-target selectivity. Some broad correlation between the targets can be seen; nevertheless, this series of compounds provides fairly uniform coverage of affinity/activity space, allowing the selection of compounds with different predicted activity profiles. For instance, let us identify the potential selective MT_1 receptor agonists. Filtering the dataset by the condition shown in Equation (5.5) yields a series of only six candidate compounds:

$$pRA_1 > -1, \ pRA_2 < -2 \qquad (5.5)$$

The distribution plots shown in Figure 5.11 confirm that they can be expected to behave as rather good MT_1 agonists with weak MT_2 partial agonist or antagonist activity.

5.3.2.2 MFTA-based Virtual Screening of Generated Structure Libraries

As mentioned above, the databases of available compounds often do not provide sufficient coverage of regions of the chemical space that may be interesting to explore during lead optimization of the compounds possessing a particular activity. A useful complementary approach involves virtual screening of artificial structure libraries obtained from the structure generator software. Even general-purpose structure generators for QSAR studies[87] can build a library containing all structures belonging to a specified chemical class or its

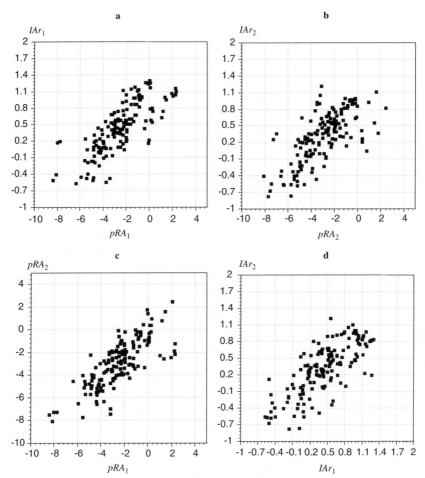

Figure 5.10 Distribution of the predicted values of affinity and intrinsic activity to the MT$_1$ and MT$_2$ melatonin receptors for the filtered virtual screening dataset of indole derivatives. (*a*) MT$_1$ affinity *vs.* MT$_1$ intrinsic activity, (*b*) MT$_2$ affinity *vs.* MT$_2$ intrinsic activity, (*c*) MT$_1$ *vs.* MT$_2$ affinity, (*d*) MT$_1$ *vs.* MT$_2$ intrinsic activity.

representative subset. Such a library could be used for the virtual screening. However, in the context of the MFTA modeling this approach has some drawbacks. In particular, it is difficult to avoid generating the structures that do not fit the molecular supergraph and thus fall outside the applicability domain of a model. On the other hand, some promising structures could be overlooked. In addition, manual definition of generation parameters based on the MFTA model may be complicated and time-consuming. Thus, a specialized generator is desirable that could take into account the integral features of an MFTA model.

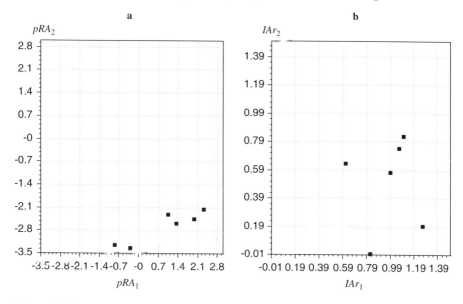

Figure 5.11 Affinity and intrinsic activity profile for the six potential selective MT_1 melatonin receptor agonists – virtual screening hit compounds: (*a*) MT_1 *vs.* MT_2 affinity and (*b*) MT_1 *vs.* MT_2 intrinsic activity.

We have developed an approach to the efficient generation of molecular graphs oriented to the MFTA models.[88] An important feature of this technique is the support for a flexible system of structural constraints. It is critical because otherwise an MFTA supergraph may in principle give rise to a very large number of generated structures, many of which are obviously unpromising, synthetically difficult or chemically unstable.

The generation algorithm can be used in two modes: deterministic and stochastic. The task of the deterministic generation involves the constructive enumeration of all possible connected molecular graphs that are subgraphs of the MFTA supergraph and do not violate the specified constraints. In stochastic generation, a representative subset of the set of all possible structures should be obtained. All generated molecular graphs must include the "central fragment" (or a scaffold) – a connected subgraph present in all the structures of a training set. Allowable valences of atoms must also be observed. In addition, the researcher can specify forbidden fragments and forbidden bonds (a fragment consisting of a pair of bonded atoms) to avoid the generation of chemically unstable compounds, toxophoric groups or structural fragments that are not represented in the training set. Constraints on the number of substituted positions of the central fragment and on the molecular weight (*e.g.*, based on the Lipinski rule[89]) make it possible to prevent the generation of overdecorated structures unfavorable for chemical synthesis and drug action.

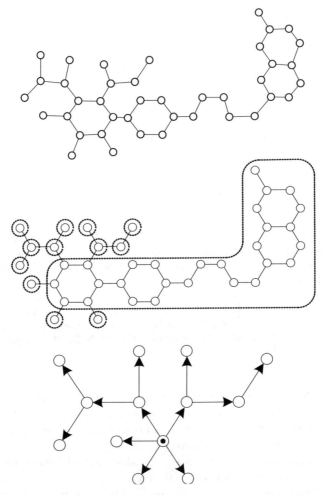

Figure 5.12 Detection of components in the MFTA supergraph and the construction of a fragmental supergraph for structure generation.
(Reproduced with permission from ref. 88, © 2007 Pleiads Publishing Ltd.)

   Since all generated structures must contain the central fragment, the problem can be reduced to the generation of substituted derivatives of a given organic compound from a set of elementary fragments.[87] In the MFTA supergraph, the following components are detected: the central fragment, cyclic and polycyclic fragments, and the vertices in acyclic fragments. Then, a so-called fragmental supergraph is constructed as a tree rooted in the central fragment (Figure 5.12). It serves as a compact representation of all possible structures. For all vertices of the fragmental supergraph, the set of possible pairs "molecular fragment – type of bond to the parent fragment" is formed and the structures are assembled.

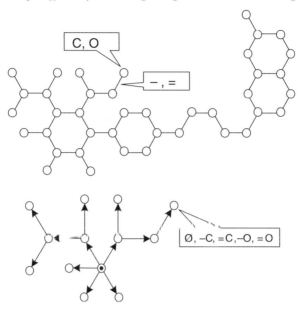

Figure 5.13 Analysis of a training set and the construction of all possible molecular fragments.
(Reproduced with permission from ref. 88, © 2007 Pleiads Publishing Ltd.)

To generate labels of vertices (atom types) and edges (bond types), the training set is examined and the lists of possible types of atom and bond are created for, respectively, each vertex and each edge of the molecular supergraph. All possible combinations of vertex and edge labels are enumerated by means of a recursive algorithm that checks for the correct valence of atoms and the absence of forbidden bonds and fragments in the generated molecular graphs (these steps are illustrated in Figure 5.13). Then, the structure is generated by stepwise extension of the central fragment using the depth-first traversal of the fragmental supergraph and the selection of fragments and bonds.

To satisfy the constraints on molecular weight, the contributions of all possible molecular fragments are precalculated. Based on the depth-first traversal order, at every step of the generation one can calculate the lower and upper bounds of the molecular weight of the resulting structures. If they are not compatible with the required molecular weight range, the addition of the current fragment is unproductive.

As an example, let us consider the application of this approach implemented in the convenient molecular generator software[90] to the virtual screening of selective AMPA receptor antagonists. Using the structure and activity data for a training set of 55 compounds,[91–93] the MFTA models with the following parameters were constructed for the affinity to the AMPA receptor [represented as $\log(1/K_{i,\text{AMPA}})$; K_i, μM] and the selectivity to AMPA over glycine/NMDA receptor [represented as $S(\text{AMPA}/\text{NMDA}) = \log(1/K_{i,\text{AMPA}}) - \log(1/K_{i,\text{NMDA}})$; K_i, μM] based on the values of the atomic charge Q and the effective van der

Waals radius of the atom's first environment *Re* as the descriptors[81] (the Q^2 values were estimated using the Stable-CV procedure):

$\log(1/K_{i,\text{AMPA}})$: $N=55$, $N_F=2$, $R^2=0.666$, RMSE$=0.488$, $Q^2=0.505$, RMSEcv$=0.600$;
S(AMPA/NMDA): $N=55$, $N_F=4$, $R^2=0.819$, RMSE$=0.292$, $Q^2=0.637$, RMSEcv$=0.417$.

Figure 5.14 gives the major contributions of the local descriptors to activity and selectivity. Similar to the previous example, the automated virtual screening approach is preferred to take into account the subtle differences in the influence of various local parameters.

Based on the MFTA model and several constraints, a set of 3000 structures was generated in the stochastic mode. For this virtual screening set, the activity and selectivity values were predicted by means of the MFTA models. Figure 5.15 illustrates the distribution of these endpoint values *versus* the outlier rating values. It can be seen that a substantial portion of the dataset has rather high activity/selectivity and low outlier rating. The structures with unreliable predictions were filtered from the screening set using the outlier rating values for both predictions (the threshold value was 20). As a result, a set of 1126 structures was obtained. Figure 5.16 shows the distribution of the predicted values. Taking into account the factors of synthetic attainability and chemical stability, one can select several compounds as the most promising

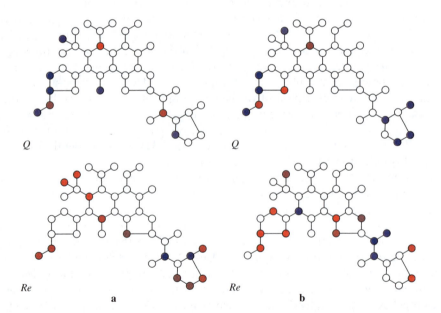

Figure 5.14   Major local descriptor contributions to the MFTA models of activity (*a*) and selectivity (*b*) for the AMPA receptor antagonists.

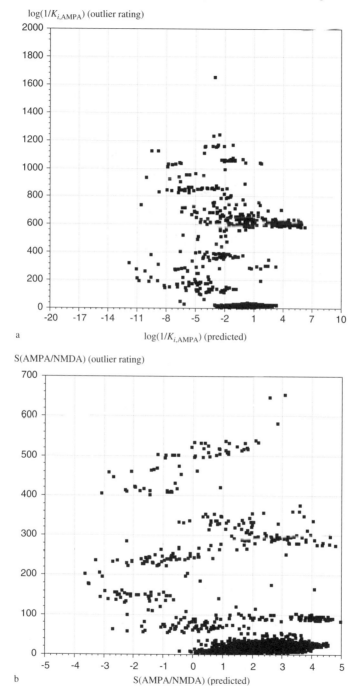

Figure 5.15 Distribution of the predicted endpoint and outlier rating values for the activity (*a*) and selectivity (*b*) of the AMPA receptor antagonists.

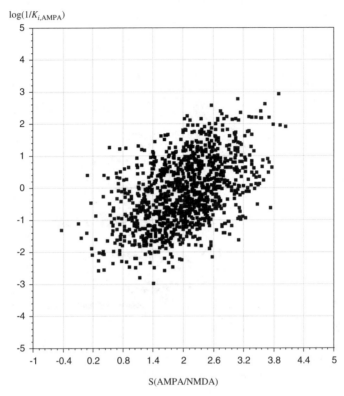

Figure 5.16 Distribution of the predicted values of selectivity and activity to the AMPA receptor for the filtered virtual screening dataset.

candidates that are predicted to provide better selectivity and activity than the training set structures.

5.4 Conclusion

The examples of application of the Molecular Field Topology Analysis presented in this chapter and other publications show that this approach provides a powerful and efficient tool for the modeling and understanding of the structure–activity relationships as well as for the virtual screening of new promising structures. It is especially useful for the series of congeneric organic compounds whose activity is based on specific interactions with one or more biotarget proteins. Despite the 2D nature of this technique, the results are consistent with the molecular models of ligand–biotarget interactions and in many cases outperform the 3D QSAR approaches in terms of model quality, computational efficiency and chemical interpretability. Further development of this approach is in progress and, with feedback from the medicinal chemistry community, we hope to make the MFTA a valuable instrument for the design and virtual screening of future drugs.

Acknowledgements

The authors thank A. A. Melnikov and M. S. Belenikin for their valuable help and discussions during the preparation of this chapter and the development of the MFTA approach in general.

References

1. H. Kubinyi, *QSAR: Hansch Analysis and Related Approaches*, VCH, Weinheim, 1993.
2. R. D. Cramer, D. E. Patterson and J. D. Bunce, *J. Am. Chem. Soc.*, 1988, **110**, 5959–5967.
3. P. Geladi and B. R. Kowalski, *Anal. Chim. Acta*, 1986, **185**, 1–17.
4. M. Wagener, J. Sadowski and J. Gasteiger, *J. Am. Chem. Soc.*, 1995, **117**, 7769–7775.
5. R. D. Cramer and S. B. Wold, *U.S. Patent No.* 5025388, 1991.
6. *SYBYL 7.3*, Tripos International, 1699 South Hanley Rd., St. Louis, Missouri, 63144.
7. U. Thibaut, in *3D QSAR in Drug Design: Theory, Methods and Applications*, ed. H. Kubinyi, ESCOM, Leiden, 1993, pp. 661–696.
8. P. J. Goodford, *J. Med. Chem.*, 1985, **28**, 849–857.
9. M. Pastor, G. Cruciani and K. A. Watson, *J. Med. Chem.*, 1997, **40**, 4089–4102.
10. U. Norinder, *J. Comp.-Aided Mol. Des.*, 1990, **4**, 381–389.
11. P.-A. Carrupt, P. Gaillard, F. Billois, P. Weber, B. Testa, C. Meyer and S. Pérez, in *Lipophilicity in Drug Action and Toxicology*, ed. R. Mannhold, VCH, Weinheim, 1996, pp. 195–217.
12. G. E. Kellogg, S. F. Semus and D. J. Abraham, *J. Comput-Aided Mol. Des.*, 1991, **5**, 545–552.
13. K. H. Kim, *Quant. Struct.- Act. Relat.*, 1993, **12**, 232–238.
14. C. L. Waller and G. R. Marshall, *J. Med. Chem.*, 1993, **36**, 2390–2403.
15. A. Poso, C. Navajas and J. Gynter, in *11th Eur. Symp. on Quantitative Structure-Activity Relationships*, September 1–6, 1996, Lausanne, Switzerland, p. P-40D.
16. M. Baroni, G. Costantino, G. Cruciani, D. Riganelli, R. Valigi and S. Clementi, *Quant. Struct.-Act. Relat.*, 1993, **12**, 9–20.
17. M. Pastor, G. Cruciani and S. Clementi, *J. Med. Chem.*, 1997, **40**, 1455–1464.
18. G. Cruciani and K. A. Watson, *J. Med. Chem.*, 1994, **37**, 2589–2601.
19. R. D. Cramer, R. D. Clark, D. E. Patterson and A. M. Ferguson, *J. Med. Chem.*, 1996, **39**, 3060–3069.

20. N. S. Zefirov, D. E. Petelin, V. A. Palyulin and J. W. McFarland, *Dokl. Akad. Nauk*, 1992, **327**, 504–508 (Russ.).
21. J. R. Hurst and T. W. Heritage, *U.S. Patent No. 6208942*, 2001.
22. M. Lipkin, D. Salt and W. Wynn, in *Computer-Assisted Lead Finding and Optimization*, eds. H. Waterbeemd, B. Testa, G. Folkers, VHCA, Basel, 1997, pp. 433–442.
23. N. J. Richmond, C. A. Abrams, P. R. N. Wolohan, E. Abrahamian, P. Willett and R. D. Clark, *J. Comp.-Aided Mol. Des.*, 2006, **20**, 567–587.
24. Y. Tominaga and I. Fujiwara, *J. Chem. Inf. Comput. Sci.*, 1997, **37**, 1152–1157.
25. T. Sulea, T. I. Oprea, S. Muresan and S. L. Chan, *J. Chem. Inf. Comput. Sci.*, 1997, **37**, 1162–1170.
26. M. Pastor, G. Cruciani, I. McLay, S. Pickett and S. Clementi, *J. Med. Chem.*, 2000, **43**, 3233–3243.
27. F. Fontaine, M. Pastor and F. Sanz, *J. Med. Chem.*, 2004, **47**, 2805–2815.
28. B. D. Silverman and D. E. Platt, *J. Med. Chem.*, 1996, **39**, 2129–2140.
29. A. P. Ferguson, T. Heritage, P. Jonathon, S. E. Pack, L. Phillips, J. Rogan and P. J. Snaith, *J. Comp.-Aided Mol. Des.*, 1997, **11**, 143–152.
30. J. W. McFarland, *J. Med. Chem.*, 1992, **35**, 2543–2550.
31. D. E. Patterson, R. D. Cramer, A. M. Ferguson, R. D. Clark and L. E. Weinberger, *J. Med. Chem.*, 1996, **39**, 3049–3059.
32. R. D. Cramer, *J. Med. Chem.*, 2003, **46**, 374–388.
33. R. D. Cramer, R. J. Jilek, S. Guessregen, S. J. Clark, B. Wendt and R. D. Clark, *J. Med. Chem.*, 2004, **47**, 6777–6791.
34. R. D. Cramer and B. Wendt, *J. Comp.-Aided. Mol. Des.*, 2007, **21**, 23–32.
35. C. Hansch, P. P. Maloney, T. Fujita and M. Muir, *Nature*, 1962, **194**, 178–180.
36. C. Hansch and T. Fujita, *J. Am. Chem. Soc.*, 1964, **86**, 1616–1626.
37. C. Hansch, *Acc. Chem. Res.*, 1969, **2**, 232–239.
38. S. M. Free and J. M. Wilson, *J. Med. Chem.*, 1964, **7**, 395–399.
39. T. Fujita and T. Ban, *J. Med. Chem.*, 1971, **14**, 148–152.
40. Chapter 1 of this book.
41. R. D. Brown, G. M. Downs and P. Willett, *J. Chem. Inf. Comput. Sci.*, 1992, **32**, 522–531.
42. R. D. Brown, G. Jones, P. Willett and R. C. Glen, *J. Chem. Inf. Comput. Sci.*, 1994, **34**, 63–70.
43. G. M. Downs, G. S. Gill, P. Willett and P. T. Walsh, *SAR QSAR Environ. Res.*, 1995, **3**, 253–264.
44. J.-E. Dubois, D. Laurent and A. Aranda, *J. Chim. Phys.*, 1973, **70**, 1616–1624.
45. Y. Sobel, P. Vizet, S. Chemtob, F. Barbieux and C. Mercier, *SAR QSAR Environ. Res.*, 1998, **9**, 83–109.
46. C. Mercier, O. Mekenyan, J.-E. Dubois and D. Bonchev, *Eur. J. Med. Chem.*, 1991, **26**, 575–592.

47. C. Mercier, S. Chemtob, P. Vizet and Y. Sobel, in *9th Intl. Workshop on Quantitative Structure-Activity Relationships in Environmental Sciences*, Bourgas, Bulgaria, September 2000, p. I.15.
48. G. K. Menon and A. Cammarata, *J. Pharm. Sci.*, 1977, **66**, 304–314.
49. A. Cammarata and G. K. Menon, *J. Med. Chem.*, 1976, **19**, 739–747.
50. P. S. Magee, *Quant. Struct.-Act. Relat.*, 1990, **9**, 202–215.
51. P. S. Magee, in *Rational Approaches to Structure, Activity, and Ecotoxicology of Agrochemicals*, ed. W. Draber, T. Fujita, CRC Press, Boca Raton, FL, 1992, pp. 79–101.
52. Z. Simon, I. Badilescu and T. Racovitan, *J. Theor. Biol.*, 1977, **66**, 485–495.
53. A. T. Balaban, A. Chiriac, I. Motoc and Z. Simon, *Steric Fit in QSAR* (Lecture notes in chemistry, v. 15), Springer, Berlin, 1980.
54. L. Kurunczi, E. Seclaman, T. I. Oprea, L. Crisan and Z. Simon, *J. Chem. Inf. Model.*, 2005, **45**, 1275–1281.
55. M. Mracec, M. Mracec, L. Kurunczi, T. Nusser, Z. Simon and G. Náray-Szabó, *J. Mol. Struct. (THEOCHEM)*, 1996, **367**, 139–149.
56. A. Bora, T. I. Oprea, L. Kurunczi and E. Seclaman, in *16th Eur. Symp. on Quantitative Structure-Activity Relationships and Molecular Modelling*, Italy, September 10–17, 2006, p. 193.
57. E. V. Radchenko, V. A. Palyulin and N. S. Zefirov, in *11th Eur. Symp. on Quantitative Structure-Activity Relationships*, Lausanne, Switzerland, September 1–6, 1996, p. P-21A.
58. N. S. Zefirov, V. A. Palyulin and E. V. Radchenko, *Doklady Chemistry*, 1997, **352**, 23–26.
59. V. A. Palyulin, E. V. Radchenko and N. S. Zefirov, *J. Chem. Inf. Comp. Sci.*, 2000, **40**, 659–667.
60. J. Gasteiger and M. Marsili, *Tetrahedron*, 1980, **36**, 3219–3228.
61. A. A. Oliferenko, V. A. Palyulin, S. A. Pisarev, A. V. Neiman and N. S. Zefirov, *J. Phys. Org. Chem.*, 2001, **14**, 355–369.
62. A. Bondi, *J. Phys. Chem.*, 1964, **68**, 441–451.
63. A. K. Ghose, A. Pritchett and G. M. Crippen, *J. Comput. Chem.*, 1988, **9**, 80–90.
64. M. H. Abraham, P. P. Duce, D. V. Prior, D. G. Barratt, J. J. Morris and P. J. Taylor, *J. Chem. Soc. Perkin Trans. 2*, 1989, **10**, 1355–1375.
65. Yu. E. Bessonov, *Vychisl. Sist.*, 1985, **112**, 3–22. (Russ.).
66. E. V. Radchenko, O. D. Baranova, V. A. Palyulin and N. S. Zefirov, in *Designing Drugs and Crop Protectants: Processes, Problems and Solutions*, ed., M. Ford, D. Livingstone, J. Dearden, H. Waterbeemd, Blackwell, Malden, 2003, p. 317–318.
67. H. Martens and T. Naes, *Multivariate Calibration*, Wiley, Chichester, 1989.
68. E. V. Radchenko, V. A. Palyulin and N. S. Zefirov, in *16th Eur. Symp. on Quantitative Structure-Activity Relationships and Molecular Modelling*, Italy, September 10–17, 2006, p. 207.
69. V. A. Palyulin, E. V. Radchenko, O. D. Baranova, A. A. Oliferenko and N. S. Zefirov, in *Designing Drugs and Crop Protectants: Processes,*

Problems and Solutions, eds. M. Ford, D. Livingstone, J. Dearden and H. Waterbeemd, Blackwell, Malden, 2003, p. 188–190.

70. L. Eriksson, E. Johansson, N. Kettaneh-Wold, J. Trygg, C. Wikström and S. Wold, *Multi- and Megavariate Data Analysis, Part I, Basic Principles and Applications*, Umetrics, Umea, 2006.
71. E. V. Radchenko, M. S. Belenikin, A. A. Sokolov, V. A. Palyulin and N. S. Zefirov, in *QSAR and Molecular Modelling in Rational Design of Bioactive Molecules*, CADD&DS, Istanbul, Turkey, 2004, p. 100–101.
72. E. V. Radchenko, V. A. Palyulin and N. S. Zefirov, *Russ. Khim. Zhurn.* (Russ.), 2006, **50**, 76–85.
73. M. J. Kukla, H. J. Breslin, R. Pauwels, C. L. Fedde, M. Miranda, M. K. Scott, R. G. Sherrill, A. Raeymaekers and J. Van Gelder, *J. Med. Chem.*, 1991, **34**, 746–751.
74. H. J. Breslin, M. J. Kukla, D. W. Ludovici, R. Mohrbacher, W. Ho, M. Miranda, J. D. Rodgers, T. K. Hitchens and G. Leo, *J. Med. Chem.*, 1995, **38**, 771–793.
75. W. Ho, M. J. Kukla, H. J. Breslin, D. W. Ludovici, P. P. Grous, C. J. Diamond, M. Miranda, J. D. Rodgers and C. Y. Ho, *J. Med. Chem.*, 1995, **38**, 794–802.
76. http://www.pdb.org/ (accessed December 2007).
77. A. Sun, A. Prussia, W. Zhan, E. E. Murray, J. Doyle, L.-T. Cheng, J.-J. Yoon, E. V. Radchenko, V. A. Palyulin, R. W. Compans, D. C. Liotta, R. K. Plemper and J. P. Snyder, *J. Med. Chem.*, 2006, **49**, 5080–5092.
78. E. V. Radchenko, G. F. Makhaeva, V. V. Malygin, V. B. Sokolov, V. A. Palyulin and N. S. Zefirov, *Doklady Biochem. Biophys.*, 2008, **418**, 47–51.
79. J. J. Irwin and B. K. Shoichet, *J. Chem. Inf. Model.*, 2005, **45**, 177–182.
80. A. A. Ivanov, V. A. Palyulin and N. S. Zefirov, *J. Mol. Graph. Mod.*, 2007, **25**, 740–754.
81. V. A. Palyulin, E. V. Radchenko, I. I. Baskin, V. I. Chupakhin, A. A. Ivanov and N. S. Zefirov, in *16th Eur. Symp. on Quantitative Structure-Activity Relationships and Molecular Modelling*, Italy, September 10–17, 2006, p. 153.
82. http://zinc.docking.org/ (accessed December 2007).
83. S. Rivara, M. Mor, C. Silva, V. Zuliani, F. Vacondio, G. Spadoni, A. Bedini, G. Tarzia, V. Lucini, M. Pannacci, F. Fraschini and P. V. Plazzi, *J. Med. Chem.*, 2003, **46**, 1429–1439.
84. G. Spadoni, C. Balsamini, G. Diamantini, A. Tontini, G. Tarzia, M. Mor, S. Rivara, P. V. Plazzi, R. Nonno, V. Lucini, M. Pannacci, F. Fraschini and B. M. Stankov, *J. Med. Chem.*, 2001, **44**, 2900–2912.
85. M.-T. Teh and D. Sugden, *Naunyn-Schmiedeberg's Arch. Pharmacol.*, 1998, **358**, 522–528.
86. D. Sugden, K. Davidson, K. A. Hough and M.-T. Teh, *Pigment Cell Res.*, 2004, **17**, 454–460.
87. A. A. Melnikov, V. A. Palyulin and N. S. Zefirov, *J. Chem. Inf. Model.*, 2007, **47**, 2077–2088.
88. A. A. Melnikov, V. A. Palyulin, E. V. Radchenko and N. S. Zefirov, *Doklady Chemistry*, 2007, **415**, 196–199.

89. C. A. Lipinski, F. Lombardo, B. W. Dominy and P. J. Feeney, *Adv. Drug Deliv. Rev.*, 1997, **23**, 3–25.
90. V. A. Palyulin, E. V. Radchenko, A. A. Melnikov and N. S. Zefirov, in *3rd German Conference on Chemoinformatics*, Goslar, Germany, November 11–13, 2007, pp. 57.
91. D. Catarzi, V. Colotta, F. Varano, G. Filacchioni, A. Galli, C. Costagli and V. Carla, *J. Med. Chem.*, 2001, **44**, 3157–3165.
92. D. Catarzi, V. Colotta, F. Varano, L. Cecchi, G. Filacchioni, A. Galli and C. Costagli, *J. Med. Chem.*, 1999, **42**, 2478–2484.
93. F. Varano, D. Catarzi, V. Colotta, G. Filacchioni, A. Galli, C. Costagli and V. Carla, *J. Med. Chem.*, 2002, **45**, 1035–1044.

CHAPTER 6

Probabilistic Approaches in Activity Prediction

DMITRY FILIMONOV AND VLADIMIR POROIKOV

Institute of Biomedical Chemistry of Russian Academy of Medical Sciences, 10, Pogodinskaya Str., Moscow, 119121, Russia

6.1 Introduction

Biological activity has a probabilistic nature, and the most appropriate approaches in activity prediction are based on the theory of probability. The statistical nature of the maximum likelihood method and the Bayesian approach is well recognized, but many other methods (multiple regression, factor analysis, pattern recognition methods such as linear discriminant analysis, linear learning machine, support vector machines *etc.*)[1–3] can also be considered as probabilistic ones.[4,5] An informational search in PubMed Central with the queries "(probabilistic approach) OR (probabilistic method)" or "(statistical approach) OR (statistical method)", will find 3477 documents and 180 475 documents, respectively. It is impossible to analyze all these publications, particularly taking into account that, despite of the presence of this term in their titles, many of them are not really probabilistic (see, for instance, refs 6–20). We propose the following definition of probabilistic approaches: "The methods that use probabilities as an essential part of the algorithm, and/or for which the results of application are presented as probability estimates." Thus, many approaches that do not correspond strictly to the definition are not considered in this chapter.

Since data on general dose–response relationships are not available in many cases, biological activity is often represented by a single quantitative or even qualitative characteristic. Therefore, many training sets are created with activity

Chemoinformatics Approaches to Virtual Screening
Edited by Alexandre Varnek and Alex Tropsha
© Royal Society of Chemistry, 2008
Published by the Royal Society of Chemistry, www.rsc.org

data presented in such mode. These probabilistic ligand-based drug design methods are further used for virtual screening. Existing training sets are not ideal, not just due to the simplified definition of biological activity but also because (i) no one activity is represented by all relevant chemical classes and (ii) no one compound has been tested against all kinds of biological activity. So, the probabilistic character of biological activity is caused not only by experimental errors of its determination but also by the incompleteness of available information.

Typically, virtual screening methods are used to select hits with a single required activity,[21–24] while the final aim of pharmaceutical R & D is to identify safe and potent leads and drug-candidates.[25–28] To overcome this problem, the authors have developed a method for prediction of many kinds of biological activity simultaneously based on the structural formula of chemical compound, which is realized in the computer program PASS (Prediction of Activity Spectra for Substances).[29,30] PASS provides the means for evaluation of general biological activity profile at the early stages of R & D, and thus its prediction can be used as a basis for the selection of compounds with the required kinds of biological activity but without unwanted ones.[31,32]

In this chapter we overview some probabilistic methods used for biological activity prediction, paying particular attention to the problems of creation of the training and evaluation sets, validation of (Q)SAR models, estimation of prediction accuracy, interpretation of the prediction results and their application in virtual screening.

6.2 Biological Activity

Biological activity is the result of a chemical compound's interaction with biological objects. It depends on the characteristics of (i) the compound (structure of molecule and its physicochemical properties), (ii) biological object (kind, sex, age, *etc.*), (iii) way of exposure (route of administration, dosage) and (iv) peculiarities of the experimental terms and conditions.

The major paradigm of the twentieth century was based on the concept "one disease – one target";[27,33] therefore, at first chemical compounds were tested against the targeted activity, and only for those leads that passed through this "filter" was a more general biological activity profile estimated. Currently, it is recognized that most pharmaceutical agents interact with several or even many targets in the organism, and thus their selectivity is rather relative. For example, by analysis of the available literature one may find that biological activity of caffeine (CAS No. 58-08-2) is described by the terms related to the following:

- ten pharmacotherapeutic effects (analeptic, antihypertensive, antihypotensive, cardiotonic, diuretic, immunosuppressant, psychostimulant, respiratory analeptic, saluretic, spasmolytic);
- 18 biochemical mechanisms of action (ATP diphosphatase inhibitor, adenosine deaminase inhibitor, cyclic AMP phosphodiesterase inhibitor, cytochrome P450 inhibitor, dATP(dGTP)-DNA purinetransferase inhibitor,

glycogen (starch) synthase inhibitor, guanylate cyclase inhibitor, hydroxy-acylglutathione hydrolase inhibitor, lactoylglutathione lyase inhibitor, nucleotide metabolism regulator, P-glycoprotein inhibitor, phosphatidylinositol kinase inhibitor, phosphodiesterase inhibitor, phosphorylase inhibitor, purine nucleosidase inhibitor, thymidine kinase inhibitor, urate oxidase inhibitor, xanthine-like agent);
- nine adverse/toxic effects [arrhythmogenic, spasmogenic, convulsant, non-mutagenic (salmonella), embryotoxic, teratogen, carcinogenic, carcino-genic (group 3), toxic];
- 16 metabolic terms (CYP1 substrate, CYP1A inhibitor, CYP1A substrate, CYP1A1 substrate, CYP1A2 inhibitor, CYP1A2 substrate, CYP2 sub-strate, CYP2B substrate, CYP2B1 substrate, CYP2B2 substrate, CYP2E substrate, CYP2E1 substrate, CYP3A substrate, CYP3A1 substrate, CYP3A4 substrate, CYP3A5 substrate).

Some apparent contradictions in terms representing the biological activity of caffeine can be explained either by its opposite effects in different doses or by peculiarities of experimental terms and conditions in the appropriate studies. A similar picture can be observed also for most well-known pharmaceuticals.

On the other hand, even acting on the same target, different chemical com-pounds can bind to them in different modes.[34] Therefore, any individual che-mical structure exhibits many biological activities, and *vice versa* a particular biological activity can be caused by many different chemical structures.[35,36]

Biological activity is tested both *in vivo* and *in vitro*. In the past 20 years, due to advances in preparative and measuring techniques, a significant part of assays is the testing of ligand binding to the macromolecular target *in vitro*. It is necessary to keep in mind that such binding can occur not with the site of macromolecule that is responsible for its biological activity or for suppressing of this biological activity. As a result, many ligands found in high-throughput assays may appear to be nonspecific or "promiscuous" inhibitors.[37] Moreover, binding is not a sufficient condition for ensuring that a beneficial function will ensue in the cell or in the organism as a whole.[38] After the deciphering of the human genome and first results in postgenomic studies it became obvious that many diseases have a complex etiology,[27] while drug action on a certain target often leads to activation/inhibition of other elements in the appropriate reg-ulatory network. As a consequence of negative feedback, expected pharma-cotherapeutic action may be significantly decreased or even completely suppressed.[39] Therefore, specially designed multi-targeted drugs may have certain advantages over single-targeted medicines.[33]

Since the final purpose of pharmaceutical studies to find hits & leads with the required, but without unwanted, properties the virtual screening should pro-vide the estimation of general biological activity profile because such experi-mental studies are highly expensive and time-consuming.

We proposed the biological activity spectrum of a substance concept, which seems to be a fundamental basis for description of biologically active sub-stances.[29,30,32,40–43] The "biological activity spectrum" of a substance is the set of different kinds of biological activity, which reflect the results of chemical substance's interaction with various biological entities. This more general concept was introduced earlier than "biospectra"[44,45] or other "activity spectra".[46] Biological activity is defined qualitatively ("yes"/"none"), suggesting that the "biological activity spectrum" represents the "intrinsic" property of a substance, depending only on its structure and physicochemical characteristics. Certainly, this is a simplified definition because the exhibition of biological activity depends on the presence and state of the corresponding targets and experimental conditions (object, route of administration, dose, *etc.*). However, such approximation provides a possibility for combining of information from many different sources, which is necessary because no one particular publication represents comprehensively different aspects of biological action of a compound. For example, to collect information on the biological activity profile of caffeine discussed above, an extensive information search was performed of the available literature and databases.

6.2.1 Dose–Effect Relationships

In the most general form the description of biological activity of a certain chemical compound can be represented as a probability of occurrence of a certain biological response, depending on the experimental conditions (object, its state, means of exposure) and "dosage" of the compound ("dosage" can be represented in many different ways, in particular a single *per os* administration or fixed amount of a substance): Pr(Doze,Test). Under the fixed experimental conditions one obtains a simple relationship "dose–effect": Pr(Doze,Test) = $P(D)$. It must be stressed that $P(D)$ is the probability of occurrence of a certain effect, which depends on a dose D as a parameter.

According to the recommendations,[47] in quantitative measurements of biological activity drug action is expressed in terms of the effect, E, produced when an agonist, A, is applied at a concentration $[A]$. The relationship between E and $[A]$ can often be described empirically by Hill's equation,[48,49] which has the form:

$$\frac{E}{E_{max}} = \frac{[A]^{nH}}{[A]^{nH} + [A]_{50}^{nH}} \tag{6.1}$$

where E_{max} is the maximal action of A, nH is the Hill coefficient and $[A]_{50}$ is the concentration that produces an effect that is 50% of E_{max}. Figure 6.1 shows an example of effect–concentration relationships estimated according to the Hill equation (Equation 6.1). Clearly, if $[A] = [A]_{50}$, all curves pass through the point at which the effect is half of its maximal value.

Unfortunately, Hill's equation (Equation 6.1) is only a convenient mathematical idealization, which can be realized for ligand binding to the pure isolated receptor *in vitro*. In an intact biological object a ligand interacts with several or even many different macromolecules,[50] and the final biological effect may dramatically differ from the simple relationship presented in Figure 6.1.

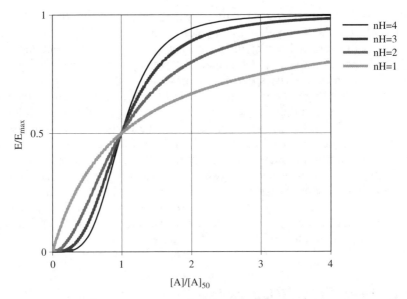

Figure 6.1 Relative values of effect depending on relative agonist concentration calculated according to the Hill equation. *nH* are the different values of the Hill coefficient.

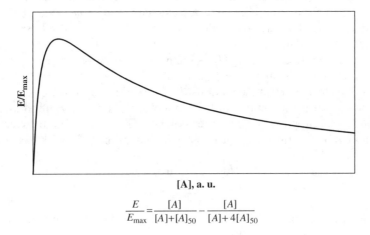

$$\frac{E}{E_{max}} = \frac{[A]}{[A]+[A]_{50}} - \frac{[A]}{[A]+4[A]_{50}}$$

Figure 6.2 Relative effect *vs.* concentration of agonist provided the agonist simultaneously acts on another target as a weak antagonist.

For example, if some effect may be caused by two mechanisms, and a ligand interacts with the appropriate receptors, both activating and inhibiting them, then either activating or suppressing of the effect E can be observed depending of the concentration $[A]$ of the ligand (Figures 6.2 and 6.3).

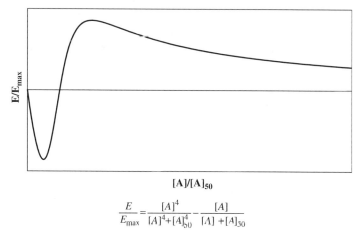

$$\frac{E}{E_{max}} = \frac{[A]^4}{[A]^4+[A]_{50}^4} - \frac{[A]}{[A] +[A]_{50}}$$

Figure 6.3 Example of relative effect dependence on the agonist's concentration provided the agonist acts on another target as antagonist with equal semi-effective concentrations and different Hill coefficients.

In experimental testing of toxicity the results are presented by the numbers of surviving (n) and dying (m) biological objects within the fixed period of time under the fixed doses of acting substance D. The conditional probability $P(m,n|D)$ of certain numbers m and n at the certain D corresponds to the Bernoulli distribution:

$$P(m,n|D) = (m + n)!\frac{P(D)^m(1 - P(D))^n}{m!n!} \tag{6.2}$$

where $P(D)$ is the probability of death of a biological object at the obtained dosage D. Based on the experimental results, $P(D)$ can be estimated only approximately by calculation of parameters for a definite parameter relationship $P(D)$, for instance Equation (6.1).

Usually, the dose–effect relationship $P(D)$ is simplified to the single quantitative or even qualitative characteristic. For example, for a certain level of probability q it is possible to determine an appropriate characteristic dose (quantile) $D_q = Arg\{P(D)=q\}$. Most often, the ED_{50} for $q=0.5$ values are used, but $q=0.16$, $q=0.75$, $q=0.84$ are also considered sometimes. However, for a non-monotonic dependence of $P(D)$, D_q can be ambiguous or even not exist if $P(D)<q$; for instance if (i) the part of population is resistant to the acting compounds and (ii) the suggested threshold q exceeds the fraction of the responsible part of population.

In accordance with the probabilistic nature of the biological activity concept, the most relevant methods for prediction of activity are those based on probabilistic theory and mathematical statistics, and the purpose of prediction is the complete relationship $Pr(Doze) = P(D)$.

Unfortunately, in practice, the application of such approaches is strongly limited by the available experimental data, which in most cases are presented by semi-effective doses and even by qualitative characteristics "active/inactive".[51,52]

6.2.2   Experimental Data

The determination of biological activity is always associated with some experimental errors, which may be caused by variability of biological objects, inaccuracy of measurements due to the limited precision of the used equipment, inaccuracy of the personnel doing manual and mental work.

If the experimental measurements have been repeated several times, the resultant data are presented as average values and standard deviations (SDs) of the measurements. In many cases numerical data in the literature and, particularly, in databases are presented without SDs even in cases where such values could be calculated on the basis of primary data. Also, the results of testing in high-throughput assays for inactive compounds typically mean that the compound does not cause the studied effect at a certain threshold, *e.g.*, at 10 µM, 1 µM, *etc.*[52]

Experimental errors associated with human error may be introduced both in experimental procedures (*e.g.*, inaccuracies of sample preparation) and in theoretical analysis of the study results (*e.g.*, errors in data drawing in publications, errors during the input of data into a computer).

As was concluded by Christoph Helma *et al.*:[53]

After summarizing our experiences with the quality assurance of chemical data in predictive toxicology, we conclude that the currently available databases and computational chemistry programs are too faulty to be trusted without further inspection. The development of reliable quality control procedures definitely needs more discussion, exchange of experience, and research activity. In this sense, we hope that we will raise some awareness in regard to data quality issues and quality assurance in predictive toxicology.

The necessity of quality control for chemical structures, particularly when the data are aggregated from different sources, was recently emphasized in another publication.[54]

However, the main source of scattering in experimental data is certainly determined by the variability of biological response. As was shown by comparison of results obtained in rodent carcinogenicity experiments, the concordance between the results taken from general literature and the results obtained from US National Toxicology Program is only about 57%.[55] Therefore, the reproducibility of biological assays may be quite poor. It is well known that LD_{50} values for rodents obtained in different laboratories may vary significantly (*e.g.*, in LD_{50} studies performed by eleven laboratories to standardize a type A botulinum toxin assay for accessing the toxin in food contaminations, up to a ten-fold difference in results was shown).[56]

Notably, in actual practice training sets are not ideal: in addition to a simplified definition and high variability of biological activity they do not contain all chemical classes relevant to a particular biological activity, and information about all kinds of biological activity that can be revealed by a particular compound is always incomplete (no one compound is tested against all kinds of biological activity, and there is no one activity for which all possible ligands are known). Consequently, the probabilistic character of biological activity is caused not only by experimental errors of its determination but also by the incompleteness of available information.

6.3 Probabilistic Ligand-based Virtual Screening Methods

Virtual screening methods are based on the modeling of the biological phenomenon of molecular recognition, either by the principle of complementarity or by the principle of similarity.[57]

Probabilistic ligand-based virtual screening methods look rather simple and fast; however, for their successful application it is necessary to have a training set of compounds with known activity. Probabilistic methods are based on the achievements of machine learning and have a long history, starting from pattern recognition methods.[4,5,58–63] Especially for the purposes of drug design, probabilistic methods were developed by Golender and Rozenblit,[64] and realized later in the expert system OREX.[65] In Section 6.4 we describe in detail the probabilistic method developed by our team, and to which the methods[7,66–74] and binary QSAR[51,52,75,76] are rather close in basic characteristics.

An important component of probabilistic ligand-based virtual screening methods is the design of the training set, which is the set of ligands available or selected to develop the virtual screening system.[77–81] The selection of this set and its usage strongly influence the overall performance of the final system.[82,83] Also, it is necessary to use the appropriate evaluation of prediction accuracy and reliability, and the representation and interpretation of biological activity prediction results is very important. Based on the probabilistic approach, it is possible to solve all these problems.

6.3.1 Preparation of Training Sets

Training sets should be representative for the compounds to be classified by the ligand-based virtual screening system.[83] Virtual screening is usually performed on a set containing a large number of ligands with a high diversity of molecular structure. For successful results, the diversity of structures from the training set must be comparable to those from the corresponding set used for virtual screening. As a rule, any training set must include sufficient active compounds as well as inactive ones.

It seems obvious that an "ideal" training set must include all tested active and inactive compounds. However, in practice it is necessary to be very careful during the design of training set because "a data set consisting of database chemical drawings and HTS assay measurements may be very misleading".[52]

There exist some other peculiarities, for instance every compound in the MDDR database (MDL® Drug Data Report[84]) has one or several records in the field "activity class", indicating that the compound is related to a certain therapeutic area. However, because of "umbrella patents", not each substance in MDDR was actually tested in biological assays. Those substances for which biological activity was studied in detail are called "principal compounds", and they have some records in the field "Action", such as experimental data on activity, LD_{50}, IC_{50}, K_i, *etc.* There are some publications, in which the training set is prepared on the basis of the MDDR database but this peculiarity is not taken into account.[7,73,74,85–87] In these publications, for each ligand from the training sets that was actually tested in biological assays there are several structurally similar molecules for which biological activity was assigned with the purpose of umbrella patenting. Therefore, unsurprisingly, structure similarity methods studied in these publications were shown to be rather successful during the validation.

In a well-designed training set the structural diversity must be as uniform as possible. It is very difficult to control such uniformity; however, the presence of closely similar compounds series in the set could (and have to) be checked, to avoid degeneracy.

In general, any ligand-based virtual screening method is based on direct or generalized similarity between the screened compound and compounds from the training set. Therefore, if such similarity is absent at all, no reasonable prediction of screened compound's properties can be made by using this training set.

6.3.2 Creation of Evaluation Sets

There are two fundamental problems in ligand-based virtual screening systems development: model selection and performance estimation. Almost invariably, all ligand-based methods have one or more adjustable parameters. To select the "optimal" parameter(s) or model for a given classification problem, it is necessary to utilize the independent evaluation set that was not used in the training procedure. Once the predictive system is developed, to estimate its performance, one must utilize the test set that was not used during the development process. To obtain the precise estimation of system performance, the test set must be large, ideally infinite. However, for a good choice of a model or its parameter(s), the number of compounds in training and evaluation sets must also be large. For theoretical analysis one can subdivide all available data into two (training and test) or three (training, evaluation and test) sets, which have to be approximately equal in size. However, to develop the actual working virtual screening system one must used all available data for the training;

therefore, nothing remains for the evaluation and test sets. To overcome this contradiction, the most suitable methods for construction of evaluation (test) sets are K-Fold Cross-Validation (KF CV) and Leave-One-Out Cross-Validation (LOO CV).[87–90]

To perform KF CV a K-fold partition of the data set is created. For each from K experiments, K-1 folds are used for training and the remaining one for testing. The true error estimate is obtained as the average of the separate K estimates. LOO CV is the degenerated case of KF CV, where K is chosen as the total number of examples. For a data set with N examples, perform N experiments, for each experiment use $N–1$ examples for training and the remaining one example for testing. The true error is estimated as the average error value on test examples – on all existing examples. Vapnik[4] proved several theorems, which stated unbiasedness and consistency of LOO CV estimation, if LOO CV is carefully performed: no information about the excluded compound is used for training and tuning the system based on a residual part of data set. Unfortunately, in the general case the computational time for LOO CV or even for KF CV will be very large due to the large number of sequential experiments. Fortunately, the probabilistic approaches usually have a small or zero number of tuned parameters and the LOO CV procedure can be performed quite easily. Thus, all available data can be used both for training and for evaluation of probabilistic ligand-based virtual screening systems. Earlier we have shown[91] that LOO CV provides a more rigorous accuracy estimation than the repeated many times 2-Fold (or jack-knife) CV.

6.3.3 Mathematical Approaches

Many different methods can be applied to virtual screening, and such methods are described in other chapters of this book and/or in the *Handbooks of Cheminformatics*.[3] Here we discuss the methods based on a probabilistic approach. Unfortunately, there are many publications in which the "probabilistic" or "statistical" approach items are farfetched. The Binary Kernel Discrimination[8–10,17,20] and the Bayesian Machine Learning Models[6] are actually special cases of Artificial Neural Networks; whereas the Probabilistic Neural Networks[14–16] are really similarity-based methods, which do not take into account the results of well-developed nonparametric regression methods.[92]

In virtual screening of the chemical structures set called the Screening Set (SS) for each compound $C \in SS$ any proposed method P should give the estimate $P(C)$, which, being compared with a certain criterion, provides the basis for decision about the advisability of further testing of the chemical compound C. In other words, it is necessary to recognize whether compound C belongs to the class of compounds in which we are interested in, *i.e.*, to solve the task of pattern recognition (PR), which is a typical problem of Machine Learning (ML). There are a lot of publications, monographs and specialized journals devoted to the problems of ML and PR; machine learning approaches are widely used in cheminformatics (see, for example, refs.

11,67,69–71,73,87,93,94). Notably, the fundamentals of machine learning were developed much earlier than the informational technologies (IT) became widely introduced. For example, Nilsson[61] noted, referring to Kanal,[59] that the engineers rediscover for themselves well-known methods of statistics. Later, in machine learning these methods were discovered for a second time, and now the same situation is observed in cheminformatics: methods well known to engineers and IT specialists are rediscovered once again.

Mathematically, the estimate $P(C)$ in many cases can be represented as:[1,61]

$$P(C) = \sum_i a_i f_i(C) \tag{6.3}$$

where $f_i(C)$ are the different functions of chemical structure of compound C, independent from the coefficients a_i. Various methods differ in the values of estimates $P(C)$, in the choice of functions $f_i(C)$, and in approaches that are used to determine the coefficients a_i. Without restriction of generality, let us suggest that the estimate $P(C)$ is a real quantity, and the decision about advisability of further testing of chemical compound C is taken if $P(C) > \theta$, where θ is a threshold value. If the functions $f_i(C)$ represent physicochemical parameters or other quantitative characteristics of molecular structure and/or every possible function of these characteristics, and coefficients a_i are determined on the basis of regression, PLS, SVM *etc.*, then the estimate $P(C)$ is the result of a QSAR method. If, at the same time, $f_i(C)$ are determined as a measure of similarity of structure of molecule C with another molecule C_i from the training set, it is a QSAR method based on similarity. If the functions $f_i(C)$ possess only the values 0 and 1, and coefficients a_i are determined on the basis of probabilistic approach, it is the method described in this chapter.

It is widely accepted that probabilistic approach was first developed and applied in expert systems MYCIN[95,96] and PROSPECTOR.[97] In these expert systems the likelihood estimates are calculated for several competitive hypothesis H on the basis of available evidences E. In the expert system MYCIN each hypothesis was estimated by a confidence factor $CF(H|E_1,E_2, \ldots)$ as a difference of estimates for the measure of belief $MB(H|E_1,E_2,\ldots)$ and the measure of distrust $MD(H|E_1,E_2,\ldots)$:

$$CF(H|E_1, E_2, \ldots) = MB(H|E_1, E_2, \ldots) - MD(H|E_1, E_2, \ldots) \tag{6.4}$$

where MB and MD were calculated by aggregation of values for separate evidences E_i $MB(H|E_i)$ and $MD(H|E_i)$ according to the theory of probability rules. In fact, these aggregation rules are piecewise-linear approximations of simpler formula:

$$CF(H|E_1, E_2, \ldots, E_m) = \frac{CF(H|E_1, E_2, \ldots, E_{m-1}) + CF(H|E_m)}{1 + CF(H|E_1, E_2, \ldots, E_{m-1})CF(H|E_m)} \tag{6.5}$$

These equations follow directly from the approach, which is very popular in recent times in Machine Learning, Data Mining, Text Mining and Knowledge

Data Discovery, bioinformatics and cheminformatics, and called "naive Bayes classifier".[7,63,66,68,98,99] Such an approach was applied for virtual screening by Labute and Gao,[51,52,75,76] and other researchers,[67,69–71,73] and also by the authors of this chapter.[91,100–104]

When applied to virtual screening the naive Bayes classifier consists in the following.

Let a molecular structure of compound C be represented by the set of descriptors $\{D_1, D_2, \ldots D_m\}$, and the probability that it belongs to a given class A is estimated by a conditional probability $P(A|C) = P(A|D_1, D_2, \ldots D_m)$.

Using Bayes' theorem, we write:

$$P(A|D_1, D_2, \ldots, D_m) = \frac{P(A) \cdot P(D_1, D_2, \ldots, D_m|A)}{P(D_1, D_2, \ldots, D_m)} \tag{6.6}$$

where $P(D_1, D_2, \ldots D_m|A)$ is the conditional probability of the descriptors set $\{D_1, D_2, \ldots D_m\}$ occurrence in a compound C from class A; $P(A)$ is the class A *prior probability*, $P(D_1, D_2, \ldots D_m)$ is the descriptors set $\{D_1, D_2, \ldots D_m\}$ *prior probability*. The "naive" conditional independence assumptions mean that each descriptor D_i is conditionally independent of every other descriptor D_j for $j \neq i$. This means that:

$$P(D_1, D_2, \ldots, D_m|A) \cong P(D_1|A)P(D_2|A) \ldots P(D_m|A) = \prod_{i=1}^{m} P(D_i|A) \tag{6.7}$$

As a result, the log-likelihood ratio of the conditional probability $P(A|D_1, D_2, \ldots D_m)$ of the class A and $P(\neg A|D_1, D_2, \ldots, D_m)$ of its complement $\neg A$ can be expressed as:

$$\ln\left[\frac{P(A|D_1, D_2, \ldots, D_m)}{P(\neg A|D_1, D_2, \ldots, D_m)}\right] = \ln\left[\frac{P(A)}{P(\neg A)}\right] + \sum_i \ln\left[\frac{P(D_i|A)}{P(D_i|\neg A)}\right] \tag{6.8}$$

Taking into account that $P(\neg A|D_1, D_2, \ldots, D_m) = 1 - P(A|D_1, D_2, \ldots, D_m)$ and using Bayes' theorem for ratio $P(D_i|A)/P(D_i|\neg A)$, we find:

$$\ln\left[\frac{P(A|D_1, D_2, \ldots, D_m)}{P(\neg A|D_1, D_2, \ldots, D_m)}\right] = \ln\left[\frac{P(A)}{1 - P(A)}\right] + \sum_i \left\{\ln\left[\frac{P(A|D_i)}{1 - P(A|D_i)}\right] - \ln\left[\frac{P(A)}{1 - P(A)}\right]\right\} \tag{6.9}$$

In terms of the general formula Equation (6.3), we can write:

$$P(C) = \ln\left[\frac{P(A|D_1, D_2, \ldots, D_m)}{P(\neg A|D_1, D_2, \ldots, D_m)}\right] \tag{6.10a}$$

$$a_0 = \ln\left[\frac{P(A)}{1 - P(A)}\right], \quad f_0(C) \equiv 1 \tag{6.10b}$$

$$a_i = \sum_i \left\{ \ln\left[\frac{P(A|D_i)}{1 - P(A|D_i)}\right] - \ln\left[\frac{P(A)}{1 - P(A)}\right] \right\} \qquad (6.10c)$$

$f_i(C) = 1$ if $D_i \in \{D_1, D_2, \ldots, D_m\}$ and $f_i(C) = 0$ if $D_i \notin \{D_1, D_2, \ldots, D_m\}$.

$$(6.10d)$$

Clearly, the constant a_0 can be included into threshold value θ, so that the function $f_0(C) \equiv 1$ is not necessary. We must stress that in such form the probabilistic approach has no tuned parameters at all. Some tuning of naive Bayes classifier can be performed by selection of the molecular structure descriptors [or $f_i(C)$] set. This is a wonderful feature in contrast to QSAR methods, especially to Artificial Neural Networks.

The describing functions $f_i(C) = 1$ if $D_i \in \{D_1, D_2, \ldots D_m\}$ [and $f_i(C) = 0$ otherwise] can be constructed on the basis of very wide approaches. In our investigations we use as descriptor sets $\{D_1, D_2, \ldots D_m\}$ substructure fragment descriptors (see below). For quantitative parameters the describing function $f_i(C)$ can be equal to 1 if molecule parameter(s) x_j satisfies the certain condition k, *e.g.*, if the value of x_j belongs to some interval or multidimensional x_j belongs to some region in appropriate space, and so on. Like this, the naive Bayes approach was proposed and developed by Labute and Gao.[51,52,74–76,105]

The naive Bayes approach has several well-known difficulties. The conditional independence of descriptors of a molecule structure is not true as a rule. The probability $P(A|D_i)$ estimations can be close or even equal to 0 or 1 and in such case coefficients a_i become too large or infinite. To overcome this problem, we have substituted the logarithms of probabilities ratios $\ln[P(A|D_i)/(1 - P(A|D_i))]$ for $\mathrm{ArcSin}(2P(A|D_i) - 1)$. The $\mathrm{ArcSin}(2P(A|D_i) - 1)$ shape coincides with the shape of $\ln[P(A|D_i)/(1 - P(A|D_i))]$ for almost all values of $P(A|D_i)$, but $\mathrm{ArcSin}(2P(A|D_i) - 1)$ values are bounded by the values $\pm\pi/2$.

Interestingly, the naive Bayes approach is "too simple", but as a rule it provides high accuracy of recognition.[7,63,68]

6.3.4 Evaluation of Prediction Accuracy

When a classifier that provides the estimation of $P(C)$ is constructed, its performance must be estimated. The most important estimation is of the prediction accuracy. To do this, an evaluation set (test set or validation set – see Section 6.3.2) must be used. The evaluation set (ES) must be relevant and include both type of examples – positive and negative ("active" and "inactive" compounds). For all compounds $C \in$ ES, estimations $P(C)$ are calculated, and obtained values are analyzed using knowledge about the "true" classification of compounds in ES. Figure 6.4 shows the main features of this task.

Let us suggest that for compounds in ES we have values of some targeted molecular property. "Expert" divides ES into two parts: positive and negative examples. Using a constructed estimator we calculate $P(C)$ values and, selecting the threshold value, divide ES into two other parts: predicted positive

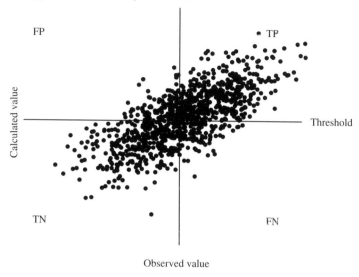

FP • TP

Calculated value

Threshold

TN • FN

Observed value

Figure 6.4 An artificially generated relationship between observed and calculated values of effect is shown as points with binormal distribution. Compounds are divided by the vertical line into actives and inactives according to the experimental values and by the horizontal line into predicted actives and inactives, at the selected threshold value. Compounds that fall into the appropriate quadrants are classified based on the test as "True Positives" (*TP*), "True Negatives" (*TN*), "False Positives" (*FP*), and "False Negatives" (*FN*).

if $P(C) > \theta$ and predicted negative if $P(C) < \theta$. We compare prediction results with known data and calculate four numbers: *TP* is the number of true positives, *FP* is the number of false positives, *TN* is the number of true negatives, and *FN* is the number of false negatives (Figure 6.4).

It is important to keep in mind that the situation illustrated in Figure 6.4 is a common case and it has symmetry in relation to errors: errors can be both in estimations $P(C)$ and in experimental values. The result like that shown in Figure 6.4 occurs, even if the classifier is ideally true but experimental values are known with finite accuracy.

For pattern recognition or classification, usually, the following characteristics of recognition accuracy are used (see, for example, refs. 66,106–108):

$$\text{Sensitivity}: \quad = \frac{TP}{TP+FN}$$

$$\text{Specificity}: \quad = \frac{TN}{TN+FP}$$

$$\text{Accuracy (concordance)}: \quad = \frac{TP+TN}{TP+FP+TN+FN}$$

$$\text{Predictive value positive}: \quad = \frac{TP}{TP+FP}$$

$$\text{Predictive value negative}: \quad = \frac{TN}{TN+FN}$$

$$\text{False negative rate}: \quad = \frac{FN}{TP+FN}$$

$$\text{False positive rate}: \quad = \frac{FP}{TN+FP}$$

and others; each of them has some disadvantages. To minimize the disadvantages, Youden's index was proposed in 1950.[109] Youden's index summarizes the test accuracy into a single numeric value, Sensitivity + Specificity − 1, or:

$$YI = \frac{TP}{TP + FN} + \frac{TN}{TN + FP} - 1 = \frac{TP \cdot TN - FP \cdot FN}{(TP + FN) \cdot (TN + FP)} \tag{6.11}$$

The recognition accuracy estimation described above faces one very important problem: what is the best choice for the threshold value θ? To solve this problem, statistical decision theory is used.[110–113] The basis for this is an analysis of the so-called the Received Operating Characteristic (ROC) curve. By tradition, ROC is plotted as a function of true positive rate $TP/(TP + FN)$ (or sensitivity) *versus* false positive rate $FP/(TN + FP)$ (or 1-Specificity) for all possible threshold values θ. Figure 6.5 presents an example of such a ROC curve for the results obtained with our computer program PASS in predicting antineoplastic activity.

Estimation of the optimal threshold value is provided by minimizing a risk function, which depends on *a priori* probabilities of positive and negative

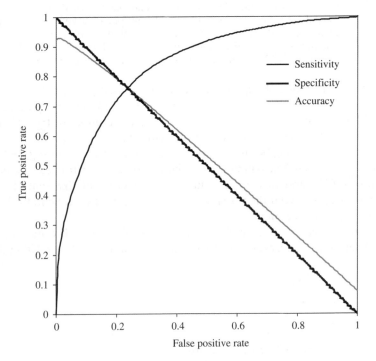

Figure 6.5   Relationships between the sensitivity $[TP/(TP + \text{FN})]$ (shown by the curve), specificity $[TN/(TN + FP)]$ and accuracy (concordance) $[(TP + TN)/(TP + FP + TN + FN)]$ as functions of False Positive Rate $[FP/(TN + FP)]$. The estimations were obtained by PASS 2007 in a leave-one-out cross-validation procedure for antineoplastic activity.

examples and loss values for all four (*TP*, *FP*, *TN* and *FN*) possible results. If *a priori* probabilities or losses are not known, the optimal choice is *MiniMax* (*Mini*mizing the *Max*imum possible loss) according to which the optimal threshold value must satisfy the condition "Sensitivity = Specificity". Another choice may be the maximum of Youden's index.

In any case, this approach uses several additional assumptions. For this reason in the last time in ML the recognition accuracy criterion of the Area Under the ROC Curve (*AUC*), which is free of additional assumptions, becomes very popular.[7,63,68–71,106–108,112–116] Mathematically, *AUC* equals the probability that the estimation $P(C)$ assigns the higher value to a randomly drawn positive example C_+ than to the randomly drawn negative example C_-:

$$AUC(P) = \text{Probability}\{P(C_+) > P(C_-)\} \qquad (6.12)$$

In our papers[91,117] we have used the Invariant Accuracy of Prediction (*IAP*) criterion, which exactly coincides with the *AUC*, and it is calculated as:

$$IAP = \frac{\text{Number of } P(C_+) > P(C_-)}{(\text{Number of } C_-) \cdot (\text{Number of } C_+)} \qquad (6.13)$$

In our computer program PASS (Section 6.4) we also use the Invariant Error of Prediction (IEP) criterion: $IEP \equiv 1 - IAP$.

Computationally, it is more convenient to calculate the estimate of prediction accuracy on ES as an Invariant Accuracy (*IA*), which equals $2AUC(P) - 1$ and can be calculated as a result of comparison of estimates $P(C_+)$ for positive and $P(C_-)$ for negative examples through all pairs (each positive example and each negative example) in a form:

$$IA = \frac{\sum_{ES} Sgn[P(C_+) - P(C_-)]}{N_+ N_-}, \quad Sgn(z) = \begin{cases} -1, & z < 0 \\ 0, & z = 0 \\ +1, & z > 0 \end{cases} \qquad (6.14)$$

which is the difference of numbers of cases of true $P(C_+) > P(C_-)$ and false $P(C_+) < P(C_-)$ divisions of pairs of positive and negative examples, divided by the number of all pairs $N_+ N_-$. These are the following general cases:

- If all objects are predicted with the same value $P(C)$, then $IA = 0$.
- If the prediction is random and the estimates $P(C_+) > P(C_-)$ and $P(C_+) < P(C_-)$ have equal probabilities, then $IA = 0$ on average of probability.
- If all outcomes $P(C_+) > P(C_-)$ or $P(C_+) < P(C_-)$, then $IA = 1$ or -1, respectively.

If inaccuracy of division of ES onto two classes exists, then:

$$IA = A \cdot \left(1 - \frac{m_-}{N_-} - \frac{m_+}{N_+}\right)$$

where $A \leq 1$ is the potential accuracy of the method, m_- is the number of compounds mistakenly described as negative examples (not found yet or not studied positive examples), and m_+ is the number of compounds mistakenly described as positive examples, for instance due to the errors in data used for creation of the sets, mistakes of personnel, *etc.* With:

$$IA = A \cdot \left(1 - \frac{m_-}{N_-} - \frac{m_+}{N_+}\right)$$

it is possible to compare the accuracy of several classifiers using ES with "errors of the teacher" correctly.

The *IA* (*IAP*, *AUC*) criterion gives a robust estimation of general classifiers performance, but in the case of virtual screening to find several ligands at a top of ranked compounds list, the minimal number of decoys may be more important.[116,118] For this purpose, Enrichment Factor[7,115,119–122] analysis of the robust initial enhancement (RIE)[116,118] and Boltzmann-Enhanced Discrimination ROC (BEDROC)[116] criteria were proposed.

6.3.5 Single-targeted *vs.* Multi-targeted Virtual Screening

Most existing virtual screening methods have been developed to be used for selection of hits with a single targeted activity.[22–24] However, most discovered pharmaceutical agents have several or even many kinds of biological activity. Some of these biological activities represent adverse/toxic effects, some others can be considered as a reason for utilization of known medicines according to new indications, which is called repositioning of drugs.[123–126]

Both new pharmacotherapeutics and adverse/toxic effect can be discovered on the basis of computer predictions with probabilistic methods. Different methods can be applied either sequentially or simultaneously. Early attempts to predict many kinds of biological activity simultaneously using such an approach were performed by Avidon and co-authors,[127] Golender and Rozenblit,[64,65] and Vassiliev and co-authors.[128]

Since the early 1990s, the authors have been developing the computer program PASS, which predicts many kinds of biological activity based on the structural formula of a compound.[29,30,32,40,41,43,100] This program, the present version of which predicts over 3000 kinds of biological activity with a mean accuracy of about 95%, is described in more detail below. Different PASS applications in virtual screening of multi-targeted ligands have been presented in several publications.[100–104,129]

The Prous Institute for Biomedical Research[130] is developing a computational method based on a wide range of molecular descriptors and binding profiles, called BioEpisteme®, which is claimed to facilitate the discovery of new medicines and new uses for existing drugs. Pre-requisites of the BioEpisteme approach are quite close to the PASS concept: "A drug may interact with multiple targets and produce more than one therapeutic response and/or adverse effect." Unfortunately, we could not find a detailed description of the method used in BioEpisteme in the available literature – only the very general scheme presented on the web-site.[130] Recently, the number of different molecular mechanisms covered by BioEpisteme was reported to be about 400.[131]

Quantum Pharmaceuticals[132] recently proposed a new method for toxicity prediction based on computation of small molecules' affinity to about 500 human proteins. The analysis of binding profiles for about 1000 known pharmaceutical agents led to establishment of a relation between the toxicological properties of a molecule and its activity against the selected representatives of approximately 50 protein families. This activity profile was further used as a "natural" set of descriptors for various toxicological endpoints predictions, including human-MRDD, human-MRTD, human-TDLo, mouse-LD_{50} (oral, intravenous, subcutaneous), rat-LD_{50} (oral, intravenous, subcutaneous, intraperitoneal), *etc.*[46]

Thus, probabilistic biological activity prediction methods can be used for both estimation of adverse/toxic effects in molecules under study and for finding the multi-targeted ligands, which might "yield drugs of superior clinical value compared with monotargeted formulations".[33]

6.4 PASS Approach

The computer program PASS was designed to predict many kinds of biological activity simultaneously based on the structural formulae of chemical compounds. Thus, PASS may estimate the biological activity profiles for virtual molecules, prior to their chemical synthesis and biological testing.

6.4.1 Biological Activities Predicted by PASS

The latest version of PASS (2007) predicts 3300 kinds of biological activity with a mean prediction accuracy of about 95%. PASS could predict about 1000 kinds of biological activity in 2004,[32] only 541 activities in 1998,[133] and 114 activities in 1996.[30]

The default list of predictable biological activities currently includes 374 pharmacotherapeutic effects (*e.g.*, antihypertensive, hepatoprotectant, nootropic, *etc.*), 2755 mechanisms of action, (*e.g.*, 5-hydroxytryptamine antagonist, acetylcholine M1 receptor agonist, cyclooxygenase inhibitor, *etc.*), 50 adverse and toxic effects (*e.g.*, carcinogenic, mutagenic, hematotoxic, *etc.*) and 121 metabolic terms (*e.g.*, CYP1A inducer, CYP1A1 inhibitor, CYP3A4 substrate, *etc.*). Information about novel activities and new compounds can be straightforwardly included into PASS.

In PASS *biological activities* are described qualitatively ("active" or "inactive"). Qualitative presentation allows integrating information concerning compounds tested under different terms and conditions and collected from many different sources, as in the general PASS training set. Any property of chemical compounds that is determined by their structural peculiarities can be used for prediction by PASS. Clearly, the applicability of PASS is broader than the prediction of biological activity spectra. For example, we used this approach to predict drug-likeness[134] and the biotransformation of drug-like compounds.[135]

6.4.2   Chemical Structure Description in PASS

The 2D structural formulae of compounds were chosen as the basis for *description of chemical structure* because this is the only information available in the early stage of research. Plenty of characteristics of chemical compounds can be calculated on the basis of structural formulae.[3,67,136–139] Earlier[29] we applied the Substructure Superposition Fragment Notation (SSFN) codes.[140] But SSFN, like many other structural descriptors, reflects rather abstraction of chemical structure by the human mind than the nature of the biological activity revealed by chemicals. The Multilevel Neighborhoods of Atoms (MNA) descriptors[91,141,142] have certain advantages over SSFN. These descriptors are based on the molecular structure representation, which includes the hydrogens according to the valences and partial charges of present atoms and does not specify the types of bonds. MNA descriptors are generated as a recursively defined sequence:

- zero-level MNA descriptor for each atom is the mark A of the atom itself;
- any next-level MNA descriptor for the atom is the sub-structure notation $A(D_1 D_2 \ldots D_i \ldots)$,

where D_i is the previous-level MNA descriptor for i-th immediate neighbors of the atom A. The mark of atom may include not only the atomic type but also any additional information about the atom. In particular, if the atom is not included into the ring it is marked by "-". The neighbor descriptors $D_1 D_2 \ldots D_i \ldots$ are arranged in uniquely, *e.g.*, in lexicographic order. Iterative process of MNA descriptors generation can be continued, covering first, second, *etc.* neighborhoods of each atom. MNA descriptors have a more general background than the descriptors,[67,137] which look like MNA.

The molecular structure is represented by the set of unique MNA descriptors of the 1st and 2nd levels. The substances are considered to be *equivalent* in PASS if they have the same set of MNA descriptors. Since MNA descriptors do not represent the stereochemical peculiarities of a molecule, substances whose structures differ only stereochemically are formally considered as equivalent.

6.4.3   SAR Base

The PASS estimations of biological activity spectra of new compounds are based on the Structure–Activity Relationships data and knowledge-base (SAR Base), which accumulates the results of the training set analysis. The in-house developed general PASS training set currently (December 2007) includes about 117000 known biologically active substances (drugs, drug-candidates, leads, and toxic compounds). Since new information about biologically active compounds is discovered regularly, we perform a special informational search and analyze the new information, which is further used for updating and correcting the PASS training set.

6.4.4 Algorithm of Activity Spectrum Estimation

The algorithm of activity spectrum estimation is based on the above-mentioned Bayesian approach, but differs in several details. For each kind of activity A_k, which can be predicted by PASS, on the basis of a molecule's structure represented by the set of MNA descriptors $\{D_1 D_2 .. D_m\}$ the following values are calculated:

$$S_{0k} = 2P(A_k) - 1 \tag{6.15a}$$

$$S_k = Sin\left[\frac{1}{m} \sum ArcSin(2P(A_k|D_i) - 1)\right] \tag{6.15b}$$

$$B_k = \frac{S_k - S_{0k}}{1 - S_k S_{0k}} \tag{6.15c}$$

where $P(A_k)$ is the *a priori* probability of finding a compound with activity of kind A_k; $P(A_k|D_i)$ is a conditional probability of activity of kind A_k if the descriptor D_i is present in a set of a molecule's descriptors. For each kind of activity, if for all descriptors of molecule $P(A_k|D_i) = 1$, then $B_k = 1$; if for all descriptors of molecule $P(A_k|D_i) = 0$, then $B_k = -1$; if the relationship between descriptors of molecule and activity A_k does not exist and $P(A_k) \approx P(A_k|D_i)$, then $B_k \approx 0$.

The simplest frequency estimations of probabilities $P(A_k)$, $P(A_k|D_i)$ are given by:

$$P(A_k) = \frac{N_k}{N}, \ P(A_k|D_i) = \frac{N_{ik}}{N_i} \tag{6.16}$$

where N is the total number of compounds in the SAR Base; N_k is the number of compounds containing the activity A_k in the activity spectrum; N_i is the number of compounds containing descriptor D_i in the structure description; N_{ik} is the number of compounds containing both the activity A_k and the descriptor D_i.

In PASS version 1.703 and later the estimations of probabilities $P(A_k)$, $P(A_k|D_i)$ are calculated as:

$$P(A_k) = \frac{\sum_n f_n(A_k) \sum_i g_n(D_i)}{\sum_n \sum_i g_n(D_i)} \tag{6.17a}$$

$$P(A_k|D_i) = \frac{\sum_n f_n(A_k)g_n(D_i)}{\sum_n g_n(D_i)} \qquad (6.17b)$$

where $f_n(A_k)$ is the generic function of compound n belonging to a set of compounds containing the activity A_k in the activity spectrum, $f_n(A_k)$ is equal to 0 or 1; $g_n(D_i)$ is the measure of compound n belonging to the set of compounds containing descriptor D_i in the structure description, now $g_n(D_i)$ is equal to 0 or $1/m_n$, where m_n is the number of descriptors for the molecule n, and $\sum_i g_n(D_i) \equiv 1$ in this case.

The estimations Equations (6.17a, b) of probabilities $P(A_k)$, $P(A_k|D_i)$ not only increase the algorithm's prediction accuracy but also open up new possibilities. For example, function $f_n(A_k)$ in the range [0,1] can be considered as a measure of molecule n belonging to a fuzzy set of molecules that reveal activity A_k. The descriptor weight $g_n(D_i)$ can be considered in the same manner, and then the molecule structure descriptors can be of arbitrary nature, *e.g.*, such as in the refs. 51 and 52.

The main purpose of PASS is the prediction of activity spectra for new, possibly not yet synthesized compounds. Therefore, the general principle of the PASS algorithm is the exclusion from SAR Base of substances that is equivalent to the substance under prediction. So, if molecule n is equivalent to the molecule under prediction then this substance is excluded from sums in (Equations 6.17a,b).

To obtain the qualitative ("Yes/No") results of prediction, it is necessary to define the threshold B_k values for each kind of activity A_k. On the basis of statistical decision theory (Section 6.3.4) it is possible using the risk functions minimization, but nobody can *a priori* determine such functions for all kinds of activity and for all possible real-world problems. Therefore the *predicted activity spectrum* is presented in PASS by the list of activities with probabilities "to be active" P_a and "to be inactive" P_i calculated for each activity. The list is arranged in descending order of P_a-P_i; thus, the more probable activities are at the top of the list. The list can be shortened at any desirable cutoff value, but $P_a > P_i$ is used by default. If the user chooses a rather high value of P_a as a cutoff for selection of probable activities, the chance to confirm the predicted activities by the experiment is high too, but many activities will be lost. For instance, if $P_a > 80\%$ is used as a threshold, about 80% of real activities will be lost; for $P_a > 70\%$, the portion of lost activities is 70%, *etc.*

An example of prediction results for sulfathiazole is shown in Figure 6.6. This substance was found in SAR Base and was excluded from the SAR Base on prediction of its activity spectrum. The known (contained in SAR Base of PASS version 2007) activity spectrum includes the following activities: antibacterial, antibiotic, dihydropteroate synthase inhibitor, iodide peroxidase inhibitor. In Figure 6.6 the predicted activity spectrum includes 65 of 374 pharmacological effects, 176 of 2755 molecular mechanisms, 7 of 50 side effects and toxicity, 11 of 121 metabolism terms at default $P_a > P_i$ cutting points. All activities included in the SAR Base are predicted with $P_a > P_i$. The activity of as

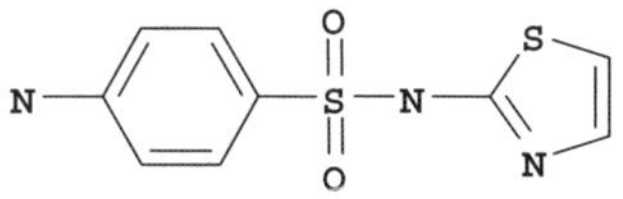

```
>  <PASS_MNA_COUNT>
   32

>  <PASS_KNOWN_ACTIVITIES>
               Antibacterial
               Antibiotic
               Dihydropteroate synthase inhibitor
               Iodide peroxidase inhibitor

>  <PASS_RESULT_COUNT>
    65 of   374 Possible Pharmacological Effects at Pa > Pi
   176 of  2755 Possible Molecular Mechanisms at Pa > Pi
     7 of    50 Possible Side Effects and Toxicity at Pa > Pi
    11 of   121 Possible Metabolism at Pa > Pi

>  <PASS_EFFECTS>
0.886  0.004  Antiobesity
0.769  0.004  Antidiabetic
0.766  0.008  Antieczematic Atopic
0.738  0.010  Antiprotozoal (Toxoplasma)
0.752  0.027  Antineoplastic (colorectal cancer)
0.721  0.002  Antiprotozoal (Coccidial)
0.651  0.043  Antineoplastic (brain cancer)
0.601  0.072  Antinephritic
0.601  0.091  Antiviral (Arbovirus)
0.578  0.083  Antineoplastic (lymphocytic leukemia)
0.578  0.083  Antineoplastic (non-Hodgkin's lymphoma)
0.418  0.005  Hypoglycemic
0.484  0.093  Allergic conjunctivitis treatment
0.408  0.019  Diuretic inhibitor
0.395  0.016  Antibacterial
0.421  0.043  Hematopoietic inhibitor
       . . .
0.253  0.059  Antiprotozoal (Trichomonas)
0.209  0.021  Antibiotic
0.267  0.093  Anticoagulant
       . . .
0.008  0.005  Histone acetylation inducer

>  <PASS_MECHANISMS>
0.732  0.004  Para amino benzoic acid antagonist
0.675  0.004  Dihydropteroate synthase inhibitor
0.661  0.028  Chloride peroxidase inhibitor
0.592  0.025  5 Hydroxytryptamine 6 agonist
0.591  0.062  Phthalate 4,5-dioxygenase inhibitor
       . . .
0.265  0.227  Pterin deaminase inhibitor
0.138  0.100  Iodide peroxidase inhibitor
0.166  0.129  Cathepsin H inhibitor
       . . .
0.141  0.140  3-Hydroxybenzoate 4-monooxygenase inhibitor

>  <PASS_TOXICITY>
0.555  0.112  Hematotoxic
0.442  0.139  Hepatotoxic
0.392  0.135  Nephrotoxic
0.275  0.066  Carcinogenic, female rats
0.205  0.114  Carcinogenic, female mice
0.341  0.269  Torsades de pointes
0.162  0.123  Carcinogenic

       . . .
```

Figure 6.6 Structure of sulfathiazole and part of its predicted activity spectrum. Activities contained in the SAR Base of PASS version 2007 are marked in bold.

a dihydropteroate synthase inhibitor is second among the 176 predicted
molecular mechanisms.

The probabilities P_a and P_i are functions of the initial estimation B_k defined
by the equations:

$$FA_k(P_a) = B_k, \quad FI_k(P_i) = B_k \tag{6.18}$$

where the functions FA_k and FI_k are obtained as the final result of the *training
procedure*, which consists of the following.

For each kind of activity and each MNA descriptor the estimations of prob-
abilities $P(A_k)$, $P(A_k|D_i)$ are calculated by Equations (6.17a,b). For each kind of
activity A_k, for each p of N_k active, and for each q of $N - N_k$ inactive compound in
SAR Base, after excluding this compound, the estimates B_{kp} and B_{kq} are calculated. The N_k estimates of B_{kp} for active compounds are sorted in ascending order;
the $N - N_k$ estimates of B_{kq} for inactive compounds are sorted in descending
order. The functions FA_k and FI_k are calculated as conditional expectations:

$$FA_k(F) = \sum_{p=1}^{N_k} C_{N_k-1}^{p-1} F^{p-1} (1-F)^{N_k-p} B_{kp} \tag{6.19a}$$

$$FI_k(F) = \sum_{q=1}^{N-N_k} C_{N-N_k-1}^{q-1} F^{q-1} (1-F)^{N-N_k-q} B_{kq} \tag{6.19b}$$

where $C_n^m F^m (1-F)^n$ is the binomial distribution, $C_n^m = n!/[m!(n-m)!]$ is the
binomial coefficient,

F is in the range [0, 1]. Clearly, FA_k and FI_k are estimations of the quantile
functions of the probability distributions of the estimations B_{kp} and B_{kq}. Thus,
the probabilities P_a and P_i are both the measures of belonging to subsets of
"active" and "inactive" compounds and the probabilities of the 1st and 2nd
kinds of prediction error, respectively. These two interpretations of the probabilities P_a and P_i are equivalent and can be used in understanding the results
of prediction.

In Figure 6.7 shows an example of probabilities $P_a(B)$ and $P_i(B)$ estimation
as functions of B value, and in terms of Sensitivity, Specificity and Youden's
index, for antihypertensive activity in the SAR Base of PASS version 2007.

Leave one out cross-validation for 3300 kinds of biological activity and
117 332 substances provides the estimate of PASS prediction accuracy during
the training procedure. The average accuracy of prediction is about 94.7%
according to the LOO CV estimation, while that for particular kinds of activity
varies from 65% [System lupus erythematosus treatment, Immunomodulator
(HIV)] to 99.9% (Allergic rhinitis treatment, histone acetylation inducer). The
estimated accuracy of prediction for all kinds of biological activity predicted by
PASS is presented at the web site.[143]

The accuracy of PASS predictions depends on several factors, of which
the quality of the training set seems to be the most important (Section 6.3.1).

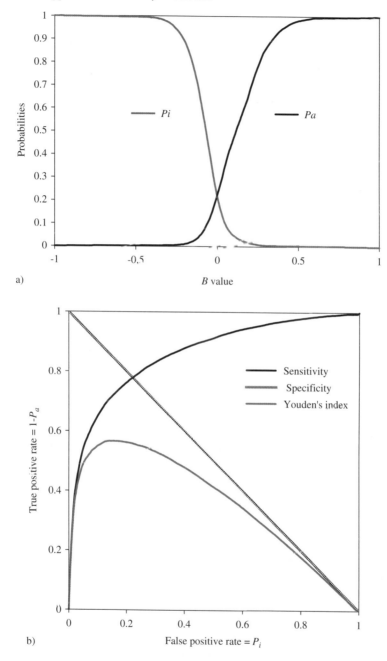

Figure 6.7 Estimations of probabilities $P_a(B)$ and $P_i(B)$ as functions of B value (*a*) and in terms of sensitivity, specificity and Youden's index (*b*). The curves are obtained for activity antihypertensive based on data presented in SAR Base PASS version 2007.

A perfect training set should include comprehensive information about biological activities known or possible for each compound. In other words, the whole *biological activity spectrum* should be thoroughly investigated for each compound included into the PASS training set. Actually, no database exists with information about biologically active compounds tested against each kind of biological activity. Therefore, information concerning known biological activities for any compound is always incomplete. We investigated the influence of the information's incompleteness on the prediction accuracy for new compounds. About 20 000 "principal compounds" from the MDDR database (Section 6.3.1) were used to create the heterogeneous training and evaluation sets. At random, 20, 40, 60, 80% of information were excluded from the training set. Either structural data or biological activity data were removed in two separate computer experiments. In both cases it was shown that even if up to 60% of information is excluded the results of prediction are still satisfactory.[91] Thus, despite the incompleteness of information in the training set, the method used in PASS is robust enough to get reasonable prediction results.

6.4.5 Interpretation of Prediction Results

Only activities with $P_a > P_i$ are considered as possible for a particular compound.

It is necessary to remember that the probability P_a first of all reflects the similarity of molecule under prediction with the structures of molecules that are the most typical in a sub-set of "actives" in the training set. Therefore, usually, there is no direct correlation between the P_a values and quantitative characteristics of activities.

Even an active and potent compound, whose structure is not typical of the structures of "actives" from the training set, may obtain a low P_a value and even $P_a < P_i$ during the prediction. This is clear from the way the functions $P_a(B)$ and $P_i(B)$ are constructed: the values P_a for "actives" and P_i for "inactives" are distributed fully uniformly. Taking this into account, the following interpretation of prediction results is possible.

If, for instance, P_a equals to 0.9, then for 90% of "actives" from the training set the B values are less than for this compound, and only for 10% of "actives" is this value higher. If we decline the suggestion that this compound is active, we will make a wrong decision with probability 0.9.

If P_a is less than 0.5, but $P_a > P_i$, then for more than half of "actives" from the training set the B values are higher than for this compound. If we decline the suggestion that this compound is active, we will make a wrong decision with a probability of <0.5. In such a case the probability of confirming this kind of activity in the experiment is small, but there is a more than 50% chance that this structure has a high degree of novelty and may become a New Chemical Entity (NCE).

If the predicted biological activity spectrum is wide, the structure of the compound is quite simple, and does not contain peculiarities that are responsible for the selectivity of its biological action.

If it appears that the structure under prediction contains a few new MNA descriptors (in comparison with the descriptors from the compounds of the training set), then the structure has low similarity with any structure from the training set, and the results of prediction should be considered as very rough estimates.

Based on these criteria, one may choose which activities have to be tested for the studied compounds on the basis of a compromise between the novelty of pharmacological action and the risk of obtaining a negative result in experimental testing. Certainly, one will also take into account a particular interest in some kinds of activity, experimental facilities, *etc.*

6.4.6 Selection of the Most Prospective Compounds

A fundamental limitation must be kept in mind: any observation, estimation or calculation has only restricted accuracy. In absolutely all cases instead of the desirable unknown intrinsic *Real* value we have only:

$$Observation = Real + Noise$$

This is critically important for (virtual) screening especially. To highlight this, Figure 6.8 presents the generated data of 1000 points with binormal distribution and correlation coefficient square $R^2 = 0.95$ and $R^2 = 0.5$ Clearly, for $R^2 = 0.5$ the relationship looks like a weak tendency only. Figures 6.9–6.11 show the results of the selection of the 100 *Bests* (with the highest *Real* values) and the 100 *Winners* (with the highest *Estimation* values) among 1 000 000 "screened" examples. Clearly, only for $R^2 = 0.95$ is coincidence of the *Winners* and the *Bests* relatively good (about 60%), while for $R^2 = 0.5$ it is practically zero.

It is possible to perform a complete analysis of such relationships, but even the presented data provide enough evidence for the following conclusion: the method for (virtual) screening must be highly accurate, and/or many different virtual screening methods must be used in combination and/or the number of selected candidates must be sufficiently large at all stages of screening (in Figures 6.9 and 6.10, the number 100 is not "sufficiently large").[99,116,144,145]

6.5 Conclusions

Since the predicted with PASS biological activity spectra contain the estimates of probabilities for the pharmacological main and side effects, molecular mechanisms of action and specific toxicity, the choice of the most prospective compounds from the available samples of chemical compounds can be realized on the basis of complex criteria. Both the presence of targeted biological effects

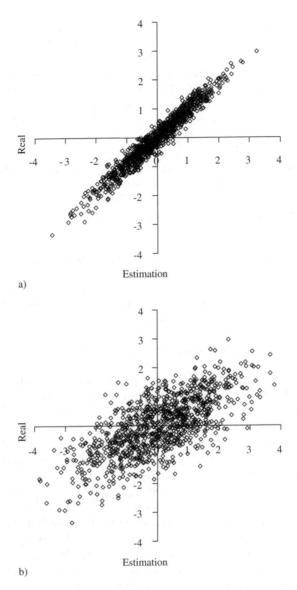

Figure 6.8 Example of relationships between the available measured values and unavailable true values. 1000 points are presented; all values have a normal distribution. Error of measurement (calculation) corresponds to the square of correlation coefficient $R^2 = 0.95$ (*a*) and 0.5 (*b*).

with desirable mechanisms of action and the absence of unwanted adverse effects and toxicity have to be taken into account. In such studies, the search for leads with the required properties and their optimization to decrease the adverse and toxic effect, usually performed sequentially, will be solved

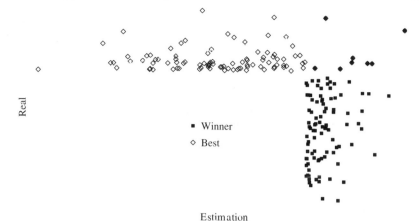

Figure 6.9 Example of relationship between the available measured (calculated) values and unavailable true values. The 100 Winners and the 100 Bests of 1 000 000 are presented. All compounds have a normal distribution; the error of measurement (calculation) corresponds to the square of correlation coefficient $R^2 = 0.5$.

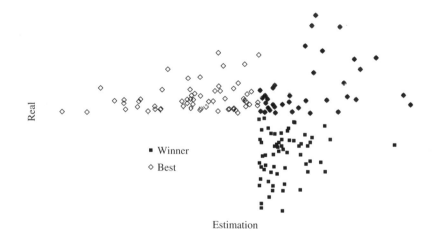

Figure 6.10 Example of relationship between the available measured (calculated) values and unavailable true values. The 100 Winners and the 100 Bests of 1 000 000 are presented. All values have a normal distribution; the error of measurement (calculation) corresponds to the correlation coefficient $R^2 = 0.8$.

simultaneously. Moreover, it was shown that the algorithms used in PASS can be successfully applied for discrimination between the so-called drug-like and drug-unlike compounds,[134] which provides the possibility for extension of the applicability of the program by "filtering" in the early stages chemical compounds, for which probability of becoming a drug is rather small.

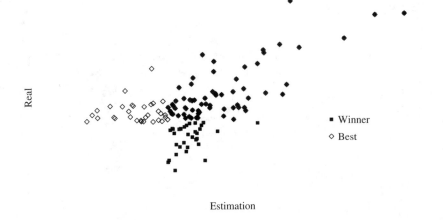

Figure 6.11 Example of relationship between the available measured (calculated) values and unavailable true values. The 100 Winners and the 100 Bests of 1 000 000 are presented. All values have a normal distribution; the error of measurement (calculation) corresponds to the correlation coefficient $R^2 = 0.95$.

The evolution of any molecule from hit to lead and from lead to drug-candidate typically is associated with the detailed evaluation of pharmaco-dynamics and pharmacokinetics of the compound. Using several different probabilistic methods for virtual screening together it might be possible to increase significantly the rate of promising substances in the selected sub-set.[101,103] A challenging task is to optimize simultaneously both pharmacody-namics and pharmacokinetics of lead compounds because it is very difficult to modify the appropriate molecular determinants that define the desired com-pound characteristics in a consistent manner. However, even this task might be solved using "an integrated software framework that monitors ligand (or library) alterations in the context of 'fitness landscape'".[26]

References

1. S. Wold and W. J. Dunn III, *J. Chem. Inf. Comput. Sci.*, 1983, **23**, 6–13.
2. D. Livingstone, *Data Analysis for Chemists: Applications to QSAR and Chemical Product Design*, Oxford Science Publications, Oxford University Press, UK, 1995.
3. J. Gasteiger, ed., *Handbooks of Cheminformatics: From Data to Know-ledge*, 4 Vols., Wiley-VCH, Weinheim, 2003.
4. V. N. Vapnik, *Estimation of Dependences Based on Empirical Data*, Springer-Verlag, New York, 1982.
5. V. Vapnik, *Statistical Learning Theory*, John Wiley and Sons, New York, 1998.

6. D. Bahler, B. Stone, C. Wellington and D. W. Bristol, *J. Chem. Inf. Comput. Sci.*, 2000, **40**, 906–914.
7. E. O. Cannon, A. Amini, A. Bender, M. J. E. Sternberg, S. H. Muggleton, R. C. Glen and J. B. O. Mitchell, *J. Comput. Aided Mol. Des.*, 2007, **21**, 269–280.
8. J. W. Godden, J. R. Furr and J. Bajorath, *J. Chem. Inf. Comput. Sci.*, 2003, **43**, 182–188.
9. J. W. Godden and J. Bajorath, *J. Chem. Inf. Model.*, 2006, **46**, 1094–1097.
10. G. Harper, J. Bradshaw, J. C. Gittins, D. V. S. Green and A. R. Leach, *J. Chem. Inf. Comput. Sci.*, 2001, **41**, 1295–1300.
11. C. Helma, T. Cramer, S. Kramer and L. De Raedt, *J. Chem. Inf. Comput. Sci.*, 2004, **44**, 1402–1411.
12. R. D. King, S. Muggleton, R. A. Lewis and M. J. E. Sternberg, *Proc. Natl. Acad. Sci. U.S.A.*, 1992, **89**, 11322–11326.
13. R. D. King, S. H. Muggleton, A. Srinivasan and M. J. E. Sternberg, *Proc. Natd. Acad. Sci. U.S.A.*, 1996, **93**, 438–442.
14. P. D. Mosier and P. C. Jurs, *J. Chem. Inf. Comput. Sci.*, 2002, **42**, 1460–1470.
15. T. Niwa, *J. Chem. Inf. Comput. Sci.*, 2003, **43**, 113–119.
16. T. Niwa, *J. Med. Chem.*, 2004, **47**, 2645–2650.
17. R. Rosipal, L. J. Trejo, B. Matthews and K. Wheeler, in *Proceedings of 3rd International Symposium on PLS and Related Methods (PLS'03)*, Lisbon, Portugal, 2003, 249–260.
18. M. J. E. Sternberg and S. H. Muggleton, *QSAR Comb. Sci.*, 2003, **22**, 527–532.
19. W. Tong, H. Hong, H. Fang, Q. Xie and R. Perkins, *J. Chem. Inf. Comput. Sci.*, 2003, **43**, 525–531.
20. D. J. Wilton, R. F. Harrison and P. Willett, *J. Chem. Inf. Model.*, 2006, **46**, 471–477.
21. B. Waszkowycz, T. D. J. Perkins, R. A. Sykes and J. Li, *IBM Systems Journal*, 2001, **40**(2), 360–376.
22. P. D. Lyne, *Drug Discov. Today*, 2002, **7**, 1047–1055.
23. A. N. Jain, *Curr. Opin. Drug Discov. Devel.*, 2004, **7**, 396–403.
24. G. Klebe, *Drug Discov Today*, 2006, **11**, 580–594
25. T. I. Oprea, *Molecules*, 2002, **7**, 51–62.
26. T. I. Oprea and H. Matter, *Curr. Opin. Chem. Biol.*, 2004, **8**, 349–358.
27. H. Kubinyi, *Nat. Rev. Drug Discov.*, 2003, **2**, 665–668.
28. H. Kubinyi, in *Computational Approaches to Structure Based Drug Design*, ed. R. M. Stroud, Royal Society of Chemistry, London, 2007, 24–45.
29. D. A. Filimonov, V. V. Poroikov and E. I. Karaicheva, *et al.*, *Exper. Clin. Pharmacol. (Rus)*, 1995, **58**, 56–62.
30. V. V. Poroikov and D. A. Filimonov, in *QSAR and Molecular Modelling Concepts, Computational Tools and Biological Applications*, eds. F. Sanz, J. Giraldo and F. Manaut, Prous Science Publishers, Barcelona, 1996, 49–50.

31. V. V. Poroikov and D. A. Filimonov, *J. Comput. Aid. Molec. Des.*, 2002, **16**, 819–824.

32. V. Poroikov and D. Filimonov, in *Predictive Toxicology*, ed. C. Helma, Taylor & Francis, New York, 2005, 459–478.

33. C. G. Wermuth, *Drug Discov. Today*, 2004, **9**, 826–827.

34. T. W. Schwartz and B. Holst, *Trends Pharmacol. Sci.*, 2007, **28**, 366–373.

35. A. C. R. Martin, C. A. Orengo, E. G. Hutchinson, S. Jones, M. Karmirantzou, R. A. Laskowski, J. B. O. Mitchell, C. Taroni and J. M. Thornton, *Structure*, 1998, **6**, 875–884.

36. M. H. V. Van Regenmortel, *J. Mol. Recognit.*, 1999, **12**, 1–2.

37. B. Y. Feng, A. Shelat, T. N. Doman, R. K. Guy and B. K. Shoichet, *Nat. Chem. Biol.*, 2005, **1**, 146–148.

38. M. H. V. Van Regenmortel, *J. Mol. Recognit.*, 2000, **13**, 1–4.

39. J. J. Hornberg, F. J. Bruggeman, H. V. Westerhoff and J. Lankelma, *BioSystems*, 2006, **83**, 81–90.

40. V. V. Poroikov, D. A. Filimonov and A. P. Boudunova, *Automat. Document. Math. Linguist.*, Allerton Press, Inc., 1993, **27**, 40–43.

41. D. A. Filimonov and V. V. Poroikov, in *Bioactive Compound Design: Possibilities for Industrial Use*, BIOS Scientific Publishers, Oxford, 1996, pp. 47–56.

42. V. Poroikov and D. Filimonov, in *Rational Approaches to Drug Design*, ed. H.-D. Holtje and W. Sippl, Prous Science Publishers, Barcelona, 2001, pp. 403–407.

43. D. A. Filimonov and V. V. Poroikov, *Rus. Chem. J.*, 2006, **50**, 66–75.

44. A. F. Fliri, W. T. Loging, P. F. Thadeio and R. A. Volkmann, *Proc. Natl. Acad. Sci. U.S.A.*, 2005, **102**(2), 261–266.

45. A. F. Fliri, W. T. Loging, P. F. Thadeio and R. A. Volkmann, *J. Med. Chem.*, 2005, **48**(22), 6918–6925.

46. P. O. Fedichev and A. A. Vinnik, in *Fourth International Symposium Computational Methods in Toxicology and Pharmacology Integrating Internet Resources (CMTPI-2007)*, Moscow, 2007, 46.

47. R. R. Neubig, *et al.*, *Pharmacol. Rev.*, 2003, **55**, 597–606.

48. A. V. Hill, *Proc. Physiol. Soc.*, 1910, **40**, 4–7.

49. E. J. Ariens, *Molecular Pharmacology*, Academic Press, New York, 1964.

50. A. Macchiarulo, I. Nobeli and J. M. Thornton, *Nature Biotechnology*, 2004, **22**(8), 1039–1045.

51. P. Labute, in *Proceedings of the Pacific Symposium on Biocomputing'99*, eds. R. B. Altman, A. K. Dunker, L. Hunter, T. E. Klein and K. Londerdale, World Scientific, New Jersey, 1999, 444–455.

52. P. Labute, S. Nilar and C. Wiiliams, *Comb. Chem. and HTS*, 2002, **5**, 135–145.

53. C. Helma, S. Kramer, B. Pfahringer and E. Gottmann, *Environ. Health Perspect.*, 2000, **108**, 1029–1033.

54. A. M. Richard, L. S. Gold and M. C. Nicklaus, *Curr. Opin. Drug Discov. Devel.*, 2006, **9**, 314–325.

55. E. Gottmann, S. Kramer, B. Pfahringer and C. Helma, *Environ. Health Perspect.*, 2001 **109**, 509–64; V. Golender and A. Rozenblit, *Logical and Combinatorial Algorithms for Drug Design*, Research Studies Pr., Letchworth, England, 1983.

56. D. Sesardic, K. McLellan, T. A. N. Ekong and D. R. Gaines, *Pharmacol. Toxicol.*, 1996, **78**, 283–288.

57. M. H. J. Seifert, K. Wolf and D. Vitt, *BioSilico*, 2003, **1**, 143–149.

58. G. Sebestyen, *Decision Making Processes in Pattern Recognition*, MacMillan, New York, 1962.

59. L. Kanal, *et al.*, in *Proceedings of National Electronics Conference*, Chicago, 1962, 279–295.

60. M. A. Aizerman, E. M. Braverman and L. I. Rozonoer, *Automation and Remote Control*, 1964, **25**, 821–837.

61. N. J. Nilsson, *Learning Machines: Foundations of Trainable Pattern-Classifying Systems*, McGraw-Hill Book, New York, 1965.

62. V. Vapnik and A. Lerner, *Automation and Remote Control*, 1963, **24**, 774–780.

63. P. Domingos and M. Pazzani, *Machine Learning*, 1997, **29**, 103–130.

64. V. Golender and A. Rozenblit, *Logical and Combinatorial Algorithms for Drug Design*, Research Studies Pr., Letchworth, England, 1983.

65. N. Veretennikova and A. Skorova, *et al.*, in *Quantitative Structure-Activity Relationships in Environmental Sciences – VII, Proceedings of QSAR 96*, SETAC Press, Elsinore, Denmark, 1996, pp. 115–131.

66. L. Eriksson, J. Jaworska, A. P. Worth, M. T. D. Cronin, R. M. McDowell and P. Gramatica, *Environmental Health Perspectives*, 2003, **111**, 1361–1375.

67. A. Bender, H. Y. Mussa and R. C. Glen, *J. Chem. Inf. Comput. Sci.*, 2004, **44**, 170–178.

68. P. A. Flach and N. Lachiche, *Machine Learning*, 2004, **57**, 233–269.

69. A. E. Klon, M. Glick and J. W. Davies, *J. Chem. Inf. Comput. Sci.*, 2004, **44**, 2216–2224.

70. A. E. Klon, M. Glick, M. Thoma, P. Acklin and J. W. Davies, *J. Med. Chem.*, 2004, **47**, 2743–2749.

71. A. E. Klon, M. Glick and J. W. Davies, *J. Med. Chem.*, 2004, **47**, 4356–4359.

72. U. Brefeld and T. Scheer, in *Proceedings of the ICML 2005 Workshop on ROC Analysis in Machine Learning*, Bonn, Germany, 2005.

73. J. Hert, P. Willett, D. J. Wilton, P. Acklin, K. Azzaoui, E. Jacoby and A. Schuffenhauer, *Org. Biomol. Chem.*, 2004, **2**, 3256–3266.

74. Nidhi, M. Glick, J. W. Davies and J. L. Jenkins, *J. Chem. Inf. Model.*, 2006, **46**(3), 1124–1133.

75. H. Gao, C. Williams, P. Labute and J. Bajorath, *J. Chem. Comput. Sci.*, 1999, **39**, 164–168.

76. H. Gao, M. S. Lajiness and J. Van Drie, *J. Mol. Graphics Modelling*, 2002, **20**, 259–268.

77. W. J. Streich, S. Dove and R. Franke, *J. Med. Chem.*, 1980, **23**, 1452–1456.

78. S. Dove, W. J. Streich and R. Franke, *J. Med. Chem.*, 1980, **23**, 1456–1459.
79. A. Golbraikh and A. Tropsha, *J. Comp.-Aided Mol. Design*, 2002, **16**(5), 357–369.
80. A. Golbraikh, M. Shen, Z. Xiao, Y.-D. Xiao, K.-H. Lee and A. Tropsha, *J. Comp.-Aided Mol. Design*, 2003, **17**(2), 241–253.
81. C. Szantai-Kis, I. Kovesdi, G. Keri and L. Orfi, *Molecular Diversity*, 2003, **7**(1), 37–43.
82. R. P. W. Duin and E. Pekalska, in *Computer Recognition Systems (Proc. of 4th Int. Conf. on Computer Recognition Systems CORES'05)*, M. Kurzynski, E. Puchala, M. Wozniak and A. Zolnierek, eds., Advances in Soft Computing, Springer Verlag, Berlin, 2005, 27–42.
83. L. D. Hughes, D. S. Palmer, F. Nigsch and J. B. O. Mitchell, *J. Chem. Inf. Model.*, 2008, **48**, 220–232.
84. http://www.mdl.com/.
85. R. P. Sheridan and J. Shpungin, *J. Chem. Inf. Comput. Sci.*, 2004, **44**(2), 727–740.
86. P. Willett, *Drug Discov. Today*, 2006, **11**, 1046–1053.
87. B. Chen, R. F. Harrison, G. Papadatos, P. Willett, D. J. Wood, X. Q. Lewell, P. Greenidge and N. Stiefl, *J. Comput. Aided Mol. Des.*, 2007, **21**, 53–62.
88. T. M. Mitchell, *Machine Learning*, McGraw-Hill, 1997.
89. U. M. Braga-Neto and E. R. Dougherty, *Bioinformatics*, 2004, **20**, 374–380.
90. A. Tropsha, in *QSAR and Molecular Modelling in Rational Design of Bioactive Molecules, EuroQSAR 2004* Proceedings, ed. E. Aki (Sener), I. Yalcin eds., Computer Aided Drug Design Development Society in Turkey, Turkey, 2004, 25–29.
91. V. V. Poroikov, D. A. Filimonov, Yu. V. Borodina, A. A. Lagunin and A. Kos, *J. Chem. Inf. Comput. Sci.*, 2000, **40**, 1349–1355.
92. W. Hardle, *Applied Nonparametric Regression*, Cambridge University Press, Cambridge, 1990.
93. J. R. Rose, in *Handbooks of Cheminformatics: From Data to Knowledge*, J. Gasteiger ed., Vols. 4, Wiley-VCH, Weinheim, 2003, pp. 1082–1097.
94. T. Ohgaru, R. Shimizu, K. Okamoto, M. Kawase, Y. Shirakuni, R. Nishikiori and T. Takagi, *J. Chem. Inf. Model.*, 2008, **48**, 207–212.
95. E. H. Shortliff, *Computer-Based Medical Consultation: MYCIN*, Elsevier, New York, 1976.
96. W. Van Melle, A. C. Scott, J. S. Benett and M. A. Peairs, *The EMYCIN manual*, Technical Report, Heuristica Programming Project, Stanford University, 1981.
97. J. Gashnig, in *Introductory Readings in Expert Systems*, D. Michie ed., Gordon and Breach Science Publishers, New York, 1982.
98. M. Vogt, J. W. Godden and J. Bajorath, *J. Chem. Inf. Model.*, 2007, **47**, 39–46.
99. M. Vogt and J. Bajorath, *J. Chem. Inf. Model.*, 2007, **47**, 337–341.
100. V. Poroikov, D. Akimov, E. Shabelnikova and D. Filimonov, *SAR & QSAR Environ. Res.*, 2001, **12**, 327–344.

101. V. V. Poroikov, D. A. Filimonov, W.-D. Ihlenfeldt, T. A. Gloriozova, A. A. Lagunin, Yu. V. Borodina, A. V. Stepanchikova and M. C. Nicklaus, *J. Chem. Inform. Comput. Sci.*, 2003, **43**, 228–236.

102. A. A. Lagunin, O. A. Gomazkov, D. A. Filimonov, T. A. Gureeva, E. A. Dilakyan, E. V. Kugaevskaya, Yu. E. Elisseeva, N. I. Solovyeva and V. V. Poroikov, *J. Med. Chem.*, 2003, **46**, 3326–3332.

103. A. Geronikaki, J. Dearden and D. Filimonov, *et al., J. Med. Chem.*, 2004, **47**(11), 2870–2876.

104. A. A. Geronikaki, A. A. Lagunin, D. I. Hadjipavlou-Litina, P. T. Elefteriou, D. A. Filimonov, V. V. Poroikov, I. Alam and A. K. Saxena, *J. Med. Chem.*, 2008, **51**(6), 1601–1609.

105. J. Hert, P. Willett and D. J. Wilton, *J. Chem. Inf. Model.*, 2006, **46**, 462–470.

106. J. Swets, *Science*, 1988, **240**, 1285–1293.

107. J. A. Swets, R. M. Dawes and J. Monahan, *Sci. Am.*, 2000, **283**, 82–87.

108. N. Triballeau, F. Acher, I. Brabet, J.-P. Pin and H.-O. Bertrand, *J. Med. Chem.*, 2005, **48**, 2534–2547.

109. W. J. Youden, *Cancer*, 1950, **3**, 32–35.

110. A. Wald, *Statistical Decision Functions*, John Wiley and Sons, New York, 1950.

111. D. Blackwell and M. A. Girshick, *Theory of Games and Statistical Decisions*, John Wiley and Sons, New York, 1954.

112. T. Fawcett, *Pattern Recognit. Lett.*, 2006, **27**, 861–874.

113. T. Fawcett, *Pattern Recognit. Lett.*, 2006, **27**, 882–891.

114. A. P. Bradley, *Pattern Recognition*, 1997, **30**(7), 1145–1159.

115. A. E. Cleves and A. N. Jain, *J. Med. Chem.*, 2006, **49**, 2921–2938.

116. J.-F. Truchon and C. I. Bayly, *J. Chem. Inf. Model.*, 2007, **47**, 488–508.

117. Yu. Borodina, D. Filimonov and V. Poroikov, *Quant. Struct.-Act. Relat.*, 1998, **17**, 459–464.

118. R. P. Sheridan, S. B. Singh, E. M. Fluder and S. K. Kearsley, *J. Chem. Inf. Comput. Sci.*, 2001, **41**, 1395–1406.

119. R. N. Jorissen and M. K. Gilson, *J. Chem. Inf. Model.*, 2005, **45**, 549–561.

120. T. A. Pham and A. N. Jain, *J. Med. Chem.*, 2006, **49**, 5856–5868.

121. M. H. J. Seifert, *J. Chem. Inf. Model.*, 2006, **46**, 1456–1465.

122. J. Kirchmair, S. Ristic, K. Eder, P. Markt, G. Wolber, C. Laggner and T. Langer, *J. Chem. Inf. Model.*, 2007, **47**, 2182–2196.

123. T. T. Ashburn and K. B. Thor, *Nat. Rev. Drug Discov.*, 2004, **3**, 673–683.

124. Y. Y. Li, J. An and S. J. Jones, *Genome Inform.*, 2006, **7**, 239–247.

125. L. A. Tartaglia, *Expert. Opin. Investig. Drugs*, 2006, **15**, 1295–1298.

126. C. G. Wermuth, *Drug Discov. Today*, 2006, **11**, 160–164.

127. V. V. Avidon, V. S. Arolovich, S. P. Kozlova and L. A. Piruzyan, *Chem. Pharm. J.* (Russian), 1978, **No. 5**, 88–92.

128. A. N. Kochetkov, P. N. Vassiliev and A. G. Breslaukhov, in *Abstr. First All-Union Conf. Theoret. Org. Chem.*, Volgograd, 1991, part 2, 500.

129. V. Poroikov, D. Filimonov, A. Lagunin, T. Gloriozova and A. Zakharov, *SAR & QSAR Environ. Res.*, 2007, **18**, 101–110.

130. http://www.prousresearch.com (accessed May 2008).
131. J. Prous and D. Aragones, in *Abstracts of National American Society Meeting*, Boston, Aug. 19–23, 2007.
132. http://www.q-pharm.com/ (accessed May 2008).
133. T. A. Gloriozova, D. A. Filimonov, A. A. Lagunin and V. V. Poroikov, *Chem-Pharm J.* (Russian), 1998, **32**, 32–39.
134. S. Anzali, C. Barnickel, B. Cezanne, M. Krug, D. Filimonov and V. Poroikov, *J. Med. Chem.*, 2001, **44**, 2432–2437.
135. Yu. Borodina, A. Sadym, D. Filimonov, V. Blinova, A. Dmitriev and V. Poroikov, *J. Chem. Inform. Comput. Sci.*, 2003, **43**, 1636–1646.
136. R. Todeschini and V. Consonni, *Handbook of Molecular Descriptors*, Wiley-VCH, Weinheim, 2000.
137. L. Xing and R. C. Glen, *J. Chem. Inf. Comput. Sci.*, 2002, **42**, 796–805.
138. W. Guba, in *Predictive Toxicology*, Christoph Helma ed., Marcel Dekker, New York, 2003, 11–35.
139. A. Varnek, D. Fourches, F. Hoonakker and V. P. Solov'ev, *J. Comp.-Aided Mol. Design*, 2005, **19**, 693–703.
140. V. V. Avidon, I. A. Pomerantsev, A. B. Rozenblit and V. E. Golender, *J. Chem. Inf. Comput. Sci.*, 1982, **22**, 207–214.
141. D. Filimonov, V. Poroikov, Yu. Borodina and T. Gloriozova, *J. Chem. Inf. Comput. Sci.*, 1999, **39**, 666–670.
142. A. Lagunin, A. Stepanchikova, D. Filimonov and V. Poroikov, *Bioinformatics*, 2000, **16**, 747–748.
143. http://www.ibmc.msk.ru/PASS (accessed May 2008).
144. R. P. Sheridan and S. K. Kearsley, *Drug Discovery Today*, 2002, **7**, 903–911.
145. J. W. Raymond, M. Jalaie and M. P. Bradley, *J. Chem. Inf. Comput. Sci.*, 2004, **44**, 601–609.

CHAPTER 7

Fragment-based De Novo Design of Drug-like Molecules

EWGENIJ PROSCHAK, YUSUF TANRIKULU
AND GISBERT SCHNEIDER

Goethe-University Frankfurt, Institute of Organic Chemistry and Chemical Biology, Siesmayerstr. 70, D-60323 Frankfurt am Main, Germany

7.1 Introduction

Automated molecular *de novo* design has been an active research area in cheminformatics since the early 1990s.[1-3] Academic researchers and molecular design groups in pharmaceutical industry alike have come up with numerous software solutions and implementations. The basic idea is to assemble molecular building-blocks *in silico* so that novel molecular structures emerge. The designed molecules can then be assessed by medicinal chemists. For the actual design process atoms (*atom-based* design) or fragments (*fragment-based* design) can be used as building blocks. Currently, we are witnessing renewed interest in fragment-based design approaches.[4] This may be attributed to several reasons:

1. Fragment-based *de novo* design methods have been proven to be successful in prospective applications leading to novel compounds with a desired activity profile.[2] Fragment hits often possess high "ligand efficiency" (binding affinity per heavy atom), which is pivotal for lead series development.[5,6] The level of trust in such methods has undoubtedly increased.
2. Fragment-based compound assembly tends to produce synthetically tractable structures.[3] Purely atom-based construction has often resulted in molecules

Chemoinformatics Approaches to Virtual Screening
Edited by Alexandre Varnek and Alex Tropsha
© Royal Society of Chemistry, 2008
Published by the Royal Society of Chemistry, www.rsc.org

that were not appealing to chemists, containing multiple stereocenters or substructure elements that are difficult to synthesize or impracticable.[7]

3. The combination or "tethering" of molecular fragments that were derived from known drugs or lead structures with known and desired bioactivity often results in chimera that also exhibit the desired function.[8,9] Linking of some few active fragments can result in multiple lead structures, as shown recently by researchers at Abbott Laboratories.[10]

4. Molecular fragments sometimes represent preferred substructure elements, e.g., benzamidine binding to the S_1 pocket of serine proteases, or hydroxamic acid for zinc ligation in metalloprotease inhibitors. It can thus be advantageous to design new molecules containing such function-enabling building-blocks ("needles") or core fragments.[11–13]

5. Fragment-based de novo design is supported by experimental small fragment screening, e.g., by high-throughput NMR and crystallography.[14–17] Such approaches render starting points for computational de novo design by providing validated base fragments.

Accurate structural information of validated drug targets (mainly protein–ligand complexes) provides the basis for *structure-based* molecular design. The large number of high-resolution protein structures available from the Protein Data Bank[18] (PDB) has opened up new opportunities for rational drug design. Still, the accurate description of ligand binding pockets and consideration of flexible-fit phenomena remains a central issue. This renders automated binding site analysis pivotal for rational drug design, including automated ligand docking and de novo design, and for finding potential binding pockets in proteins that lack a known ligand. These methods require exact structural information of the binding site in a ligand-bound state as a starting point. This prerequisite currently limits the application domain of receptor-based de novo design to globular proteins (mainly enzymes) and other targets for which a high-resolution (< 3 Å) structural model is available.

Various computational methods exist for the location of possible ligand binding sites. Most of these pocket detection algorithms rely on geometric criteria to find clefts and surface depressions. Empirical studies have actually shown that quite often ligand binding sites usually coincide with the largest pocket of a protein's surface.[19] Figure 7.1(a) shows an example of an automatically identified protein pocket in thrombin with an inhibitor bound. The lactam-based inhibitor is shape-complementary to the pocket surface and fills the whole pocket. Prediction and experiment are in perfect agreement. The ligand completely fits into the protein pocket, leaving almost no unoccupied space. This is an ideal scenario for application of receptor-based de novo design.

Notably, automated pocket identification does not always deliver correct results. Figure 7.1(b) shows a crystal structure of Factor Xa with an inhibitor binding-mode that accesses a shallow part of the pocket area. Although automated pocket detection software was used to find the ligand binding site and the overall prediction was correct, the experimentally determined ligand binding mode differs from the predicted pocket. The ligand occupies a non-detected sub-pocket.

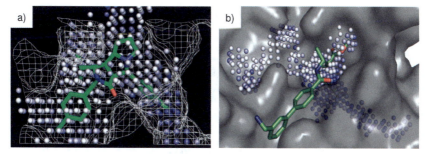

Figure 7.1 Examples of automatically identified ligand-binding pockets with inhibitors bound. For pocket detection, a grid-based approach was used (Pocket-Picker).[19] Dots represent surface cavities identified by PocketPicker, colored by "buriedness". Solvent-accessible pocket surfaces are indicated by a mesh (left) or as hard surface (right). Darker shading of the grid dots indicates greater buriedness. (*a*) Thrombin active site (PDB identifier: 2cf8, 1.3 Å resolution; with a lactam inhibitor), (*b*) co-crystal structure of Factor Xa (PDB entry 1ezq, 2.2 Å resolution; with inhibitor RPR128515). The automatically extracted pocket does not match with the surface-exposed parts of the actual inhibitor binding pocket. (Adapted from ref. 3.)

It would have been impossible to identify the ligand in Figure 7.1(b) with the pocket shape suggested by the pocket extraction software. This example illustrates some of the potential pitfalls of automated receptor-based techniques.

In contrast to receptor-based molecular design, *ligand-based* design of compound libraries requires known bioactive molecules as a starting point. These template molecules are often referred to as "seed structures", "queries" or "reference compounds". Novel structures are constructed so that they exhibit some kind of similarity to the templates. Focused compound libraries can be designed *in silico*, synthesized and tested for activity *in vitro*. We stress that the choice of an appropriate similarity measure depends on the drug discovery project and is context-dependent (*cf.* Chapter 5). In this chapter, we focus on ligand-based approaches, as these are applicable irrespective of the availability of a receptor model, and for a large group of drug targets, namely G-protein coupled receptors (GPCR) and other integral membrane proteins like ion channels, reliable atomistic models are rare or unavailable. Table 7.1 lists selected examples of recent advances in the field of ligand-based *de novo* design methods. For an overview of earlier developments and receptor-based design, the reader is referred to the literature.[1–3]

De novo design software produces molecular structures with desired properties from scratch. In this attempt we are confronted with a virtually infinite search space. In theory, the search space for *de novo* design is given by all drug-like compounds – a number estimated to be in the range of 10^{20} to 10^{100}.[25] The fragment-based design concept drastically reduces this search space as it limits the numbers of molecules. Furthermore, instead of the systematic construction and evaluation of each individual compound (as in experimental high-throughput screening), the *de novo* design process relies on the Principle of

Table 7.1   Recently published software solutions for automated ligand-based
                 de novo design.

Software	Description
Molecule Evoluator[20]	Atom-based design by evolutionary operators
FluX[21,22]	Adaptive fragment-based design by evolutionary operators
COREGEN[23]	Fragment-based ring-linker-based assembly of scaffold libraries
Evolutionary Algorithm Inventor[24]	Atom- and fragment-based evolutionary algorithm
Ftrees[25]	Fragment combination by connection rules an Ftrees analysis
Skelgen[26]	Employment of active reference ligands and a *pseudo*-receptor approach
Fragment Trees[27]	Exhaustive enumeration of fragment-based chemical space with target-specific constraints
Combinatorial library enumeration[28]	Construction of large combinatorial libraries by use of "real reactions" and a heuristic backtracking algorithm
Compound Generator (CoG) and the median molecules concept[29,30]	Multi-objective *de novo* design by Pareto-ranking and genetic operators on molecular graph-based chromosomes

Local Optimization.[31] This means that only a fraction of all potential screening candidates are actually constructed and evaluated. Since it is impossible to enumerate all possible virtual molecules in advance, only those candidate molecules are considered at a time, which represent a local neighborhood of the actually best solutions. Notably, most software implementations are non-deterministic, and design processes rarely converge on the global optimum (the one "super-molecule") but on a local or "practical" optimum.[32] Multiple runs of *de novo* design software will therefore produce different "promising" candidates rather than perfect solutions due to the nature of the search. Basically, three questions must be addressed by any *de novo* design program:

1. How to assemble candidate compounds?
2. How to evaluate molecule quality ("fitness")?
3. How to systematically navigate in search space?

One can further differentiate between "positive" and "negative" design.[3] *Positive design* restricts virtual compound assembly to regions of chemical space that have a higher probability to find drug-like molecules, while *negative design* criteria help prevent adverse properties (*e.g.*, very high lipophilicity) and unwanted substructures (*e.g.*, reactive groups, stereocenters). It is essential to understand that *de novo* design will rarely yield novel lead structures with nanomolar activity, high target selectivity, and an acceptable pharmacokinetic profile in the first place. Despite all efforts to produce synthetically tractable molecules with a good property profile there is no guarantee that a designed molecule finds immediate appraisal by a synthetic chemist! Quite the opposite: *de novo* generated structures often represent molecules that require significant further optimization. What can be expected is an increased hit rate in a focused compound library compared to screening of an arbitrary compound collection. This directly leads to the need for synthetically accessible structures. This issue was recognized early in the history of *de novo* design, but has been addressed only recently. A common pattern of the few *de novo* design programs that consider ease of synthesis is to assemble molecular building blocks by virtual reaction schemes. For example, suitable building blocks can be obtained by virtual *retro*-synthesis of drug molecules, using simplistic fragmentation schemes like RECAP.[33] The same set of reaction schemes is then employed to assemble candidate compounds.[21,34] It is reasonable to assume that the designed compounds will have some degree of "drug-likeness" and contain only few awkward structural elements. Ideally, virtual structure assembly is guided by simulated organic synthesis steps so that a synthesis route can be proposed for every generated structure. Some of the advanced software implementations actually made a step toward this goal. These programs automatically analyze generalized synthetic routes and pick potential synthons from databases of available compounds.[28,35,36]

In the following we highlight aspects of fragment-based design with an emphasis on ligand-based approaches, and the combination of shape-matching and pharmacophore-based scoring functions.

7.2 From Molecules to Fragments

The goal of *de novo* design is the generation of compounds that were not synthesized before. Synthesis of such compounds should be feasible, with few synthetic steps, high yields, and cheap starting material. There are different approaches to estimate the synthetic accessibility of *de novo* designed molecules. One of the recently developed methods introduced by Boda and co-workers[35] takes into account structural complexity, similarity to available starting material, and assessment of strategic bonds where a structure can be decomposed into retrosynthetic fragments. This concept incorporates a scoring scheme derived from the knowledge of medicinal chemists. Another sophisticated approach to obtain synthesizable molecules was conceived by Vinkers *et al.* in their program SYNOPSIS (SYNthesize and OPtimize System In Silico):[36] virtual synthesis starts from commercially available compounds and employs 70 reaction types to connect the fragments. In a prospective application to HIV inhibitors the authors demonstrated that SYNOPSIS is not only able to deliver active compounds but also a synthesis route that could be followed.

Fragment-based programs employ molecular building-blocks from a fragment database that contains virtual synthons from commercially or synthetically available compounds or from virtual retrosynthetic decomposition of a compound database. The bonds cleaved during the retrosynthetic dissection process are not necessarily the ones that will be made during the actual synthesis. For example, the formation of a tertiary amine in the reaction shown in Figure 7.2 can be performed in different ways. Typically, retrosynthesis

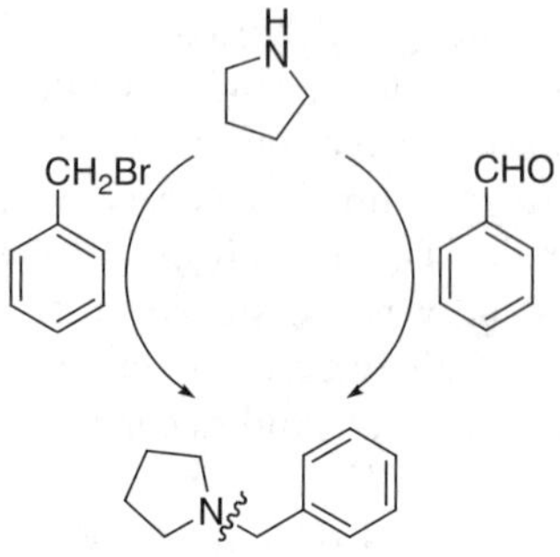

Figure 7.2 Two synthetic routes to form an amine bond.

software does not consider the effectiveness of the formation of such a bond. Another critical issue is the consideration of stereochemistry, because stereoselective reactions are not always readily applicable.

There are two main strategies to obtain a fragment library from which the *de novo* compounds can be assembled. A straightforward idea is to screen commercially available libraries for suitable fragments. The definition of a fragment has been widely discussed in literature.[37] In analogy to Lipinski's Rule-of-5 for drugs,[38] a Rule-of-3 has been suggested to define suitable fragments.[39] Accordingly, molecular weight < 300 Da, number of hydrogen bond acceptors ≤ 3, number of hydrogen bond donors ≤ 3, and clog $P \leq 3$ might be useful criteria for fragment selection. The number of rotatable bonds (NROT) of ≤ 3 and a polar surface area (PSA) of $\leq 60 \text{ Å}^2$ represent additional criteria for the selection of fragments for *in silico* reactions. Elimination of fragments with a known liability for toxicity alerts can be helpful to avoid their occurrence in the designed structures.

7.2.1 Pseudo-retrosynthesis

Another way to obtain a library of fragments is to decompose a known database of drug- or lead-like molecules by *pseudo*-retrosynthetic rules like RECAP (Retrosynthetic Combinatorial Analysis Procedure).[33] The RECAP bond cleavage procedure consists of eleven bond cleavage types shown in Figure 7.3. This approach has several advantages:

- Fragments derived from drug-like molecules are often correlated with biological activity and a good pharmacokinetic and -dynamic profile. Novel compounds designed from these building-blocks are expected to behave in the same way.
- The synthetic accessibility of the linked fragments is considered *implicitly*.
- Several approaches have been made to design targeted fragment libraries from actives for a target class (*e.g.*, kinases)[23] or even a specific target.[27] A meaningful pre-selection of compounds ensures the similarity of novel compounds to the ones with confirmed activity, often resulting in so-called "chimera". The BREED approach widely employs this technique.[40]

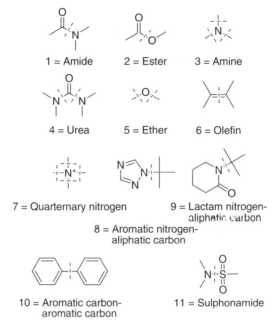

Figure 7.3 RECAP bond cleavage types.[33]

RECAP bond cleavage types were derived from common reactions that can be performed in a laboratory. The bond cleavage type is neither linked to a special reaction type nor to specific starting material. For example, the amine bond (RECAP bond cleavage type 3) can be effectively built by a nucleophilic substitution or by reductive amination (Figure 7.2). The fragments obtained from RECAP decomposition are not real synthetic units. Rather, they can be used to virtually assemble novel compounds that should be synthetically accessible. As an example, Figure 7.4 shows the RECAP dissection of a Factor Xa inhibitor **7.1**, with the three resulting fragments **7.2 7.4**.

RECAP dissection of DrugBank,[41] a database containing 877 approved drugs, yields 860 fragments. The molecular weight of the fragments is below 300 Da, and the other parameters also obey the Rule-of-3 (*vide supra*). Lipophilicity, expressed as clog *P*, is also comparable to the 62 174 fragments of the "fragment-like" ZINC collection,[42] which is a property-filtered subset of the ZINC database containing approximately 4.6 million commercially available compounds (Figure 7.5).

7.2.2 Shape-derived Fragment Definition

An important step according to several *de novo* design concepts is fragment matching on a query structure. This requires fragment representations that are

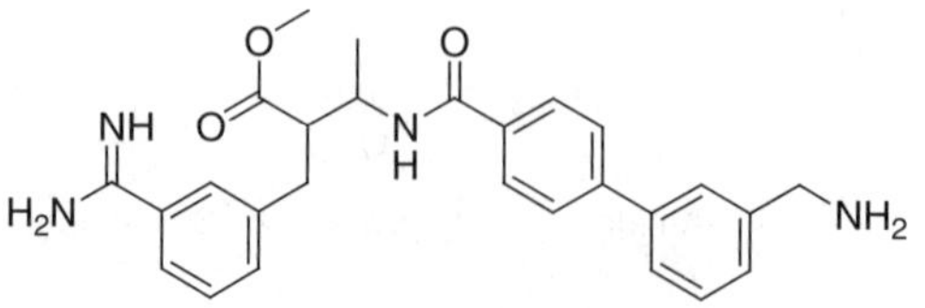

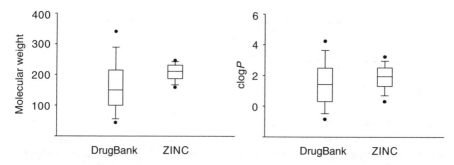

Figure 7.4   Fragment decomposition of Factor Xa inhibitor RPR128515 (**1**). Numbered atoms indicate cleavage sites, which can be used for virtual forward synthesis of variant molecules.

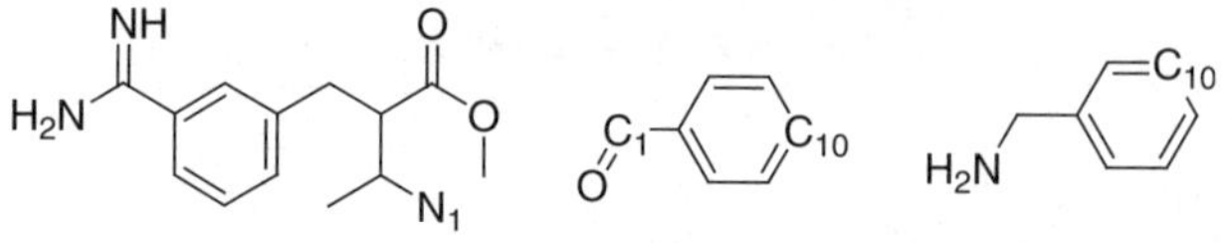

Figure 7.5   Distributions of molecular weight (Da) and lipophilicity (clog *P*) for 860 fragments from DrugBank (obtained by application of RECAP rules) and 62 174 fragments from the ZINC fragment-like collection (property-filtered).

independent of the molecular graph. One such approach is the generation of a surface-based fragment representation as implemented in our SQUIRRELnovo approach (E. Proschak *et al.*, unpublished).[43] First, the solvent-accessible surface of the molecule is calculated. Then, the surface is decomposed into hyperbolic paraboloids (called *Shapelets*).[44] These are surface patches with constant local curvature. When fragments are generated by a RECAP-like bond cleavage procedure, the surface area of the fragment and the referring hyperbolic paraboloids represent the fragment surface and can be used for matching (Figure 7.6).

While *de novo* design methods can produce fully assembled structures, a chemist's expertise is often decisive for generation of novel leads that are synthetically accessible. Step-by-step fragment assembly under supervision of the human expert is implemented in SQUIRRELnovo, which can be used as an idea generator for medicinal chemistry. Fragments that are superposed with the query

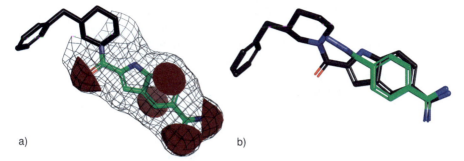

a)   b)

Figure 7.6 Fragment generation and matching using the *Shapelets* approach. (*a*) A molecular fragment is described by its *Shapelets* (paraboloids) decomposition of the solvent-accessible surface (mesh). (*b*) Result of fragments matching to a reference structure. A benzamidine building-block was matched to a thrombin inhibitor by *Shapelets*-matching.

structure are the starting points for bioisosteric replacements. SQUIRRELnovo uses *Shapelets* of the fragment and of the query molecule for partial matching. It aligns the surface of the fragment to parts of the query surface. This alignment is then assessed by manual inspection, since human experts select the most promising fragments for upcoming assembly steps. The main advantage of fragment representation by surface-derived *Shapelets* is the independence from the underlying chemotype. This abstraction allows for scaffold-hopping and suggests bioisosteric groups with similar shape.

7.3 From Fragments to Molecules

Arbitrary combination of molecular fragments creates an immense set of possibilities, which is claimed to be virtually infinite. By implementing chemical information into the molecule assembly process (virtual synthesis) one can gain (i) synthetically tractable molecules and (ii) a reduction of search space to a manageable size. There are three main ways to navigate through chemical space by assembling candidate compounds: full enumeration, deterministic and non-deterministic techniques. Obviously, full enumeration and evaluation of every single virtual molecule (brute-force approach) is straightforward. This certainly represents a method of choice for the automated construction of small libraries of up to a few million compounds. Larger libraries or "chemical spaces" demand approximate solutions, and often *de novo* design is implemented as a local optimization process. Compound optimization can be achieved in a deterministic or non-deterministic manner. The latter typically involves a stochastic sampling part. We will now briefly describe three algorithms that are representative for either one of these strategies.

Full enumeration algorithms are confronted with a prevalent problem in computer science theory: the naïve assembly of molecular fragments to form virtual candidate compounds results in a combinatorial explosion. Owing to the enormous number of different molecular fragments and the way they can be linked, we are not able to solve this problem, including our current knowledge in suitable time and quality constraints. Therefore, some programs soften one of these two constraints in order to satisfy the remaining criterion. This is achieved either by calculation of "good" candidate compounds in long computation times or by fast combinatorial algorithms yielding large candidate compound libraries that may contain some high quality compounds.

SMILIB implements an exemplary algorithm for the latter case.[45,46] On the one hand, it offers rapid construction of very large virtual molecule sets. On the other hand, for extremely large libraries one will not find a fitness function to compute the quality of all resultant virtual molecules in reasonable time. A way to direct the design process (in terms of compound quality) is given through the molecule composition theme (Figure 7.7). The combinatorial principle behind the algorithm is that virtual molecules are a composition of three substructure types: scaffolds, linkers and building-blocks. Three sets containing fragments of

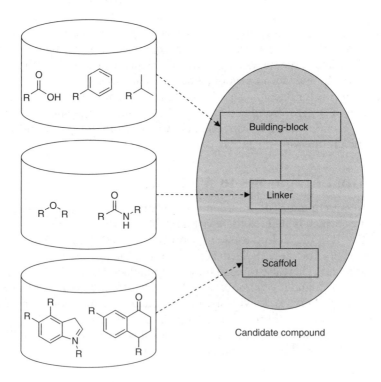

Figure 7.7 Combinatorial molecule assembly scheme. Resultant candidate molecules are objects that have building-blocks attached to scaffolds *via* linkers. Three user-definable fragment sets are needed to feed the algorithm with fragments for each domain. A few example fragments of each type are shown for clarity. For virtual library enumeration, each database fragment gets linked to each other fragment following a combinatorial assembly scheme.

only one substructure type represent the basis for combinatorial design. Each fragment of each type is combined with each fragment of the other types using predefined linkage rules. To obtain molecules exhibiting desired affinity to a target, one can feed the algorithm with "privileged" fragments.[23] Moreover, it is possible to emulate real chemistry by selecting linker fragments that are known to be bridging patterns of organic reactions. Although it is not possible to model ring formation reactions such as the Diels–Alder reaction using the scaffold/linker/building-block scheme, synthesis rules can be implemented that may bring the results closer to synthetic feasible compounds.

Other algorithms exploit the combinatorial nature of *de novo* design processes by sequentially growing molecules, *e.g.*, FlexNovo.[47,48] This software is based on an incremental construction algorithm exhibiting dynamic programming principles. The advantage over the previous presented strategy is that construction of every possible virtual molecule is avoided. Instead, the compositions are scored on the fly – in the case of FlexNovo by docking scores. Note that this is a receptor-based method, hence it is an example on how to reduce the number of enumerated candidate compounds by integrating binding site information. In the initial step, all available fragments are docked into the target binding pocket. Therefore, this method is not suitable in the absence of receptor structure information, which is the case for many integral membrane proteins. Figure 7.8

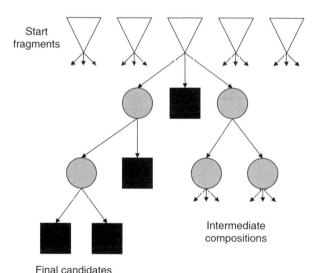

Figure 7.8 Build-up scheme of a deterministic molecule design approach. The first row shows start fragments (triangles), which are docked into the target binding pocket. Actual *de novo* design starts from the second row, which represents the first extension cycle (depth = 1). In an iterative process, fragments are assembled to yield candidate compounds (squares) or, if the resultant molecules can be further extended, composed fragments (circles). By evaluating the intermediate compositions during the design process, branches of the search tree are dropped using heuristics. (Adapted from ref. 48.)

shows an outline of the *de novo* design process. The triangle row (depth zero) represents the solutions of the initial step, *i.e.*, the pre-docked fragments. The construction phase with a user-defined number (*k*-greedy) of extension cycles starts from these initial partial solutions. This part of the algorithm is responsible for evolving virtual molecules. In the first construction cycle (depth = 1), every fragment is linked to compatible ones which may result in final molecules if dummy atoms representing "bridges" are no longer present. After evaluation of these intermediate compositions, the construction continues (depth = 2) by extending only the best compositions from the previous cycle. This iterative process continues until the *k*-th extension step is reached.

Non-deterministic approaches can solve the problem of chemical space navigation by means of stochastic algorithms. Here, the algorithm of FLUX[21,22] is exemplary (Figure 7.9). In contrast to deterministic algorithms, where unique endpoints (final candidate molecules) are defined, stochastic processes typically result in different solutions with each run. The algorithm generates offspring from the start molecule by exchanging single fragments *via* genetic operators. As known from evolutionary algorithms, each variant molecule is evaluated until a candidate compound is found. Molecule optimization is achieved by implementing chemical information about reference compounds which will route the design process to the respective "activity

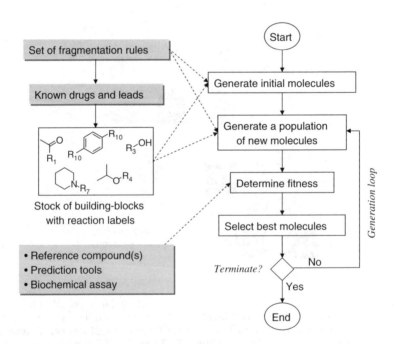

Figure 7.9   Non-deterministic *de novo* design pipeline including an evolutionary algorithm for optimization. This scheme contains a population-based algorithm that mimicks biological evolution through mutation and recombination during the breeding process.

island" in chemical space.[49] Notably, the map of chemical space indicating regions of higher and lower quality is compiled *en passant* during the design process. In terms of positive design, the process is directed directs toward active compounds, whereas negative (inactive, non-selective) reference compounds define "tabu zones" in search space.

7.4 Scoring the Design

Evaluation of candidate compounds is the most critical step in *de novo* design. It is the duty of the fitness function to decide whether a composed virtual molecule is kept or discarded in deterministic algorithms. One must always be aware that it is the used fitness function that defines the search space for novel molecules.

Many different fitness function categories exist. In ligand based *de novo* design, scoring is often reduced to the task of a similarity analysis between query molecules and virtual molecules. Receptor-based methods employ more sophisticated scoring functions combined with automated docking algorithms,[50] receptor-based pharmacophores[51] or molecular dynamics and free-energy perturbation (FEP) methods.[52] The latter has been implemented in the software BOMB, which was conceived by Jorgensen and co-workers.[53] We highlight this particular approach because it represents one of the most advanced receptor-based techniques. Most ligand-based *de novo* design methods, in contrast, are more simplistic but remarkably faster.

BOMB first places core fragments within the binding site of the target pocket using common substructures found in known drugs. Thousands of these virtual constructs are scored using multiple fragment conformations in combination with force field methods. The top-scoring candidate fragments are then optimized by the Monte Carlo/FEP technique. FEP calculations yield relative free energies of binding. Briefly, the FEP procedure can be used to "morph" one structure into another in incremental small steps. Typically, approximately 20 such steps are performed. In a recent BOMB design study, nanomolar HIV-RT inhibitors like compound **7.5** were obtained (Figure 7.10).

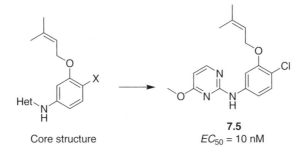

Core structure	**7.5**
	$EC_{50} = 10$ nM

Figure 7.10 *De novo* design of HIV-1 reverse transcriptase (RT) inhibitors using the software BOMB for core fragment placing and growing, and free energy perturbation (FEP) calculations for optimization of the heterocycle (Het) and the substituent X.[53]

Pharmacophore methods have demonstrated their usefulness as fitness functions because the constitutive definition of a pharmacophore allows the analysis of compounds in terms of a generalized interaction profile.[54] A pharmacophore description of a molecule is an abstract view on the underlying chemotype, because reference compounds are reduced to hypothetical interaction sites (potential pharmacophoric points, PPP) that may be responsible for ligand binding. As information on the molecular graph is abandoned, this concept enables searching for novel structures with prevalent binding modes but different molecular substructures. This is often a desired result. Pharmacophore-based methods like Recore allow for re-scaffolding.[55] Although Recore was originally implemented to work with docking poses, it also allows for definition of an anchor pharmacophore that can be matched by alternative scaffolds. When pharmacophore-based descriptors are used to evaluate the difference of virtual molecules and bioactive reference compounds, structural exchange is of course not limited to ligand core structures. For example, FLUX employs a topological pharmacophore descriptor (CATS descriptor)[56] for the design of molecules with bioisosteric replacements which appear in the core and in peripherals of the molecular graph.

Topological methods are not able to address conformational aspects in *de novo* design because one topology can yield different spatial conformations. Recent *de novo* design approaches like SQUIRRELnovo satisfy the requisition for shape-based matching of fragment surface patches into the mantle of one or multiple reference molecules. Shape matching is achieved by spatial alignment of *Shapelets* (Figure 7.6). Subsequent assessment of candidate compounds is based on the overlay of the fragment's pharmacophore points and a pharmacophore model of the reference molecule(s). Figure 7.11 shows this idea for the example of a thrombin inhibitor and a benzamidine fragment after molecular superposition by *Shapelets*. A three-dimensional

Figure 7.11   Pharmacophore-based scoring of the alignment shown in Figure 7.6(b). The reference molecule, its Gaussian field-based pharmacophore model, and the matched fragment are shown. Green fields characterize lipophilic interaction sites, red and blue ones show hydrogen-bond acceptor and donor sites. Hydrophobic atoms of the matched fragment are shown as little green balls. Small blue balls indicate hydrogen-bond donor points. Scoring of the fragment is achieved by summing up probability values of pharmacophoric points of the fragment in the respective interaction field of the reference pharmacophore model. The fragment has a dummy atom for linking (orange).

pharmacophore model of the reference molecule using trivariate Gaussian mixture models is computed (LIQUID approach).[57] Gaussian mixture models are suited for pharmacophore modeling because they allow calculation of the size and orientation of pharmacophore interaction fields in three-dimensional space. Another advantage is that one can construct a scoring function based on probability distributions.

Figure 7.12 show the results of a validation study. The task was to identify bioisosteric replacements for fragments in known PPAR (peroxisome proliferator-activated receptor) ligands. Fibrates are therapeutic agents for the treatment of metabolic disorders and activate PPARα, a member of the PPAR family.[58] It has been demonstrated that the 2-methyl-propionic acid moiety **7.6** is responsible for the selectivity of fibrates toward PPARα.[57] SQUIRRELnovo suggests bioisosteric replacement for this group. These groups have been patented for action on PPARα.[59 62]

To illustrate ligand-based *de novo* design with drug-derived molecular fragments, we present a recent application of Skelgen[26] published by Roche and *et al.*, who designed novel antagonists for the constitutively active histamine H₃ receptor.[63] The H₃ receptor is mainly known for its modulating activity on histamine production and release. It is responsible for regulation of other

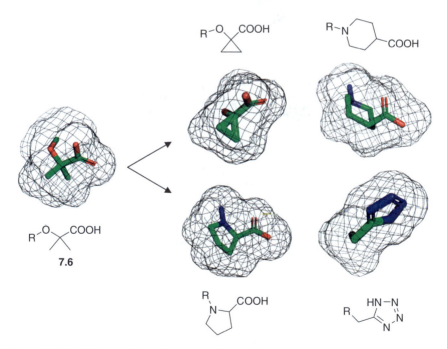

Figure 7.12 Bioisosteric replacement of 2-methylpropionic acid in **7.6**. These fragments were suggested by SQUIRRELnovo based on shape matching (mesh) and pharmacophore point scoring (LIQUID "fuzzy" pharmacophore method). All bioisosteres have been proven to exhibit the desired bioactivity as building-blocks for PPAR agonists.

neurotransmitters in both the central and peripheral nervous system.[64,65] The starting point for the design was given by two previously known H_3 ligand pharmacophore hypotheses, model 1 and model 2.[66,67] Literature and patent analyses led to a merged pharmacophore model, which was used by the software Skelgen to assemble novel candidate structures fulfilling the three-dimensional pharmacophore constraints (Figure 7.13). Several compounds were suggested, synthesized and tested. Potent inverse agonists were obtained, exhibiting an EC_{50} of 0.2 nm and $K_i = 0.3$ nm (human H_3 receptor).

Another successful ligand-based study has recently been performed with our software FLUX to generate novel ligands of the trans-activation response element (TAR) of the human immunodeficiency virus (HIV)-1 mRNA, which is a potential drug target in the treatment of AIDS.[68,69] Specific binding of the Tat protein to TAR is essential for viral replication. Inhibitors blocking this protein–RNA interaction are lead structure candidates for drug development. Owing to its remarkable flexibility, TAR RNA represents a comparably challenging drug target.[70] It had been shown in NMR experiments that a central bulge region of this small RNA element can accommodate intercalating ligands

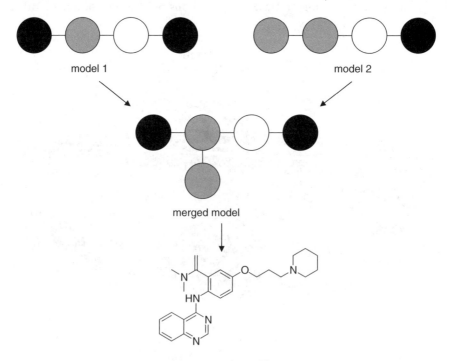

Figure 7.13 The *de novo* design software Skelgen[26] was used by Roche to fit fragments into a pharmacophore model that was obtained by merging the information of models 1 and 2. The best designed molecule is shown ($EC_{50} = 0.2$ nM; $K_i = 0.3$ nM). The pharmacophore features are displayed as black, grey and white circles, representing basic amines, electron-rich regions and aromatic interaction centers, respectively.

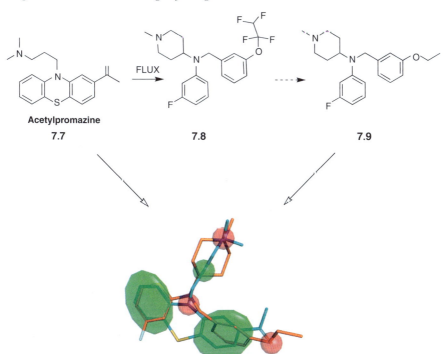

Figure 7.14 Using the TAR RNA ligand acetylpromazine as a template structure, FLUX suggested compound **7.8** as a result of automated fragment-based *de novo* design. Compound **7.9** was derived from this precursor, synthesized and successfully tested in a binding assay. The distribution of pharmacophoric points is very similar in **7.7** and **7.9**.

like, for example, acetylpromazine (**7.7**, Figure 7.14).[71] The task for fragment-based *de novo* design was to find novel TAR RNA binding agents mimicking the reference structure acetylpromazine.

For *de novo* design with FLUX, Schüller *et al.* employed a fragment database containing approximately 6000 unique building blocks, which had been compiled from a drug and lead database.[72] Both virtual *retro*-synthesis and virtual forward synthesis were performed using RECAP reactions. Then, new structures were designed by applying the evolutionary optimization concept of FLUX. The population size was set to 100 individuals, and optimization was terminated after 75 generations. One-hundred runs were performed to obtain multiple suggestions. The topological pharmacophore descriptor CATS served as a basis for "fitness" calculation, *i.e.*, Euclidian distances between the 150-dimensional CATS descriptor vectors of **7.7** and the virtual structures were computed. The design objective was to obtain a virtual compound with a minimal distance to the template. In addition, the molecular weight of designed candidate compounds was limited to 200–750 Da.

The best design was compound **7.8**. Inspection of this compound by experienced chemists resulted in a variant structure (**7.9**), because the original design **7.8** contains a tetrafluoroethoxy group. For ease of synthesis, and because the CATS descriptor does not explicitly consider fluorine moieties as pharmacophoric, molecule **7.8** was modified to obtain **7.9**, where an ethoxy group substitutes for the 1,1,2,2-tetrafluoro-1-ethoxy moiety. Figure 7.14 shows the almost perfect fit of all pharmacophoric features in a superposition of **7.9** with the reference acetylpromazine (**7.7**). Chemical synthesis was straightforward, which is a consequence of using drug-derived molecular fragments and a *pseudo* retro-synthesis approach for *de novo* design. In a binding assay the desired activity of the compound was conformed, namely disruption of the Tat–TAR interaction.[72]

These example show that it is possible to perform scaffold-hopping by fragment-based *de novo* design. One might argue that by using fragments as building-blocks instead of atoms, no "true" *de novo* design is performed. Also, it has long been unclear whether fragment-based approaches actually suggest truly novel molecular structures. Only recently, Krüger *et al.* addressed this question systematically and demonstrated that, for one or multiple templates of a given chemotype, alternative replacements are reached during *de novo* compound generation, thus indicating successful scaffold-hops.[73] In an extensive design study using 73 known inhibitors of angiotensin converting enzyme (ACE) as template structures, the FLUX algorithm produced approximately 20% known scaffolds (*i.e.*, identical chemotypes as the templates) and 80% new scaffolds, which were not contained in the set of 73 template structures. In total, over 9000 designed compounds were analyzed. Figure 7.15 shows the most prominent scaffold classes. Notably, the benzene ring is by far the most popular scaffold among these designs. This analysis clearly shows that fragment-based *de novo* design can come up with novel compound suggestions.

7.5 Conclusions and Outlook

We currently witness renewed strong interest in automated *de novo* design methods after their introduction in the 1990s. Recent successful applications of fragment-based design techniques have demonstrated their applicability and usefulness to drug discovery projects. Although *de novo* designed molecules are usually treated as suggestions only, some of the original computer-generated designs have actually been synthesized and successfully tested in various assays. It is safe to say that fragment-based *de novo* design has been proven to work. The strong interest in this technique is motivated by several additional facts: Drug discovery teams are confronted with the task of finding novel, patent-free lead structures for optimization, which is often hampered by high attrition rates during later stages of the drug discovery pipeline.[74,75] As a consequence, only a few new chemical entities have been approved by the authorities in recent years.[76] In this context, *de novo* design might provide a valuable source of

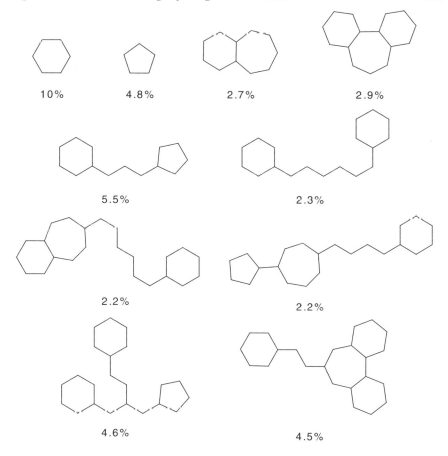

Figure 7.15 The ten most prominent frameworks produced by FLUX from 73 reference structures (known ACE inhibitors).[66]

inspiration in the early stages of hit and lead identification. This concept is further supported by the successful application of experimental fragment-based approaches like high-throughput X-ray or NMR screening. As a consequence of recent new software developments, *de novo* design methods have become available for a broader group of users. Limited access to such software tools might have hindered its wide-spread use in the past. Several academic and commercial software suites now contain a "*de novo* design module". However, such tools are not always readily applicable. Improved software with comfortable and versatile user interfaces is needed. This brings us to the limitations of the available methods. We think that there still is much room for future development, *e.g.*, by inclusion of realistic chemical reactions instead of *pseudo*-reaction schemes, and multidimensional optimization taking secondary constraints like ADMETox issues into account. First software implementations already include such options. Certainly, more sophisticated scoring functions are required that allow for proper consideration of enthalpic and entropic contributions of individual molecular fragments to ligand–receptor complex formation. Experimental data from spectroscopic and calorimetric measurements of fragment–receptor interaction could provide a valuable basis for tuning adjustment of scoring functions. The use of shape descriptors and pharmacophoric properties might represent a more coarse-grained straight-forward heuristic approach toward this goal and seems to be particularly suited for rapid first-pass molecular design. A further possibility for improving the impact of fragment-based *de novo* design is provided in the form of natural-product derived fragments and scaffolds.[77,78] The influence of natural product structures has been marked in past decades.[79] Therefore, the automated generation of both individual compounds as well as whole compound libraries will benefit from natural product-derived fragment collections. Finally, the assessment of intermediate design suggestions by a human expert has been proven to be helpful if not essential for the final success. In our view, there is no real need for increased computer power for molecular *de novo* design. Instead, smart interfaces for human–machine interaction are required that enable "chemical intelligence" be transferred to the actual software implementation.

Acknowledgements

The authors are grateful to B. Krüger, A. Schüller, Dr P. Schneider, and Dr K.-H. Baringhaus for helpful discussion of *de novo* design concepts. B. Krüger performed the calculations shown in Figure 7.15. This work was supported by the Deutsche Forschungsgemeinschaft (SFB 579, project A11).

References

1. P. M. Dean, D. G. Lloyd and N. P. Todorovs, *Curr. Opin. Drug Discov. Devel.*, 2004, **7**, 348.
2. G. Schneider and U. Fechner, *Nat. Rev. Drug Discov.*, 2005, **4**, 649.
3. G. Schneider and K.-H. Baringhaus, *Molecular Design – Concepts and Applications*, Wiley-VCH, Weinheim, New York, 2008.
4. P. J. Hajduk and J. Greer, *Nat. Rev. Drug Discov.*, 2007, **6**, 211.
5. P. J. Hadjuk, *J. Med. Chem.*, 2006, **49**, 6972.
6. R. A. Carr, M. Congreve, C. W. Murray and D. C. Rees, *Drug Discov. Today.*, 2005, **10**, 988.
7. W. L. Jorgensen, *Science*, 2004, **303**, 1813.
8. D. A. Erlanson, J. A. Wells and A. C. Braisted, *Annu. Rev. Biophys. Biomol. Struct.*, 2004, **33**, 199.
9. M. G. Siegel and M. Vieth, *Drug Discov. Today*, 2007, **12**, 71.
10. J. R. Huth, C. Park, A. M. Petros, A. R. Kunzer, M. D. Wendt, X. Wang, C. L. Lynch, J. C. Mack, K. M. Swift, R. A. Judge, J. Chen, P. L. Richardson, S. Jin, S. K. Tahir, E. D. Matayoshi, S. A. Dorwin, U. S. Ladror, J. M. Severin, K. A. Walter, D. M. Bartley, S. W. Fesik, S. W. Elmore and P. J. Hajduk, *Chem. Biol. Drug Des.*, 2007, **70**, 1.

11. A. R. Leach, R. A. Bryce and A. J. Robinson, *J. Mol. Graph. Model.*, 2000, **18**, 358–526.
12. E. W. Lameijer, J. N. Kok, T. Bäck and A. P. Ijzerman, *J. Chem. Inf. Model.*, 2006, **46**, 553.
13. J. Batista and J. Bajorath, *J. Chem. Inf. Model.*, 2007, **47**, 1405.
14. S. Bartoli, C. I. Fincham and D. Fattori, *Curr. Opin. Drug Discov. Devel.*, 2007, **10**, 422.
15. R. E. Hubbard, B. Davis, I. Chen and M. J. Drysdale, *Curr. Top. Med. Chem.*, 2007, **7**, 1568.
16. H. O. Villar and M. R. Hansen MR, *Curr. Top. Med. Chem.*, 2007, **7**, 1509.
17. H. Jhoti, A. Cleasby, M. Verdonk and G. Williams, *Curr. Opin. Chem. Biol.*, 2007, **11**, 485.
18. H. M. Berman, J. Westbrook, Z. Feng, G. Gilliland, T. N. Bhat, H. Weissig, I. N. Shindyalov and P. E. Bourne, *Nucleic Acids Research*, 2000, **28**, 235.
19. M. Weisel, E. Proschak and G. Schneider, *Chemistry Central J.*, 2007, **1**, 8.
20. E. W. Lameijer, J. N. Kok, T. Bäck and A. P. Ijzerma, *J. Chem. Inf. Model.*, 2006, **46**, 545.
21. U. Fechner and G. Schneider, *J. Chem. Inf. Model.*, 2006, **46**, 699.
22. U. Fechner and G. Schneider, *J. Chem. Inf. Model.*, 2007, **47**, 656.
23. A. M. Aronov and G. W. Bemis, *Proteins*, 2004, **57**, 36.
24. M. Feher, Y. Gao, J. C. Baber and W. A. Shirley, *J. Saunders, Bioorg. Med. Chem.*, 2007, **16**, 422.
25. H. Mauser and M. Stahl, *J. Chem. Inf. Model.*, 2007, **47**, 318.
26. D. G. Lloyd, C. L. Buenemann, N. P. Todorov, D. T. Manallack and P. M. Dean, *J. Med. Chem.*, 2004, **47**, 493.
27. J. Parn, J. Degen and M. Rarey, *J. Comput. Aided Mol. Des.*, 2007, **21**, 328.
28. S. Nikitin, N. Zaitseva, O. Demina, V. Solovieva, E. Mazin, S. Mikhalev, M. Smolov, A. Rubinov, P. Vlasov, D. Lepikhin, D. Khachko, V. Fokin, C. Queen and V. Zosimov, *J. Comput. Aided Mol. Des.*, 2005, **19**, 48.
29. N. Brown, B. McKay and J. Gasteiger, *J. Comput. Aided Mol. Des.*, 2004, **18**, 761.
30. N. Brown, B. McKay, F. Gilardoni and J. Gasteiger, *J. Chem. Inf. Comput. Sci.*, 2004, **44**, 1079.
31. I. Rechenberg, *Evolutionsstrategie – Optimierung technischer Systeme nach Prinzipien der biologischen Evolution*, Fommann-Holzboog, Stuttgart, 1973.
32. G. Schneider and S.-S. So, *Adaptive Systems in Drug Design*, Landes Bioscience, Austin, 2002.
33. X. Q. Lewell, D. B. Judd, S. P. Watson and M. M. Hann, *J. Chem. Inf. Comput. Sci.*, 1998, **38**, 511.
34. G. Schneider, M.-L. Lee, M. Stahl and P. Schneider, *Comput. Aided Mol. Des.*, 2000, **14**, 488.
35. K. Boda, T. Seidel and J. Gasteiger, *J. Comput. Aided Mol. Des.*, 2007, **21**, 311.
36. H. M. Vinkers, M. R. de Jonge, F. F. Daeyaert, J. Heeres, L. M. Koymans, J. H. van Lenthe, P. J. Lewi, H. Timmerman, K. Van Aken and P. A. Janssen, *J. Med. Chem.*, 2003, **46**, 2765.

37. E. R. Zartler and M. J. Shapiro, *Curr. Opin. Chem. Biol.*, 2005, **9**, 366.
38. C. A. Lipinski, F. Lombardo, B. W. Dominy and P. J. Feeney, *Adv. Drug Deliv. Rev.*, 2001, **46**, 3.
39. M. Congreve, R. Carr, C. Murray and H. Jhoti, *Drug Discov. Today*, 2003, **8**, 876.
40. A. C. Pierce, G. Rao and G. W. Bemis, *J. Med. Chem.*, 2004, **47**, 2768.
41. DrugBank, v. February 2006, http://redpoll.pharmacy.ualberta.ca/drugbank/index.html.
42. J. J. Irwin and B. K. Shoichet, *J. Chem. Inf. Model.*, 2005, **45**, 178.
43. E. Proschak, Y. Tanrikulu, B. Hofmann, O. Rau, H. Zettl, M. Weisel, J. Kriegl, D. Steinhilber, M. Schubert-Zsilavecz and G. Schneider, Presentation at the 3rd German Conference on Chemoinformatics, Goslar, Germany, 2007.
44. E. Proschak, M. Rupp, S. Derksen and G. Schneider, *J. Comput. Chem.*, 2008, **29**, 108.
45. A. Schüller, G. Schneider and E. Byvatov, *QSAR Comb. Sci.*, 2003, **23**, 719.
46. A. Schüller, V. Hähnke and G. Schneider, *QSAR Comb. Sci.*, 2007, **6**, 408.
47. J. Degen and M. Rarey, *ChemMedChem*, 2006, **1**, 854.
48. J. Degen and M. Rarey, http://www.zbh.uni-hamburg.de/flexnovo/, 2008.
49. G. Schneider and P. Schneider, In: *Chemogenomics in Drug Discovery* (H. Kubinyi, G. Müller Eds), Wiley-VCH, Weinheim, 2004, 341.
50. H. J. Böhm, *J. Comput. Aided Mol. Des.*, 1992, **6**, 61.
51. V. J. Gillet, G. Myatt, Z. Zsoldos and A. P. Johnson, *Perspect. Drug Discov. Des.*, 1995, **3**, 34.
52. D. A. Pearlman and M. A. Murcko, *J. Med. Chem.*, 1996, **39**, 1651.
53. W. L. Jorgensen, J. Ruiz-Caro, J. Tirado-Rives, A. Basavapathruni, K. S. Anderson and A. D. Hamilton, *Bioorg. Med. Chem. Lett.*, 2006, **16**, 663.
54. C. Wermuth, C. R. Ganellin, P. Lindberg and L. Mitscher, *Pure Appl. Chem.*, 1998, **70**, 1129.
55. P. Maass, T. Schulz-Gasch, M. Stahl and M. Rarey, *J. Chem. Inf. Model.*, 2007, **47**, 390.
56. G. Schneider, W. Neidhart, T. Giller and G. Schmid, *Angew. Chem. Int. Ed.*, 1999, **38**, 2894.
57. Y. Tanrikulu, M. Nietert, U. Scheffer, E. Proschak, K. Grabowski, P. Schneider, M. Weidlich, M. Karas, M. Göbel and G. Schneider, *ChemBioChem*, 2007, **8**, 1932.
58. M. L. Sierra, V. Beneton, A. B. Boullay, T. Boyer, A. G. Brewster, F. Donche, M. C. Forest, M. H. Fouchet, F. J. Gellibert, D. A. Grillot, M. H. Lambert, A. Laroze, C. Le Grumelec, J. M. Linget, V. G. Montana, V. L. Nguyen, E. Nicodème, V. Patel, A. Penfornis, O. Pineau, D. Pohin, F. Potvain, G. Poulain, C. B. Ruault, M. Saunders, J. Toum, H. E. Xu, R. X. Xu and P. M. Pianetti, *J. Med. Chem.*, 2007, **50**, 685.
59. Patent: JP 2004-202918, 20040709.
60. P. S. Humphries, S. Bailey, J. V. Almaden, S. J. Barnum, T. J. Carlson, L. C. Christie, Q. Q. T. Do, J. D. Fraser, M. Hess, J. Kellum, Y. H. Kim,

G. A. McClellan, K. M. Ogilvie, B. H. Simmons, D. Skalitzky, S. Sun, D. Wilhite and L. R. Zehnder, *Bioorg. Med. Chem. Lett.*, 2006, **16**, 6120.
61. Patent: US 2007191371, A1 20070816.
62. Patent: WO 2004069793.
63. O. Roche and R. M. Rodriguez Sarmiento, *Bioorg. Med. Chem. Lett.*, 2007, **17**, 3670.
64. H. Stark, *Expert Opin. Ther. Pat.*, 2003, **13**, 851.
65. R. Leurs, P. Blandina, C. Tedford and H. Timmerman, *Trends Pharmacol. Sci.*, 1998, **19**, 178.
66. R. Apodaca, C. A. Dvorak, W. Xiao, A. J. Barier, J. D. Boggs, S. J. Wilson, T. W. Lovenberg and N. I. Carruthers, *J. Med. Chem.*, 2003, **46**, 3938.
67. R. Faghih, W. Dwight, J. Bao Pan, G. B. Fox, K. M. Krueger, T. A. Esbenshade, J. M. McVey, K. Marsch, Y. L. Bennani and A. Hanckock, *Bioorg. Med. Chem. Lett.*, 2003, **13**, 1325.
68. J. Karn, *J. Mol. Biol.*, 1999, **293**, 235.
69. S. Bannwarth and A. Gatignol, *Curr. HIV Res.*, 2005, **3**, 61.
70. Z. Du, K. E. Lind and T. L. James, *Chem. Biol.*, 2002, **9**, 708.
71. Z. Du, K. E. Lind and T. L. James, *Chem. Biol.*, 2002, **9**, 185.
72. A. Schüller, M. Suhartono, U. Fechner, Y. Tanrikulu, S. Breitung, U. Scheffer, M. W. Göbel and G. Schneider, *J. Comp. Aided Mol. Des.*, 2008, **22**, 59.
73. B. Krüger, A. Dietrich, K.-H. Baringhaus and G. Schneider, *Comb. Chem. High-Throughput Screen.*, 2008, in press.
74. D. K. Walker, *Br. J. Clin. Pharmacol.*, 2004, **58**, 601.
75. S. S. Singh, *Curr. Drug Metab.*, 2006, **7**, 165.
76. R. L. Lalonde, K. G. Kowalski, M. M. Hutmacher, W. Ewy, D. J. Nichols, P. A. Milligan, B. W. Corrigan, P. A. Lockwood, S. A. Marshall, L. J. Benincosa, T. G. Tensfeldt, K. Parivar, M. Amantea, P. Glue, H. Koide and R. Miller, *Clin. Pharmacol. Ther.*, 2007, **82**, 21.
77. M.-L. Lee and G. Schneider, *J. Comb. Chem.*, 2001, **3**, 284.
78. K. Grabowski and G. Schneider, *Curr. Chem. Biol.*, 2007, **1**, 115.
79. D. J. Newman and G. M. Cragg, *J. Nat. Prod.*, 2007, **70**, 461.

CHAPTER 8

Early ADME/T Predictions: Toy or Tool?

IGOR V. TETKO[a] AND TUDOR I. OPREA[b]

[a] Helmholtz Zentrum München-German Research Center for Environmental Health (GmbH), Institute of Bioinformatics and Systems Biology, Neuherberg, 85764, Germany; [b] Division of Biocomputing, Department of Biochemistry and Molecular Biology, University of New Mexico School of Medicine, MSC 11 6145, Albuquerque NM 87131, USA

8.1 Introduction

The successful development of new drugs critically depends on the ability of researchers to predict the ADME/T (absorption, distribution, metabolism, excretion, and toxicity) properties of chemical compounds. These properties are important to narrow the search for New promising Chemical Entities (NCEs) in the early phases of drug discovery. Despite significant increases in R&D funds in the top 50 major pharmaceutical companies over the last decade, the number of NCEs remains practically unchanged. Indeed, a rigorous analysis indicates that the total cost of developing new drugs increased from $350 million in 1991 to over $800 million in 2003 (normalized for US$ in 2000).[1] When analyzed by disease class, new drugs cost more for respiratory disorders ($1.134 billion) and cancer ($1.042 billion) than for, e.g., HIV/AIDS ($504 million). Some of the low success rates can be attributed to the failure of drug candidates in clinical studies due to poor ADME/T properties. Thus, frontloading the risk by utilizing reliable in silico ADME/T tools may become a cost-saving endeavor.

A further increase in the interest for predictive ADME/T methods is due to the development of high-throughput screening and synthesis methods. The possibility to design, make and test millions of compounds has increased the

Chemoinformatics Approaches to Virtual Screening
Edited by Alexandre Varnek and Alex Tropsha
© Royal Society of Chemistry, 2008
Published by the Royal Society of Chemistry, www.rsc.org

risk of following poor leads – as pointed out by Lipinski and colleagues in 1997.[2] Shortly thereafter, Lipinski identified poor permeability and poor solubility as critical issues related to the poor success of developing orally available drugs.[3] Strategies to select better leads[4] and the lead-like concept[5,6] emerged at the same time. Tools for rapid *in silico* evaluation emerged in response to this pressure, *i.e.*, computational chemists and cheminformaticians were tasked to evaluate molecules faster than the experimentalists could make and evaluate them,[7] which led to significant attempts to integrate virtual screening and ADME/T evaluation.[8] The past decade has seen increased interest in the accurate estimation of ADME/T properties. The industrial sector has given stronger preference for the progression of candidates with the most favorable physicochemical and biological profile to clinical studies, in an effort to minimize the risk of failure in later stages of drug development. The *in silico* profiling of virtual libraries as a means of focusing on the most promising compounds for pharmaceutical development, places an increased emphasis on developing computational tools that reliably predict ADME/T properties.

Each year, a growing number of publications report on computational methods for the development of predictive ADME/T models. However, currently available methods are not reliable enough and are limited in their application,[9] despite the recognition of their importance in the drug discovery process.[10] Are we able to generate such reliable models, considering the severe limitations related to the intrinsic chemical diversity, the quantity and quality of the data? In this chapter, we critically review data and approaches used to develop physicochemical and biological ADME/T models, in an attempt to address this question.

8.2 Which Properties are Important for Early Drug Discovery?

Which computational properties are considered relevant during the early stages of drug discovery? Several reports from leading pharmaceutical companies provide a comprehensive review of these properties and the methods deployed in early stages.

8.2.1 Pfizer

The "rule of 5" (RO5) by Lipinski *et al.* (molecular weight, MW, ≤ 500, Clog $P \leq 5$, the number of hydrogen bond donor atoms ≤ 5 and the count of nitrogen and oxygen atoms [accounting for hydrogen bond acceptor atoms] ≤ 10) is the most implemented four-parameter system. Its implementation as an early alert system in industrial medicinal chemistry research has significantly altered the way early drug discovery has been carried out in the past decade. Indeed, over 2000 papers cite this publication[2] and its reprint[11] according to Scopus (http://www.scopus.com) and ISI knowledge (http://isiknowledge.com/). Violation of two or more of these R05 criteria is used to indicate that low absorption and thus low permeability are likely (exception: biological transporter substrates). One R05 criterion, Clog P, is in fact a QSAR (fitted) descriptor, while the others can (un)ambiguously be assigned from molecular structure; (Un)ambiguous relates to the relatively simple task of counting hydrogen bond donors and acceptors. Consider the case of water (H_2O), and the ambiguity caused by cheminformatics tools that count only one donor and one acceptor. This system does not allow the formation of the tetrahedral ice structure, where each water molecule donates, and accepts, two hydrogen bonds. Aiming to evaluate a relatively large number of small molecules at Pfizer with the R05 criteria may explain why simplicity (sum of N, O) was preferred over scientific accuracy.

Not only due to the R05 approach, but also to numerous previous papers in the QSAR field, pioneered by Corwin Hansch,[12] Clog P has become one of the most relevant computed properties when small molecules are initially evaluated and prioritized. In the words of Lombardo *et al.*:[13]

At any rate, due to its wide popularity and the vast availability of literature and in-house data, whether computed or experimentally determined, octanol–water partition or distribution data, remain by far the most utilized single parameter for ADME, QSPR and QSAR predictions.

8.2.2 Abbot

Figure 8.1 shows that the properties that Abbot scientists predicted most frequently for their compounds, using their WWW portal, are structural alerts for toxicity and mutagenicity.[14] The second largest number of predictions is for log P followed by bioavailability scores (rule of five[2] and internal Abbot score[15]), pK_a, and solubility prediction as well as prediction of some other more complex properties, such as Blood–Brain Barrier (BBB) permeability, binding energy, polar/non-polar surface area, and log D.

8.2.3 Novartis

Novartis uses the *In Silico Profiling* web tool.[16] Available properties include the octanol–water partition coefficient log P, molar refractivity, flexibility index, hydrogen bonding characteristics and molecular polar surface area. Various drug properties, such as intestinal absorption, BBB permeability or Plasma–Protein Binding (PPB) are calculated based on in-house models.

8.2.4 Bayer

Scientists at Bayer implemented "traffic lights" for the prioritization of molecules from the HTS hits.[17] Their score is based on solubility, lipophilicity, molecular weight, Polar Surface Area (PSA), and number of rotatable bonds.

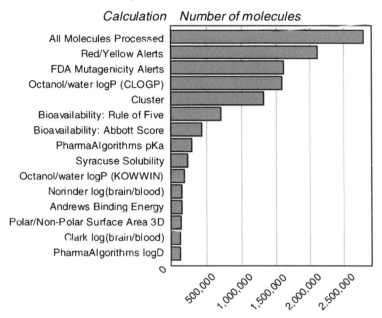

Figure 8.1 Illustration of the number of molecules processed at the Abbot site through the various algorithms available on the property calculation web page. (Reproduced from ref. 14 with permission of Wiley-VCH Verlag GmbH.)

These descriptors are similar to R05 criteria, since PSA directly correlates with the sum of hydrogen-bond donors and acceptors. Water solubility is also critically related to permeability and bioavailability.

8.2.5 Inpharmatica

The Admensa Interactive tool[18] developed by scientists from Inpharmatica incorporates several ADME properties, namely log P, log $D_{7.4}$ (log P at pH 7.4), aqueous solubility, human intestinal absorption, BBB penetration, cytochrome P450 (CYP) affinities, P-glycoprotein transport, hERG inhibition and PPB to score and prioritize their molecules.

The above examples highlight the early frontloading of ADME/T properties, which includes physicochemical properties such as log P, aqueous solubility, pK_a and pH-dependent variants of these properties. These are deemed relevant by medicinal chemists, when prioritizing small molecules for early drug discovery. Other criteria, such as those used in R05, as well as PSA and the numbers of non-terminal rotatable bonds are also deemed relevant for early lead profiling. Although these are important by themselves, complex ADME/T properties such as BBB permeability and CYP affinities are also estimated using such descriptors. Therefore, the accuracy of physicochemical property estimation, *e.g.* Clog *P*, may ultimately determine the prediction accuracy of complex ADME/T properties.

Considering the data quantity and quality limitations related to ADME/T measurements, we raise a concern related to the use of "complex" ADME/T models – which in turn rely on "simple" models such as Clog *P*, and the untested assumption that using such "complex" models is more advantageous than using "simple" properties on which complex models are built. This concern holds particularly true in the very early stages of drug discovery, where a large number of molecules are profiled (only) *in silico*, without the influence of a proper feedback loop, whereby the cycle prediction → experiment → error correction → model improvement → (better) prediction is absent.

We start this chapter with an analysis of methods to predict log *P* and aqueous solubility. In this context, we discuss the issue of applicability domain for QSAR models and the accuracy of prediction. Data available for simple physicochemical and ADME/T properties are compared by discussing the limitations of prediction of biological ADME/T properties. We restrict ourselves to several absorption and distribution properties, without discussing ME/T models. The interested reader is referred to the relevant sections in *Comprehensive Medicinal Chemistry II* (>1100 pages).[19]

8.3   Physicochemical Profiling

8.3.1   Lipophilicity

The partition coefficient is defined as the ratio of the concentration of a solute in the organic phase to its concentration in the water phase. This definition applies to the same neutral microspecies. However, many small molecules of pharmaceutical, agricultural and environmental interest may assume different protomeric and tautomeric forms, which increases the complexity of the above (simple) definition. Indeed, many small molecules contain moieties that ionize in water, thus contributing to a decrease in lipophilicity. The distribution coefficient, log D_{xy}, measures the pH-dependent distribution of drug in octanol–water phases at pH *xy*.

Practically, the determination of log *D* at fixed pH is simpler than the log *P* measurement, which may require multiple titration experiments and/or extrapolations to the neutral state of the compound. Moreover, by using some specific ranges of pH (*e.g.*, pH 1–2 for stomach or neutral pH 6.5 for jejunum), one can better simulate the medium in the gastrointestinal tract.

8.3.1.1   Data Availability

Data quality and quantity are important issues when addressing the limitations of the existing calculated log *P* models. The amount of data for log *P* prediction is one of the largest in the field. The MedChem database[20] contains the largest commercially available collection, with over 60 000 measurements of log *P* and log *D* (http://www.biobyte.com). The PHYSPROP database[21] of the Syracuse Research Inc. provides experimental log *P* values for 13 058 compounds. LOGKOW database (http://logkow.cisti.nrc.ca), which is supported by the Sangster Research Laboratories, provides on-line access to about 20 000 molecules, including log *P*, log *D* and pK_a values. This database is updated quarterly and is the largest publicly available collection of octanol–water partition coefficients in the field. The inter-laboratory variation of log *P* values were estimated as $\sigma = 0.45$ (MAE = 0.26) log-units.[22] The larger number of experimental measurements is available within the industry, particular for log *D* measurements. For example, more than 96 000 molecules were used to benchmark several methods in Pfizer and in Nycomed.[23] HQSAR Tripos descriptors were used by Bayer AG to develop log *D* models at pH 2.3 using 70 000 compounds.[24] The log *D* values for 11 283 measurements performed using the shake flask method at pH 7.4 were used by Merck.[25]

8.3.1.2 Models

Many models for log *P* prediction have been developed and published in the literature, which is a consequence of the availability of a large amount of log *P* data. There are much fewer log *D* models and, moreover, models built with large datasets are almost exclusively from large companies.[13] One can distinguish two main groups for log *P* calculation: fragmental and based on descriptors calculated for the whole molecule.[23,26]

The log *P* is to a large extent an additive property. Thus, not surprisingly, a considerable number of fragmental methods to predict this property have been published. The general equation for this group of methods can be represented as:

$$\log P = a + \sum_{i=1}^{N} b_i G_i + \sum_{j=1}^{K} c_j F_j \tag{8.1}$$

where G_i is the number of occurrences of the group *i*, F_j are the correction factors and *a*, b_i and c_j are the regression coefficients. Several popular methods, *e.g.* Clog *P*,[27,28] ACD/log *P*,[29] Σf-SYBYL,[30,31] Klog *P*,[32] Hlog *P*,[33] AB/log *P*,[34,35] use the fragmental representation of molecules to correlate lipophilicity of molecules with their structures.

The second group includes methods that use the 3D structure representation of molecules, such as CLIP,[36] QikProp,[37] COSMOlogP.[38,39] Methods based on atom-type or topological indices calculated for the whole molecules, such as AUTOLOGP,[40] KOWWIN,[41,42] VLOGP,[43,44] XLOGP[45] and ALOGPS[46,47] were also proposed. Several reviews describe and compare the advantages and features of these approaches.[23,48,49]

One may think that the main difference of the second group compared to fragmental methods is the absence of the "missed" fragments, a problem that can seriously hamper prediction ability of fragment-based methods. However, it appears that the use of methods based on whole molecule descriptors can also have very low prediction ability for molecules that were not covered by the training set (Section 8.4).

8.3.2 Solubility

Permeability and solubility/dissolution are two major determinants of gastro-intestinal drug absorption. The prediction of solubility of molecules is more difficult than for lipophilicity. Solubility critically depends on the solid-state properties of compounds. The same compound can exist in amorphous or in several crystalline states[50] and this can result in very different solubility of molecules. The prediction of crystalline properties, represented, for example, by the melting point, is one of the most difficult problems of physical chemistry.[51] Like the octanol–water partition coefficient, water solubility critically depends on the pH and ionization state of molecules.

8.3.2.1 Data Availability

The largest commercially available datasets are the Physical Properties (PHYSPROP)[21] and AQUASOL databases (*ca.* 6000 compounds in each database). The AQUASOL database has been published as a book.[52] Fur-thermore, two relatively large sets of aqueous solubility data models were used in many other studies.[53,54] Data from the AQUASOL database had an inter-laboratory variation of about $\sigma = 0.49$ log-units (as estimated for $N = 1031$ molecules).[55] Moreover, large inter-laboratory errors mask the influence of temperature, and differences as large as $\Delta T = 30\,^\circ\text{C}$ do not increase this error. In-house models developed at pharmaceutical companies could be based on similar or even larger numbers of measurements. For example, about 5000 molecules were used to develop a model at Bayer Healthcare AG.[24]

8.3.2.2 Models

The prediction of aqueous solubility is more complex compared to lipophilicity. Frequently, solubility models incorporate log P as one of the descriptors. Yalkowsky[56,57] considered a physical model of the solubilization and provided a theoretical basis of the link between log P and solubility of molecules in his General Solubility Equation (GSE). The 2001 version[58] of GSE is amazingly simple:

$$\log S = 0.5 - 0.01(\text{MP} - 25) - \log P \tag{8.2}$$

where MP is the melting point – fixed at 25 °C for liquid compounds. Despite its simplicity, the equation has good predictive ability. Indeed, the accuracy of this equation[59–61] is similar to that of the Monte Carlo simulation of Jorgenson and

Duffy,[62] the group contribution approach of Klopman[63] and the neural network model of Huuskonen.[54] The PCCHEM program used at the US Environmental Protection Agency (EPA) incorporates three different equations. All of them are similar to GSE but have different coefficients to predict aqueous solubility depending on the range of log P values.[64] Meylan and Howard used a database of 817 (RMSE = 0.62) compounds to derive a heuristic equation:

$$\log S = 0.69 - 0.96 \log P - 0.0031 \text{MW} - 0.0092(\text{MP} - 25) + \sum f_i \quad (8.3)$$

which contains 15 additional correction factors.[41]

Given that MP as a property is difficult to predict, some groups do not make such predictions. For example, a median melting point value of 125 °C is commonly used at Syngenta to apply the GSE in the absence of experimental values.[65] In the early stages of drug R&D, for example when evaluating HTS libraries, this omission makes practical sense: compounds can exist in different forms, ranging from amorphous (less pure, thus lower MP) to crystalline (high purity, higher MP). In the later stages, *e.g.*, in drug development, molecules reach a higher purity, which in turn may result in decreased solubility. Thus, a use of median MP values may lead to perhaps significant errors, given the changes in solubility caused by the purification of substances that occurs during the development process.

Log P can be used as an additional parameter, in combination with other descriptors. For example, neural network models developed by Liu and So[66] and Goller *et al.*[24] use log P in combination with topological and quantum-chemical descriptors. Many methods do not use log P as a descriptor. These methods have been described in several reviews.[55,65,67] However, there is a clear relationship between these two physicochemical properties, namely log P and aqueous solubility.

Analysis of *in-house* data in pharma companies frequently demonstrates a low prediction ability of current models for both lipophilicity[68–70] and aqueous solubility.[71,72] The calculated errors of these models are often around or higher than 1 log-unit, which is not sufficient for screening purposes. Thus, despite relatively large amounts of data for physicochemical properties and their simplicity compared to more complex ADME/T properties, the accuracy of prediction remains low. Let us consider factors that limit the prediction accuracy of models.

8.4 Why Predictions Fail: The Applicability Domain Challenge

The failure of models to yield accurate predictions is a consequence of either experimental design or differences in the chemical spaces used to develop and test the models.[73] The identification of model Applicability Domain (AD) can differentiate reliable from non-reliable predictions. There are, indeed, many different approaches to determine AD. These methods can be classified in two categories: methods that explore the similarity of molecules in the descriptor space[74] and methods based on the predicted property.[75]

8.4.1 AD Based on Similarity in the Descriptor Space

These methods can be categorized as (i) range-based methods, (ii) geometric methods, (iii) distance-based methods and (iv) probability-density distribution range methods.[76] The range-based methods introduce upper and lower limits (which can also be infinity or no limit) for some descriptors. Those molecules with descriptor values above or below these limits are placed outside the AD. Descriptors can be either structural features (*e.g.*, number of fragments of a particular type) or properties (*e.g.*, lipophilicity, aqueous solubility, *etc.*). Range-based methods have simple interpretation, and are thus popular in cheminformatics. Often, range-based methods are implicitly present in a model. For example, R05 criteria are in fact a range-based AD for a model, which can be stated as "all molecules are orally permeable". Any violation of two of these rules corresponds to an "out of the applicability domain" prediction, which implies a decreased probability that such molecules have good oral permeability.

Another known case of implicit range-based AD methods is that of fragment-based log P methods that evaluate missing fragments. For example, with the CLOGP software, versions 4 or earlier denied the user the ability to obtain a numerical estimate for lipophilicity when the input molecules contained "missing fragments". Quite possibly due to market ("evolutionary") pressure, later versions of CLOGP include an "*ab initio*" estimation of the contribution for missed fragments.[27] This "*ab initio*" calculation, however, may lead to less reliable log P predictions. For example, about 67% of the molecules (376 out of 558) with large CLOGP errors (>1.5 log-units) in the PHYSPROP dataset contained fragment values calculated by the "*ab initio*" method.[47] The ALOGPS program,[47] which predicts lipophilicity and aqueous solubility of chemical compounds, flags unreliable predictions when the input molecule contains one or more E-state atom or bond types that were not represented in the training set. This simple AD flag makes it possible to identify 90% of the outliers (357/394) with large prediction errors (>1.5 log-units) for the same dataset.

Range-based cut-offs are used to determine whether the input molecule is inside or outside the space defined by the training set, to determine the Optimal Prediction Space (OPS) used in the TOPKAT package.[43,44] An example of the geometric methods is a convex envelope that is the smallest convex region enclosing all points from the training set.[74] This method provides a nice visual interpretation for models with few variables, but this feature is lost in higher dimension spaces. Distance-based methods include different metrics, such as Euclidian, city block, as well as three other interrelated measures such as Mahalanobis, hoteling T^2 and leverage, to assess quality of predictions.[76–80] Leverage is defined as:

$$h = x^T (X^T X)^{-1} x \tag{8.4}$$

where x is the vector of descriptors of a query compound and X is the matrix formed with descriptors from the training set. High h values indicate that the input molecule stands out from the training set and may involve considerable extrapolation, rather than interpolation. Leverage was recommended for assessing AD in several studies.[80,81] The probability density function can also be used to determine the distance to the descriptor space.[74]

8.4.2 AD Based on Similarity in the Property-based Space

ADs in the descriptor space actually ignore the most important parameter, the predicted property itself. Indeed, the target property is implicitly included in similarity measures, since it guides the selection of descriptor sets to find the optimal target property model(s).[82] One way to account for the influence of the target property during AD determination is to weigh variables for similarity measures using, *e.g.*, the importance of a given descriptor for the model.[74] Several methods explore variations in the model residuals. In such methods, not one but a set (ensemble) of models is usually generated (*e.g.*, using different subsets of the data,[46] different variables[83] *etc.*). Residuals and/or confidence values for predictions are analyzed to derive model ADs.

The decision forest method builds multiple models by combining, in one predictor, results of multiple Decision Trees (DTs).[83] DTs are constructed to be as heterogeneous as possible, using each variable maximum one time in all models. Using the example of the analysis of estrogen receptor binding, the prediction accuracy was demonstrated to increase as the confidence level of the prediction increased.[74,84] A similar result was observed in methods developed to discriminate soluble from poorly soluble molecules.[85] The authors applied an ensemble of neural network models and demonstrated that molecules with small standard deviations of predictions (<0.01) had 2–3 times lower errors than the rest of the dataset. Thus, predictions with high standard deviations were outside the AD of models. The standard deviations were also used to predict the vapor pressure and solubility of chemical models.[24,86,87] In another study,[88] the standard deviation of predictions originating from an ensemble of Bayesian Regularized Neural Nets was shown to be positively correlated to the distance to the model in the descriptor space; both metrics were used to estimate prediction errors.

The Gaussian Process could be also used to estimate prediction accuracy based on the variance of different models derived within this approach. The usefulness of this approach for confidence intervals prediction of aqueous solubility and lipophilicity was shown.[89,90]

The Associative Neural Networks (ANNs)[46] method uses residuals calculated from an ensemble of models to categorize new input molecules. The property-based similarity, R, of a given molecule to the training dataset is defined as the square of maximum correlation of a vector of residuals of the query molecule to vectors of residuals of all molecules in the training set.[91,92] The log P prediction analysis from a PHYSPROP dataset using the ALOGPS program[46,47] showed that molecules with $R > 0.8$ and $R < 0.3$ had Mean

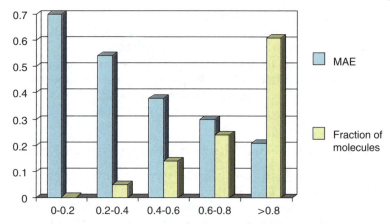

Figure 8.2 Prediction accuracy of ALOGPS[46] program as a function of a property-based similarity of a molecule to the training set compounds (MAE = mean absolute error).[91,92]

Absolute Errors (MAEs) of approximately 0.3 and 0.7 log-units,[91,92] respectively (Figure 8.2). Property-based similarity was used to differentiate reliable *vs.* non-reliable log *P* predictions, showing excellent agreement of actual *vs.* predicted values for 7498 and 8750 neutral molecules from AstraZeneca and Pfizer datasets, respectively.[75] Moreover, for >50% of the in-house Pfizer compounds (as characterized by property-based similarity values of >0.8), predicted log *P* values had an accuracy of 0.35 log-units, which is similar to that of experimental measurements.[93] The same approach was successful in estimating the accuracy of log *P* predictions for Pt(II) complexes.[94]

8.4.3 How Reliable are Physicochemical Property Predictions?

AD approaches can be used to estimate prediction errors for new molecules. For example, using the error-correlation dependencies from Figure 8.2 it was possible to estimate that only less than 20% of molecules from the NCI database will have predicted errors with MAE <0.5 log-units using a log *P* model based on the PHYSPROP database.[91] For chemical vendor databases, such as Asinex or Ambinter, the same accuracy could be expected for only ~3% of compounds (Figure 8.3).[91] The use of in-house data does not dramatically change the coverage and accuracy of predictions. Indeed, the expected accuracy of the ALOGPS program increased only by 0.06–0.07 log-units on average, when predicting the 10 million compounds from the iResearch Library, after improving ALOGPS with in-house AstraZeneca and Pfizer data.[75]

Conceivably, by using different descriptors, one may develop models with improved applicability domains. However, we noticed that the prediction accuracy of methods developed to predict lipophilicity and aqueous solubility was governed by the diversity of molecules in the training set, not by the choice of

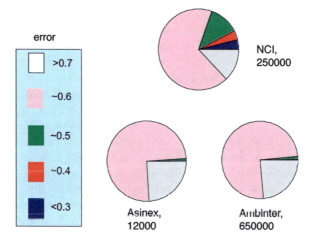

Figure 8.3 MAE and percentage of compounds predicted with ALOGPS program for the National Cancer Institute and two commercial providers.[91,92]

machine learning methods.[47,55,95] This observation is in agreement with similar conclusions, based on 20 diverse datasets (log *D*, aqueous solubility, pK_a and biological activities), made by Sheridan *et al.*[25] Benchmarking of 14 different distances to model for prediction of toxicity against *T. pyriformis* also indicates that the diversity and distribution of data the training set, not the choice of computational approaches and molecular descriptors, were the limiting factors determining the accuracy of predictions and applicability domain of the models.[96]

Thus, within the current set of descriptors, datasets and machine learning tools, one is unlikely to develop models that work like "magic" to provide accurate predictions for all possible chemicals, even for relatively simple physicochemical properties such as log *P* and aqueous solubility. Indeed, the quantity and quality of data, as well as the diversity of the training set, less so the machine learning method and the choice of molecular descriptors, are likely to determine model quality with respect to prediction accuracy. With this conclusion in mind, let us examine the prediction of more complex, biological ADME/T properties.

8.5 Available Data for ADME/T Biological Properties

8.5.1 Absorption

Several parameters relate to the prediction of absorption. One of the parameters used to quantify Human Intestinal Absorption (HIA) is the fraction absorbed (%F), which is defined as the mass absorbed divided by the total mass of the given drug dose:

$$\%FA = m_a/m_t \tag{8.5}$$

where m_a and m_t are absorbed and total dose, respectively. In most cases (and always specified), %FA relates to *oral* absorption. Oral %FA is governed by, among other factors, the area of the absorbing surface and the residence time of the drug at the absorbing surface – which are likely to directly influence the fraction of the drug dose that is absorbed, while several other factors (*e.g.*, complexation with bile salts, metabolic reactions in the intestine, *etc.*) are likely to result in decreased %FA.[97,98] Oral %FA estimates include specific regions of the gastro-intestinal tract that may absorb (or not) drugs at different rates, which further increase the biological complexity of this parameter.

Furthermore, there are drug delivery routes that may lead to systemic availability, among which are topical administration routes, *e.g.*, intra-nasal, trans-dermal, aerosol inhalation, sublingual tablets, pellet implants and ophthalmic, as well as injectable formulations (other than intra-venous, intra-arterial and intra-cardiac), *e.g.*, intra-muscular, subcutaneous, subdural, epidural, intra-amniotic, intra-cerebral, in the cerebro-spinal fluid and intra-cavernosal. Therefore, the %FA for non-oral administration routes is relevant, in particular for those drugs that have significant side-effects and long plasma half-life; unfortunately, it is also subject to significantly fewer experimental and modeling studies. This is, in no small part, because an overwhelming majority of drugs are developed for oral formulation – which, in turn, is dictated by patient preference as the option of choice for drug delivery. Therefore, unsurprisingly, RO5 and most ADME/T modeling tools are focused on oral permeability and on those properties that influence oral delivery. Consequently, for the remainder of this chapter, we discuss ADME/T in the context of oral drug delivery.

For lead molecules with poor %FA, one is usually interested in under-standing the underlying molecular determinants and modifications of the chemical structure that are required to improve it. Intestinal permeability can be guided either by passive or active transport. Caco-2 cells derived from col-orectal carcinoma cells[99,100] or Madin-Darby Canin Kidney (MDCK) cells have been used to evaluate the intestinal permeability of drugs. Active trans-port includes absorptive carriers (such as peptide and amino acid transporters), while ATP-Binding Cassette (ABC) transporters such as ABCB1 (also known as P-gp, P-glycoprotein, or MDR1, multi-drug resistance protein 1) are responsible for drug efflux.

8.5.1.1   Data

Bioavailability, Fraction Absorbed and Human Intestinal Absorption (HIA). The dataset of Abraham and colleagues,[101] which contains data for 241 com-pounds, is one of the most used. The authors not only collected the data but also classified the drugs in several categories, *i.e.*, GOOD, OK, Questionable, and Dose-Dependent (DP), according to the original work and protocols used. The dataset of Klopman *et al* contained 467 drugs[102] while a larger set of 647 molecules were collected by Ho *et al.*[103] The latter database is available from http://modem.ucsd.edu/adme. A bigger dataset of 1290 compounds, consisting of structures and their human oral pharmacokinetic data, including bioavailability and/or absorption and/or radio labeled studies, was used in ref. 104.

Human Jejune Permeability. Lennernäs[105] has reported data on 38 molecules, some of which were collected from the literature. A larger set of 51 molecules (45 drugs and six amino acids) was used by SimulationsPlus to develop their commercially available model from the previous publication of the same author[106] and other literature sources.

Caco-2/MDCK. The dataset of Yazdanian *et al.*[107] of 38 structurally diverse compounds is frequently used by other authors to develop methods or/and to test their approaches. Yamashita *et al.*[108,109] collected a larger set (87 compounds). This dataset was extended to 100 molecules by Hou *et al.*[110] The amount of data for this *in vitro* model of intestinal absorption remains low. The number of measurements for MDCK cells, 55, is even lower.[111]

8.5.1.2 Models

The QSAR paradigm for structure–permeability correlations, used to evaluate oral absorption, was summarized by Van der Waterbeemd *et al.*[112] as follows:

$$\text{oral permeability} = f(\log \boldsymbol{D_{7.4}}, \text{ molecular size, H-bond capacity}) \qquad (8.6)$$

where $\log D_{7.4}$ is the octanol–water partition coefficient at pH 7.4, molecular size is a measure related (among other properties) to the mass, volume and surface of the input molecule, and H-bond capacity relates to the number and strength of the hydrogen bonds that can be donated or accepted by the same molecule. The seminal R05 paper from Pfizer selected four criteria governing oral permeability for drugs according to the QSAR paradigm (Equation 8.6), which are easily calculated, and which have a direct physicochemical interpretation. However, R05 criteria do not necessarily allow one to rapidly evaluate oral *bioavailability*. More to the point, oral *permeability* reflects the ability of a molecule to pass the intestinal barrier, as estimated by %FA (Equation 8.5), or by the HIA (human intestinal absorption) parameter. These estimates look at the concentration of the parent drug in the portal vein, *i.e.*, before the drug is exposed to metabolizing enzymes in the liver, and typically ignoring the metabolism that may occur in the intestine. Indeed, Caco-2 models for permeability where the apparent apical to basal (A→B) drug permeability is approximately equal to the reverse (B→A; basal to apical) permeability serve as *in vitro* surrogates for this property.[99] However, oral *bioavailability* (%oral) reflects both the fraction of the parent drug that is absorbed intact through the intestinal barrier (%FA) *and* the fraction of the parent drug that survives first-pass metabolism. In other words, it evaluates how much of the parent drug is *systemically* available, *i.e.*, able to reach its intended drug target(s). Systemic availability is further influenced by other ADME parameters, *e.g.*, volume of distribution at steady-state (VD_{ss}), systemic clearance (CL), plasma protein binding (PPB) and plasma half-life ($t_{1/2}$), among others. Some drugs have good %FA but are significantly metabolized during first-pass: felodipine (100% HIA, 15.5% oral) and labetalol (95% HIA, 20% oral) are such examples. These drugs are sometimes "first in class", and can be replaced by "best in class" drugs, *e.g.*, amlodipine (74% oral) enjoys a higher financial success than felodipine, for the same therapeutic indications. In rarer cases, drugs have lower %FA but higher %oral, due to transporter effects: Cefuroxime axetil (38% HIA, >60% oral) is such an example. In Caco-2 models, transporter effects may be inferred when A→B is significantly different to B→A. Of course, from an ADME/T perspective, it is more important to capture %oral, not %HIA.

Many models for absorption parameters include log P as one of the important parameters – as reflected in Equation (8.6). For example, TPSA, log $D_{6.5}$, the number of R05 violations and the number of H-bond donors were used by Hou *et al.*[103] to model simple hierarchical rules that classify compounds into "poor" (%FA <30%) and "good" (%FA >30%) intestinal absorption categories. These authors used the largest database, 579 molecules, which were considered by them as transported by passive phenomena. Quantitative predictions were also developed by Hall *et al.*,[113] using more complex methods, such as neural networks and E-state indices; again PSA and log P were found as the dominant properties. The high relevance of TPSA for HIA was confirmed by its selection (out of 2929 descriptors) using Support Vector Machines (SVM).[114] Konovalov *et al.* found that only one descriptor, log P predicted using ALOGP program, was sufficient to predict HIA using Monte Carlo variable selection.[115] The use of other descriptors did not improve the results. A training set of 77 structurally diverse organic molecules was used to construct significant QSAR models[110] for Caco-2 cell permeation: cellular permeation was found to depend primarily upon the experimental distribution coefficient log $D_{7.4}$, the High Charged Polar Surface Area (HCPSA) – a factor related to H-bond capacity, and the radius of gyration (rgyr) – a size-related parameter, for 77 structurally diverse drugs. Other studies found that measured or computed log P are useful for estimating this property using smaller datasets.[116,117] Finally, one study modeled two Caco-2 permeability measurements, as well as HIA, simultaneously.[118] This nonlinear model, built on a training set of 16 drugs, was based on H-bond capacity (donors, acceptors, and polar surface area), hydrophobic transferability (multiple log P and log $D_{7.4}$ estimates), and less so on size (total surface area). This model had good external predictivity: 11 out of 16 compounds (68.7%) in the Caco-2 permeability external set were predicted within 0.6 log-units error, whereas 46 out of the 69 drugs (66.7%) in the HIA external test were predicted within 23% HIA unit error. In summary, most of the surveyed models reflect the QSAR permeability paradigm (Equation 8.6), and all depend critically on log P.

8.5.1.3 Prediction of Active Transport and Efflux

Datasets for the modeling of ABCB1 (P-gp) compounds in the literature typically include 200 substrates[119–121] and up to 400 inhibitors.[122] These sets are extensions of the Seelig[123] dataset of 100 molecules. The purpose of these studies is usually to predict whether the input drugs will be ABCB1 substrates. Efflux pumps such as ABCB1 can be important for the therapeutic effect of CNS drugs and for the blood–brain barrier permeability. Other efflux pumps, *e.g.*, ABCC1 (MRP1), ABCC2 (MRP2, cMOAT) and ABCG2 (BCRP, MXR, ABCP), have been shown to influence the oral absorption and disposition of a wide variety of drugs. Naturally occurring polymorphisms of drug transporters are also responsible for individual differences in response to drug regimens. In one study, eleven different SNP (Single Nucleotide Polymorphism) ABCB1 variants were compared to wild-type, with respect to substrate specificity for 40 drugs; it was found that the nonsynonymous polymorphisms of 2677G > T, A or C in the ABCB1 gene, corresponding to amino acid 893, Ala > Ser, Thr or Pro, respectively, in P-gp, greatly impact ABCB1 substrate specificity and activity.[124]

Furthermore, elevated expression levels of ABCB1, ABCC1, ABCG2 and perhaps other ABC efflux transporters (48 known members) in human cancer cells have been found to lead to multi-drug resistance,[125] which in turn correlates with patient outcome in several cancers.[126] The appropriate study of ABC transporters allows us not only to better understand drug absorption, but also to evaluate cancer patients with respect to their responsiveness to chemotherapy, and their susceptibility to certain side effects. High-throughput assays that simultaneously evaluate drug transporter inhibitors for, *e.g.*, ABCB1, ABCC1 and ABCG2, are beginning to emerge.[127] Co-administration of ABCB1 inhibitors such as mometasone furoate with other chemotherapeutic agents is expected to lead to improved anti-cancer drug regimens.[128]

8.5.2 Distribution

Distribution parameters that are subject to computational methods include the BBB permeability, PPB, tissue partitioning (K_p) and the volume of distribution at steady state (VD$_{ss}$). These parameters are critically important for understanding drug pharmacokinetics. For example, BBB permeability is an essential property for those drugs targeting the CNS (Central Nervous System), such as antidepressants or anxiolytics. Conversely, CNS penetration (and positive BBB permeability) is, for example, not desired for anti-allergic drugs that target the peripheral histaminic H_1 receptor (which may cause drowsiness when binding to the central H_1 receptors). The volume of distribution defined as:

$$VD_{ss} = \text{dose}/C_0 \qquad (8.7)$$

where "dose" is the administered dose and C_0 is the initial concentration of drug in plasma, which increases with tissue partitioning (high K_p) and decreases with plasma binding (high PPB). The binding of small molecules to the Human Serum Albumin (HSA), which is the most abundant protein in human plasma (600 mM), determines to the largest extent the plasma binding.[129]

8.5.2.1 Data

Blood–Brain Barrier. The largest dataset, 328 molecules, was discussed by Abraham *et al.*,[130] who collected data for *in vivo* distribution of drugs from blood, plasma, or serum to rat brain. The authors showed that these data could be effectively combined into the single data set with enhanced accuracy. The experimental accuracy of BBB data was estimated to be 0.3 log-units.[130] Qualitative data for human BBB permeability (+ or –) is available for 278 (BBB+) and 172 (BBB–) drugs in WOMBAT-PK.[131]

Plasma Protein Binding (PPB). The fraction bound to plasma proteins (%PPB) can be easily calculated from binding affinity to HSA under the assumption that binding occurs mainly to the Human Serum Albumin (HSA). Thus we can ignore differentiation of both these properties. A dataset of 94 molecules with binding to HSA was collected by Colmenarejo.[132,133] Larger datasets of 138 and 154 molecules were collected by Kratochwil[134] (HSA) and by Saiakhov[135] (%PPB), respectively. The dataset of Yamazaki[136] (%PPB) contained the protein binding percentages of 346 drugs. For comparison, commercial SimulationsPlus and ChemSilico predictors were developed with data for 388 and 345 molecules, respectively.

Volume of Distribution. A dataset of 120 compounds with VDss data was published by Lombardo *et al.*[137] Later on the same authors collected a larger set of 384 compounds from the literature.[138]

8.5.2.2 Models

Again, log *P* was one of the critical descriptors to describe these properties. For example, Lombardo *et al.* showed that successful prediction of the VDss depended on two experimentally determined physicochemical parameters, log $D_{7.4}$ and the fraction of compound ionized at pH 7.4, as well as on the fraction of free drug in plasma.[137] A more advanced model by the authors had more descriptors, but also included lipophilicity as one of the major descriptors.[138] Saiakhov *et al.*[135] did not find a reliable correlation between log *P* and %PPB of all 154 analyzed molecules. However, following a structural analysis they identified eight main biophores, *i.e.*, structures appearing with high frequency, in compounds with different degrees of binding activity. The local QSAR models were developed for seven of eight of these biophores and the partition coefficients were found to be important for almost all these local models.

Several authors including, for example, Colmenarejo,[133] Kratochwil[134] or Valko[139] found log *P* as one of the important parameters for their models. However, additional structure-based parameters, such as PSA, number of hydrogen bonds, topological pharmacophores, *etc.*, were required to derive statistically significant models. We reevaluated the importance of these descriptors to predict %PPB. Based on 851 drugs with known %PPB from WOMBAT-PK, we used several simple descriptors, log *P*, number of hydrogen donors and acceptors, number of Rotatable Bonds (RBONDS), Topological Surface Area (TPSA), molecular weight in multiple linear regression. The regression equation was as follows:

$$\%\text{PPB} = 34 + 0.15(0.02)\text{TPSA} + 12.6(0.6)\log P - 1.3(0.3)\text{RTB} \qquad (8.8)$$

$$R^2 = 0.41, \ \text{RMSE} = 25.5\%, \ \text{MAE} = 20.6\%, \ N = 851, \ F = 85$$

Equation (8.8) confirms the strong dependency of %PPB on both log *P* and TPSA; surprisingly, it shows an inverse relationship between %PPB and RTB, the number of non-terminal rotatable (single) bonds.

Many models estimate BBB permeability, as reviewed by Clark[140] and Luco and Marchevsky.[141] These studies highlight lipophilicity, as measured by the octanol–water or other partition coefficients (*e.g.*, oil–water,[142] air–water,[143]) or their differences (*e.g.*, $\Delta\log P$ between octanol–water and cyclohexane–water[144,145]) or by the distribution coefficient $\log D_{7.4}$,[146] as important in predicting this property. Using the Monte Carlo variable selection, only one descriptor, TPSA(NO), which is the topological polar surface area using the N,O polar contributions, was selected by Konovalov *et al.*[115] as significant for the prediction of the log BB from a set of more than 3000 descriptors calculated using PCLIENT.[147] The simplest, most efficient rules for BBB permeability were formulated by Norinder and Haeberlein, in a R05-like manner:[148]

- *Rule 1:* if the sum of nitrogen and oxygen atoms (N + O) in a molecule is less than or equal to five that molecule has a high chance of entering the brain.
- *Rule 2:* if Clog *P* > (N + O), then the log BB is positive, *i.e.*, the concentration of the molecule in the brain is higher than in the blood.

This simple two-rule approach can be used in particular when one is selecting scaffolds following primary screening. More exact methods should be used when deciding whether to test the compound's ability to pass the blood–brain barrier.

Based on 450 drugs with *qualitative* BBB data from WOMBAT-PK, we find that 83% of molecules with (N + O) < 5 and only 30% of molecules with (N + O) > 5 were annotated in the database as passing the blood–brain barrier. The percentage of drugs passing the barrier linearly decreased with sum of (N + O) atoms up to 6 atoms followed by drastic decrease in percentage of

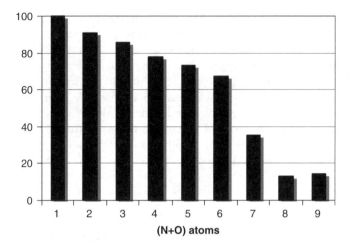

Figure 8.4   Percentage of molecules passing through the blood–brain barrier as a
function of the number of (N + O) atoms (using data from the WOMBAT-
PK database[131]).

molecules for (N + O) ≥ 7 atoms (Figure 8.4). Thus, it seems logical to increase
the limit of atoms in this rule by one. The modified Norinder–Haeberlein rule,
(N + O) < 7, correctly predicts 81% of all molecules as passing barrier. Given
our limited dataset, we conclude that molecules with (N + O) ≥ 7 have a one-in-
five probability (19%) to pass the BBB by passive phenomena.

8.6   The Usefulness of ADME/T Models is Limited by the Available Data

The amount of experimental data available to develop ADME/T models for
biological properties is rarely above 1000 molecules. Indeed, the number of
publicly available measurements is 1–2 orders of magnitude, compared to the
(much larger) number of measurements, 10^4–10^5, that are available for physicochemical ADME/T properties, *i.e.*, log P and solubility. All the above estimates of available experimental data are significantly smaller than any of the
numbers of molecules estimated by the Virtual/Tangible/Real (VTR) description framework for compounds;[149] indeed, the *Tangible* collection (small
molecules one could buy) is already above 30 million, and is already significantly
under-sampled with respect to accurate predictions (see also Figure 8.3). By the
same token, the *Virtual* collection is a number in excess of 10^{60}, and thus is
already prohibitively undersampled. Consequently, when selecting new molecules by means of virtual screening, and estimating their ADME/T properties,
we will always face the issue of extrapolation by extending our knowledge from
hundreds to millions of molecules. Perhaps global models, such as ChemGPS,[150]
which has been validated *in extenso* on various ADME/T properties[151] could

become more relevant if the applicability domain indicates these models are well validated. Indeed, simultaneous optimization of receptor binding affinity and ADME/T properties seems feasible[152] using the VolSurf paradigm.[153] VolSurf has been established as suitable for ADME/T property prediction, and appears to perform reasonably well on various complex (biological) ADME/T properties.

Considering that even simple physicochemical properties, such as log *P*, can reliably be predicted (at the level of experimental accuracy) for just a small fraction of the compounds found in commercial databases, our expectations for accuracy with respect to complex ADME/T property prediction is indeed less than optimistic. Assuming that the chemical diversity sampling per novel molecule is similar to that observed in estimating log *P*, we anticipate that only 0.020–0.2% of the compounds within large, diverse VTR sets will be predicted at the level of experimental accuracy. It is perhaps better to rely on relatively simple rule-based systems rooted in physicochemical property filters when scanning large chemical spaces for the ADME/T friendly regions, as discussed elsewhere.[154]

The process of drug discovery amounts to the search for optimality in a hyper-dimensional, multi-response surface area, and thus is a complex process. As discussed earlier, log *P* and the other R05 parameters, together with other properties such as PSA, and the number of flexible bonds, are significant contributors to a large number of models for different ADME/T properties. The open question is whether the use of such models provides an added value, compared to the simple and easily interpretable R05 criteria or the recent trends for the drug-likeness formulated by Gleeson.[155] But can these models provide added value?

Firstly, such models are developed with fewer training set molecules than are used to derive simple qualitative rules. Secondly, when predictions are insufficiently accurate even for simple properties such as log *P* and aqueous solubility for over 95% of the molecules, why should one attempt to apply more complex models that in turn rely explicitly, or perhaps implicitly, on these simple properties? Naturally, some of these predictions will prove accurate simply by chance, but most will fail. Thus, we suggest that given the present paucity of data there is no rational reason to trust that complex ADME/T models can lead to any improvement compared to simple qualitative rules. Indeed, those ADME/T models that lack a clear indication of their AD are a toy suitable for Horrobin's internally self-consistent universe, Castalia.[156] We posit that such ADME/T models are not a scientific tool for virtual screening, since they do not provide a metric for the accuracy of predictions. However, even using such models, human reasoning may prevail and decisions made by experienced medicinal chemists may lead to adequate progress. Last but not least, most ADME/T modeling tools continue to ignore the influence of ATP-binding cassette transporters (Section 8.5.1.3), as well as other carriers, and their impact on pharmacodynamics.

On the other hand, the adequate use of applicability domains will enable the user to identify those scaffolds for which reliable predictions have been made, which can further be progressed in lead optimization. Additional experiments could then be used to enhance or replace the existing ADME/T models, to increase their reliability in guiding the lead optimization process. This strategy known as "*in combo*" is gaining more attention in drug discovery.[75]

The development of models by data integration represents another potentially useful approach for ADME/T studies. Indeed, considering that some of the ADME/T properties are interrelated, and hence depend on the same physico-chemical properties, the simultaneous development of models for multiple properties can be advantageous for each of the analyzed properties – an aspect that is clearly used in the VolSurf paradigm.[8] The practical advantages of such approaches were also demonstrated for the prediction of different physico-chemical properties of alkanes by Baskin *et al.*,[157] log *P*/aqueous solubility prediction by Livingstone *et al.*[158] and for air–tissue prediction by Gaudin *et al.*[159]

8.7   Conclusions

The development of good models for ADME/T properties is rendered difficult by the paucity of reliable experimental data. Without proper consideration of the applicability domain, the developed models are of limited practical value for virtual screening, in particular for large and undersampled VTR spaces.[149] The success of such limited models critically depends on, and is limited by, the depth of expert knowledge of the human users, who may in fact call upon their past experience to decide whether the results of these models can be trusted. By reporting the accuracy of prediction for ADME/T models, by constantly refining the models with novel, complementary experimental data, as well by performing simultaneous data integration, better approaches with clearly defined applicability domains can be established. Furthermore, only by understanding and adequately modeling transporter-mediated phenomena are we going to witness dramatic improvements in the ADME/T property prediction sector. These advancements are likely to transform such methods from toys into scientific tools for drug discovery.

Acknowledgements

This study was partially supported (I.V.T.) by the Go-Bio BMBF grant 0313883 (AZ-31P4556), by the New Mexico Tobacco Settlement Fund (T.I.O.) and by data from Sunset Molecular Discovery LLC. This chapter is dedicated to Corwin Hansch, founder of the QSAR field, on his 90th birthday.

References

1. C. P. Adams and V. V. Brantner, *Health Aff.*, 2006, **25**, 420–428.
2. C. A. Lipinski, F. Lombardo, B. W. Dominy and P. J. Feeney, *Adv. Drug. Deliv. Rev.*, 1997, **23**, 3–25.
3. C. A. Lipinski, *J. Pharmacol. Toxicol. Methods*, 2000, **44**, 235–249.

4. S. J. Teague, A. M. Davis, P. D. Leeson and T. Oprea, *Angew. Chem. Int. Ed. Engl.*, 1999, **38**, 3743–3748.

5. M. M. Hann, A. R. Leach and G. Harper, *J. Chem. Inf. Comput. Sci.*, 2001, **41**, 856–864.

6. T. I. Oprea, A. M. Davis, S. J. Teague and P. D. Leeson, *J. Chem. Inf. Comput. Sci.*, 2001, **41**, 1308–1315.

7. T. I. Oprea, *Molecules*, 2002, **7**, 51–62.

8. T. I. Oprea and H. Matter, *Curr. Opin. Chem. Biol.*, 2004, **8**, 349–358.

9. G. Wess, *Drug Discov. Today*, 2002, **7**, 533–535.

10. H. van de Waterbeemd and E. Gifford, *Nat. Rev. Drug Discov.*, 2003, **2**, 192–204.

11. C. A. Lipinski, F. Lombardo, B. W. Dominy and P. J. Feeney, *Adv. Drug. Deliv. Rev.*, 2001, **46**, 3–26.

12. C. Hansch, P. P. Maloney, T. Fujita and R. M. Muir, *Nature*, 1962, **194**, 178–180.

13. F. Lombardo, B. Faller, M. Shalaeva, I. Tetko and S. Tilton, in *Drug Properties: Measurement and Computation*, R. Mannhold ed., Wiely-VCH, Weinheim, 2007, pp. 407–437.

14. Y. C. Martin, *QSAR Comb. Sci.*, 2006, **25**, 1192–1200.

15. Y. C. Martin, *J. Med. Chem.*, 2005, **48**, 3164–3170.

16. R. Lewis, P. Ertl, E. Jacoby, M. Tintelnot-Blomley, P. Gedeck, R. M. Wolf and M. C. Peitsch, *Chimia*, 2005, **59**, 545–549.

17. M. Lobell, M. Hendrix, B. Hinzen, J. Keldenich, H. Meier, C. Schmeck, R. Schohe-Loop, T. Wunberg and A. Hillisch, *ChemMedChem*, 2006, **1**, 1229–1236.

18. M. D. Segall, A. P. Beresford, J. M. Gola, D. Hawksley and M. H. Tarbit, *Expert Opin. Drug Metab. Toxicol.*, 2006, **2**, 325–337.

19. H. Van de Waterbeemd and B. Testa, in *Comprehensive Medicinal Chemistry II: In Silico Tools in ADMET*, B. Testa and H. van de Waterbeemd eds., Elsevier, Oxford, edn., 2007, vol. 5, p. 1152.

20. C. Hansch, A. Leo and D. Hoekman, *Hydrophobic, Electronic, and Steric Constants*, American Chemical Society, Washington, DC, 1995.

21. *The Physical Properties Database (PHYSPROP), Syracuse Research Corporation*, http://www.syrres.com/, Accessed April 1, 2008.

22. I. V. Tetko, in *Computational Toxicology: Risk Assessment for Pharmaceutical and Environmental Chemicals*, ed. S. Ekins, John Wiley & Sons, Inc, New Jersey, edn., 2007, vol. 1, pp. 241–275.

23. R. Mannhold, C. Ostermann, G. I. Poda and I. V. Tetko, *J. Pharm. Sci.*, 2008, in press.

24. A. H. Goller, M. Hennemann, J. Keldenich and T. Clark, *J. Chem. Inf. Model.*, 2006, **46**, 648–658.

25. R. P. Sheridan, B. P. Feuston, V. N. Maiorov and S. K. Kearsley, *J. Chem. Inf. Comput. Sci.*, 2004, **44**, 1912–1928.

26. R. Mannhold and A. Petrauskas, *QSAR Comb. Sci.*, 2003, **22**, 466–475.

27. A. J. Leo and D. Hoekman, *Persp. Drug Discov. Des.*, 2000, **18**, 19–38.

28. A. J. Leo, *Chem. Rev.*, 1993, **93**, 1281–1306.

29. A. A. Petrauskas and E. A. Kolovanov, *Persp. Drug Discov. Des.*, 2000, **19**, 99–116.
30. R. Mannhold and R. F. Rekker, *Persp. Drug Discov. Des.*, 2000, **18**, 1–18.
31. R. Mannhold, R. F. Rekker, K. Dross, G. Bijloo and G. de Vries, *Quant. Struct.-Activ. Rel.*, 1998, **17**, 517–536.
32. G. Klopman, J.-Y. Li, S. Wang and M. Dimayuga, *J. Chem. Inf. Comput. Sci.*, 1994, **34**, 752–781.
33. V. N. Viswanadhan, A. K. Ghose and J. J. Wendoloski, *Persp. Drug Discov. Des.*, 2000, **19**, 85–98.
34. P. Japertas, R. Didziapetris and A. Petrauskas, *Quant. Struct.-Activ. Rel.*, 2002, **21**, 23–37.
35. P. Japertas, R. Didziapetris and A. Petrauskas, *Mini Rev. Med. Chem.*, 2003, **3**, 797–808.
36. P.-A. Carrupt, P. Gaillard, F. Billois, P. Weber, B. Testa, C. Meyer and S. Perez, in *Lipophilicity in Drug Action and Toxicology*, eds. V. Pliska, B. Testa and H. van de Waterbeemd, VCH Publishers, Weinheim, edn., 1996, pp. 195–217.
37. E. M. Duffy and W. L. Jorgensen, *J. Am. Chem. Soc.*, 2000, **122**, 2878–2888.
38. A. Klamt, F. Eckert, M. Hornig, M. E. Beck and T. Burger, *J. Comput. Chem.*, 2002, **23**, 275–281.
39. M. Hornig and A. Klamt, *J. Chem. Inf. Model.*, 2005, **45**, 1169–1177.
40. J. Devillers, D. Domine, C. Guillon, S. Bintein and W. Karcher, *SAR QSAR Environ. Res.*, 1997, **7**, 151–172.
41. W. M. Meylan and P. H. Howard, *Persp. Drug Discov. Design*, 2000, **19**, 67–84.
42. W. M. Meylan and P. H. Howard, *J. Pharm. Sci.*, 1995, **84**, 83–92.
43. V. K. Gombar and K. Enslein, *J. Chem. Inf. Comput. Sci.*, 1996, **36**, 1127–1134.
44. V. K. Gombar, *SAR QSAR Environ. Res.*, 1999, **10**, 371–380.
45. R. X. Wang, Y. Gao and L. H. Lai, *Persp. Drug Discov. Des.*, 2000, **19**, 47–66.
46. I. V. Tetko and V. Y. Tanchuk, *J. Chem. Inf. Comput. Sci.*, 2002, **42**, 1136–1145.
47. I. V. Tetko, V. Y. Tanchuk and A. E. Villa, *J. Chem. Inf. Comput. Sci.*, 2001, **41**, 1407–1421.
48. I. V. Tetko and D. J. Livingstone, in *Comprehensive Medicinal Chemistry II: In Silico Tools in ADMET*, eds. B. Testa and H. van de Waterbeemd, Elsevier, edn., 2006, vol. 5, pp. 649–668.
49. R. Mannhold and H. van de Waterbeemd, *J. Comput.-Aided Mol. Des.*, 2001, **15**, 337–354.
50. S. Datta and D. J. Grant, *Nat. Rev. Drug Discov.*, 2004, **3**, 42–57.
51. A. Gavezzotti, *Acc. Chem. Res.*, 1994, **27**, 309–314.
52. S. H. Yalkowsky and Y. He, *Handbook of Aqueous Solubility Data*, CRC Press, Boca Raton, 2003.
53. R. Kuhne, R.-U. Ebert, F. Kleint, G. Scmidt and G. Schuurmann, *Chemosphere*, 1995, **30**, 2061–2077.

54. J. Huuskonen, *J. Chem. Inf. Comput. Sci.*, 2000, **40**, 773–777.
55. K. V. Balakin, N. P. Savchuk and I. V. Tetko, *Curr. Med. Chem.*, 2006, **13**, 223–241.
56. S. H. Yalkowsky and S. C. Valvani, *J. Pharm. Sci.*, 1980, **69**, 912–922.
57. S. H. Yalkowsky, S. C. Valvani and T. J. Roseman, *J. Pharm. Sci.*, 1983, **72**, 866–870.
58. N. Jain and S. H. Yalkowsky, *J. Pharm. Sci.*, 2001, **90**, 234–252.
59. Y. Ran, N. Jain and S. H. Yalkowsky, *J. Chem. Inf. Comput. Sci.*, 2001, **41**, 1208–1217.
60. Y. Ran and S. H. Yalkowsky, *J. Chem. Inf. Comput. Sci.*, 2001, **41**, 354–357.
61. D. L. Peterson and S. H. Yalkowsky, *J. Chem. Inf. Comput. Sci.*, 2001, **41**, 1531–1534.
62. W. L. Jorgensen and E. M. Duffy, *Bioorg. Med. Chem. Lett.*, 2000, **10**, 1155–1158.
63. G. Klopman and H. Zhu, *J. Chem. Inf. Comput. Sci.*, 2001, **41**, 439–445.
64. W. M. Meylan, P. H. Howard and R. S. Boethling, *Environ. Sci. Technol.*, 1996, **15**, 100–106.
65. J. S. Delaney, *Drug Discov. Today*, 2005, **10**, 289–295.
66. R. Liu and S. S. So, *J. Chem. Inf. Comput. Sci.*, 2001, **41**, 1633–1639.
67. I. V. Tetko and G. I. Poda, in *Drug Properties: Measurement and Computation*, R. Mannhold ed., Wiley-VCH, Weinheim, 2007, pp. 381–406.
68. I. V. Tetko and G. I. Poda, *J. Med. Chem.*, 2004, **47**, 5601–5604.
69. I. V. Tetko and P. Bruneau, *J. Pharm. Sci.*, 2004, **93**, 3103–3110.
70. M. J. Walker, *QSAR Comb. Sci.*, 2004, **23**, 515–520.
71. J. J. Morris and P. Bruneau, in *Virtual Screening for Bioactive Molecules*, H. G. Bohm and G. Schneider eds., Wiley-VCH, Weinheim, 2000, vol. **10**, pp. 33–58.
72. G. I. Poda, I. V. Tetko and D. C. Rohrer, in *229th American Chemical Society National Meeting & Exposition*, ACS, San Diego, CA, edn., 2005, p. MEDI 514.
73. T. R. Stouch, J. R. Kenyon, S. R. Johnson, X. Q. Chen, A. Doweyko and Y. Li, *J. Comput. Aided. Mol. Des.*, 2003, **17**, 83–92.
74. T. I. Netzeva, A. Worth, T. Aldenberg, R. Benigni, M. T. Cronin, P. Gramatica, J. S Jaworska, S. Kahn, G. Klopman, C. A. Marchant, G. Myatt, N. Nikolova-Jeliazkova, G. Y. Patlewicz, R. Perkins, D. Roberts, T. Schultz, D. W. Stanton, J. J. van de Sandt, W. Tong, G. Veith and C. Yang, *Altern. Lab. Anim.*, 2005, **33**, 155–173.
75. I. V. Tetko, P. Bruneau, H. W. Mewes, D. C. Rohrer and G. I. Poda, *Drug Discov. Today*, 2006, **11**, 700–707.
76. J. Jaworska, N. Nikolova-Jeliazkova and T. Aldenberg, *Altern. Lab. Anim.*, 2005, **33**, 445–459.
77. M. Shen, Y. Xiao, A. Golbraikh, V. K. Gombar and A. Tropsha, *J. Med. Chem.*, 2003, **46**, 3013–3020.
78. E. Papa, F. Villa and P. Gramatica, *J. Chem. Inf. Model.*, 2005, **45**, 1256–1266.
79. P. Gramatica, P. Pilutti and E. Papa, *J. Chem. Inf. Comput. Sci.*, 2004, **44**, 1794–1802.

80. A. Tropsha, P. Gramatica and V. K. Gombar, *QSAR Comb. Sci.*, 2003, **22**, 69–77.
81. L. Eriksson, J. Jaworska, A. P. Worth, M. T. Cronin, R. M. McDowell and P. Gramatica, *Environ. Health Perspect.*, 2003, **111**, 1361–1375.
82. S. C. Basak and G. D. Grunwald, *Chemosphere*, 1995, **31**, 2529–2546.
83. W. Tong, H. Hong, H. Fang, Q. Xie and R. Perkins, *J. Chem. Inf. Comput. Sci.*, 2003, **43**, 525–531.
84. W. Tong, Q. Xie, H. Hong, L. Shi, H. Fang and R. Perkins, *Environ. Health Perspect.*, 2004, **112**, 1249–1254.
85. D. T. Manallack, B. G. Tehan, E. Gancia, B. D. Hudson, M. G. Ford, D. J. Livingstone, D. C. Whitley and W. R. Pitt, *J. Chem. Inf. Comput. Sci.*, 2003, **43**, 674–679.
86. A. J. Chalk, B. Beck and T. Clark, *J. Chem. Inf. Comput. Sci.*, 2001, **41**, 457–462.
87. A. Breindl, B. Beck, T. Clark and R. C. Glen, *J. Mol. Model.*, 1997, **3**, 142–155.
88. P. Bruneau and N. R. McElroy, *J. Chem. Inf. Model.*, 2006, **46**, 1379–1387.
89. A. Schwaighofer, T. Schroeter, S. Mika, J. Laub, A. ter Laak, D. Sulzle, U. Ganzer, N. Heinrich and K. R. Muller, *J. Chem. Inf. Model.*, 2007, **47**, 407–424.
90. T. S. Schroeter, A. Schwaighofer, S. Mika, A. Ter Laak, D. Suelzle, U. Ganzer, N. Heinrich and K. R. Muller, *J. Comput.-Aided Mol. Des.*, 2007, **21**, 651–664.
91. I. V. Tetko, in *229th American Chemical Society National Meeting & Exposition*, San Diego, CA, edn., 2005, vol. 229, pp. U602-U602.
92. I. V. Tetko and V. Y. Tanchuk, in *229th American Chemical Society National Meeting & Exposition*, San Diego,edn., CA, 2005, vol. 229, pp. U608-U608.
93. I. V. Tetko, P. Bruneau, H. W. Mewes, D. C. Rohrer and G. I. Poda, in *232th ACS National Meeting*, San Francisco, CA, edn., 2006.
94. I. V. Tetko, I. Jaroszewicz, J. A. Platts and J. Kuduk-Jaworska, *J. Inorg. Biochem.*, 2008.
95. I. V. Tetko, V. Y. Tanchuk, T. N. Kasheva and A. E. Villa, *J. Chem. Inf. Comput. Sci.*, 2001, **41**, 1488–1493.
96. I. V. Tetko, I. Sushko, A. K. Pandey, A. Tropsha, H. Zhu, E. Papa, T. Öberg, R. Todeschini, D. Fourches and A. Varnek, *J. Chem. Inf. Model.*, 2008, submitted.
97. U. Fagerholm, *J. Pharm. Pharmacol.*, 2007, **59**, 905–916.
98. U. Fagerholm, *J. Pharm. Pharmacol.*, 2007, **59**, 1463–1471.
99. P. Artursson and J. Karlsson, *Biochem. Biophys. Res. Commun.*, 1991, **175**, 880–885.
100. P. Artursson, K. Palm and K. Luthman, *Adv. Drug. Deliv. Rev.*, 2001, **46**, 27–43.
101. Y. H. Zhao, J. Le, M. H. Abraham, A. Hersey, P. J. Eddershaw, C. N. Luscombe, D. Butina, G. Beck, B. Sherborne, I. Cooper and J. A. Platts, *J. Pharm. Sci.*, 2001, **90**, 749–784.

102. G. Klopman, L. R. Stefan and R. D. Saiakhov, *Eur. J. Pharm. Sci.*, 2002, **17**, 253–263.
103. T. Hou, J. Wang, W. Zhang and X. Xu, *J. Chem. Inf. Model.*, 2007, **47**, 208–218.
104. J. P. Bai, A. Utis, G. Crippen, H. D. He, V. Fischer, R. Tullman, H. Q. Yin, C. P. Hsu, L. Jiang and K. K. Hwang, *J. Chem. Inf. Comput. Sci.*, 2004, **44**, 2061–2069.
105. H. Lennernäs, *Xenobiotica*, 2007, **37**, 1015–1051.
106. H. Lennernäs, *J. Pharm. Sci.*, 1998, **87**, 403–410.
107. M. Yazdanian, S. L. Glynn, J. L. Wright and A. Hawi, *Pharm. Res.*, 1998, **15**, 1490–1494.
108. F. Yamashita, S. Wanchana and M. Hashida, *J. Pharm. Sci.*, 2002, **91**, 2230–2239.
109. S. Fujiwara, F. Yamashita and M. Hashida, *Int. J. Pharm.*, 2002, **237**, 95–105.
110. T. J. Hou, W. Zhang, K. Xia, X. B. Qiao and X. J. Xu, *J. Chem. Inf. Comput. Sci.*, 2004, **44**, 1585–1600.
111. J. D. Irvine, L. Takahashi, K. Lockhart, J. Cheong, J. W. Tolan, H. E. Selick and J. R. Grove, *J. Pharm. Sci.*, 1999, **88**, 28–33.
112. H. van de Waterbeemd, G. Camenisch, G. Folkers and O. A. Raevsky, *Quant. Struct.-Activ. Rel.*, 1996, **15**, 480–490.
113. J. R. Votano, M. Parham, L. H. Hall and L. B. Kier, *Mol Divers*, 2004, **8**, 379–391.
114. J. K. Wegner, H. Frohlich and A. Zell, *J. Chem. Inf. Comput. Sci.*, 2004, **44**, 931–939.
115. D. A. Konovalov, N. Sim, E. Deconinck, Y. V. Heyden and D. Coomans, *J. Chem. Inf. Model.*, 2008.
116. C. A. Bergstrom, M. Strafford, L. Lazorova, A. Avdeef, K. Luthman and P. Artursson, *J. Med. Chem.*, 2003, **46**, 558–570.
117. P. Stenberg, U. Norinder, K. Luthman and P. Artursson, *J. Med. Chem.*, 2001, **44**, 1927–1937.
118. T. I. Oprea and J. Gottfries, *J. Mol. Graph. Model.*, 1999, 17, 261–274, 329.
119. J. E. Penzotti, M. L. Lamb, E. Evensen and P. D. Grootenhuis, *J. Med. Chem.*, 2002, **45**, 1737–1740.
120. J. Huang, G. Ma, I. Muhammad and Y. Cheng, *J. Chem. Inf. Model.*, 2007, **47**, 1638–1647.
121. P. deCerqueiraLima, A. Golbraikh, S. Oloff, Y. Xiao and A. Tropsha, *J. Chem. Inf. Model.*, 2006, **46**, 1245–1254.
122. B. Zdrazil, D. Kaiser, S. Kopp, P. Chiba and G. F. Ecker, *QSAR Comb. Sci.*, 2007, **26**, 669–678.
123. A. Seelig, *Eur. J. Biochem.*, 1998, **251**, 252–261.
124. A. Sakurai, Y. Onishi, H. Hirano, M. Seigneuret, K. Obanayama, G. Kim, E. L. Liew, T. Sakaeda, K. Yoshiura, N. Niikawa, M. Sakurai and T. Ishikawa, *Biochemistry*, 2007, **46**, 7678–7693.
125. G. D. Leonard, T. Fojo and S. E. Bates, *Oncologist*, 2003, **8**, 411–424.

126. M. Wada, *Cancer. Lett.*, 2006, **234**, 40–50.
127. I. Ivnitski-Steele, R. S. Larson, D. M. Lovato, H. M. Khawaja, S. S. Winter, T. I. Oprea, L. A. Sklar and B. S. Edwards, *Assay. Drug Dev. Technol.*, 2008, **6**, 263–276.
128. S. S. Winter, D. M. Lovato, H. M. Khawaja, B. S. Edwards, I. D. Steele, S. M. Young, T. I. Oprea, L. A. Sklar and R. S. Larson, *J. Biomol. Screen.*, 2008, **13**, 185–193.
129. S. Curry, P. Brick and N. P. Franks, *Biochim. Biophys. Acta*, 1999, **1441**, 131–140.
130. M. H. Abraham, A. Ibrahim, Y. Zhao and W. E. Acree Jr., *J. Pharm. Sci.*, 2006, **95**, 2091–2100.
131. M. Olah, R. Rad, L. Ostopovici, A. Bora, N. Hadaruga, D. Hadaruga, R. Moldovan, A. Fulias, M. Mracec and O. TI, in *Chemical Biology: From Small Molecules to Systems Biology and Drug Design*, S. L. Schreiber, T. M. Kapoor and G. Wess eds., Wiley-VCH, Weinheim, 2007, pp. 760–786.
132. G. Colmenarejo, *Med. Res. Rev.*, 2003, **23**, 275–301.
133. G. Colmenarejo, A. Alvarez-Pedraglio and J. L. Lavandera, *J. Med. Chem.*, 2001, **44**, 4370–4378.
134. N. A. Kratochwil, W. Huber, F. Muller, M. Kansy and P. R. Gerber, *Biochem. Pharmacol.*, 2002, **64**, 1355–1374.
135. R. D. Saiakhov, L. R. Stefan and G. Klopman, *Perspect. Drug Discov. Des.*, 2000, **19**, 133–155.
136. K. Yamazaki and M. Kanaoka, *J. Pharm. Sci.*, 2004, **93**, 1480–1494.
137. F. Lombardo, R. S. Obach, M. Y. Shalaeva and F. Gao, *J. Med. Chem.*, 2004, **47**, 1242–1250.
138. F. Lombardo, R. S. Obach, F. M. Dicapua, G. A. Bakken, J. Lu, D. M. Potter, F. Gao, M. D. Miller and Y. Zhang, *J. Med. Chem.*, 2006, **49**, 2262–2267.
139. K. Valko, S. Nunhuck, C. Bevan, M. H. Abraham and D. P. Reynolds, *J. Pharm. Sci.*, 2003, **92**, 2236–2248.
140. D. E. Clark, *Drug Discov. Today*, 2003, **8**, 927–933.
141. J. M. Luco and E. Marchevsky, *Curr. Comput.-Aided Drug Des.*, 2006, **2**, 31–55.
142. E. J. Lien, *Annu. Rev. Pharmacol. Toxicol.*, 1981, **21**, 31–61.
143. H. Fischer, R. Gottschlich and A. Seelig, *J. Membr. Biol.*, 1998, **165**, 201–211.
144. P. Seiler, *Eur. J. Med. Chem.*, 1974, **9**, 473–479.
145. R. C. Young, R. C. Mitchell, T. H. Brown, C. R. Ganellin, R. Griffiths, M. Jones, K. K. Rana, D. Saunders, I. R. Smith, N. E. Sore and T. J. Wilks, *J. Med. Chem.*, 1988, **31**, 656–671.
146. X. Liu, M. Tu, R. S. Kelly, C. Chen and B. J. Smith, *Drug Metab. Dispos.*, 2004, **32**, 132–139.
147. I. V. Tetko, J. Gasteiger, R. Todeschini, A. Mauri, D. Livingstone, P. Ertl, V. A. Palyulin, E. V. Radchenko, N. S. Zefirov, A. S. Makarenko, V. Y. Tanchuk and V. V. Prokopenko, *J. Comput.-Aided Mol. Des.*, 2005, **19**, 453–463.

148. U. Norinder and M. Haeberlein, *Adv. Drug. Deliv. Rev.*, 2002, **54**, 291–313.
149. M. M. Hann and T. I. Oprea, *Curr. Opin. Chem. Biol.*, 2004, **8**, 255–263.
150. T. I. Oprea and J. Gottfries, *J. Comb. Chem.*, 2001, **3**, 157–166.
151. T. I. Oprea, I. Zamora and A. L. Ungell, *J. Comb. Chem.*, 2002, **4**, 258–266.
152. I. Zamora, T. Oprea, G. Cruciani, M. Pastor and A. L. Ungell, *J. Med. Chem.*, 2003, **46**, 25–33.
153. C. Cruciani, P. Crivori, P. A. Carrupt and B. Testa, *J. Molec. Struct.-Theochem*, 2000, **503**, 17–30.
154. T. I. Oprea, P. Benedetti, G. Berellini, M. Olah, K. Fejgin and S. Boyer, in *Molecular Interaction Fields*, ed. G. Cruciani, Wiley-VCH, New York, 2007, pp. 249–272.
155. M. P. Gleeson, *J. Med. Chem.*, 2008, **51**, 817–834.
156. D. F. Horrobin, *Nat. Rev. Drug. Discov.*, 2003, **2**, 151–154.
157. I. I. Baskin, V. A. Palyulin and N. S. Zefirov, *Doklady Akademii Nauk*, 1993, **332**, 713–716.
158. D. J. Livingstone, M. G. Ford, J. J. Huuskonen and D. W. Salt, *J. Comput.-Aided Mol. Des.*, 2001, **15**, 741–752.
159. C. Gaudin, G. Marcou, A. Varnek, I. I. Baskin, A. K. Pandey and I. V. Tetko, *J. Chem. Inf. Model.*, 2008, submitted.

CHAPTER 9

Compound Library Design – Principles and Applications

WEIFAN ZHENG[a] AND STEPHEN R. JOHNSON[b]

[a] Department of Pharmaceutical Sciences, BRITE, North Carolina Central University, 1801 Fayetteville Street, Durham, NC 27707, USA; [b] Computer-Assisted Drug Design, Bristol-Myers Squibb Research and Development, NJ, USA

9.1 Introduction

Despite the development of HTS (High-throughput Screening) capabilities and readily available chemical libraries, making all possible compounds and screening them all in biological assays are still very costly and difficult. Thus, it is of critical importance to be able to rationally select chemicals for specific biological targets to increase the efficiency of drug discovery and chemical genomics research. This process of rationally selecting compounds for chemical synthesis or biological screening is called *compound library design*.

Many factors need to be considered in designing compound libraries. These include factors related to the biological activity against a target or a family of targets and the factors related to drug developability, drug-likeness and other ADMET (Absorption, Distribution, Metabolism, Elimination and Toxicity) properties. In this chapter we discuss various computational approaches to compound library design, from similarity-guided methods, to diversity-based compound selection and compound collection comparison, to pharmacophore and QSAR model guided methods as well as protein structure-based methods. We also review how ADMET alerts, models and filters are being used in compound library design. Finally, we discuss the importance and broad applicability of combinatorial optimization methods in next-generation multi-objective and multi-target library design.

Chemoinformatics Approaches to Virtual Screening
Edited by Alexandre Varnek and Alex Tropsha
© Royal Society of Chemistry, 2008
Published by the Royal Society of Chemistry, www.rsc.org

9.1.1 Compound Library Design

Drug discovery and chemical genomics research involve making compounds and testing them in specific biological assays. Over the past 20 years, the cycle time for making and testing compounds has been dramatically reduced owing to the development of High-Throughput Screening (HTS) and combinatorial/parallel synthesis technologies. It is now common practice for many research laboratories to test compounds in 384-well and higher density plate format under robotic control. Chemicals from natural sources, combinatorial/parallel synthesis as well as historic synthetic efforts are becoming readily available due to the rise of chemical vendors and Contract Research Organizations (CRO), such as Asinex, ChemDiv and WuXi Pharma. Combinatorial synthetic technologies are now powerful tools in the daily life of the medicinal chemist. Despite these HTS capabilities and readily available chemical libraries, it is still very costly and ineffective to make all possible compounds and screen them all against biological assays. Consequently, to increase the efficiency of drug discovery and chemical genomics research, it is of critical importance to be able to rationally select chemicals for specific biological targets.

This chapter defines the process of rationally selecting compounds for chemical synthesis or biological screening as *compound library design*. The whole purpose of compound library design is to obtain a maximum amount of information while making and testing a minimum number of compounds. For designing and making new chemical libraries, this means the rational selection of a subset of building blocks (or reagents) from available reagent pools to obtain the best chemical libraries for a particular design goal. For compound acquisition or screening set design, this means the rational selection of a subset of compounds from commercially available or proprietary databases under some constraints or criteria. Thus, we broadly use the term compound library design to mean compound collection design or screening set design for a particular project, or the design of new chemical libraries for synthesis (Figure 9.1).

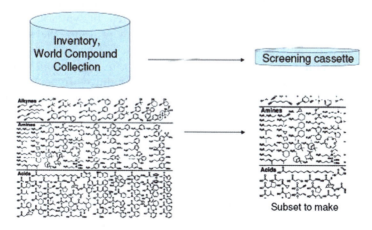

Figure 9.1 Compound library design.

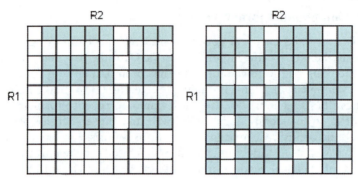

Figure 9.2   Combinatorial and cherry-picking libraries.

When designing new chemical libraries that will be made using combinatorial or parallel synthetic methods, we note two possible design scenarios. The first is to make the new library in a combinatorial fashion, where each selected building block will react with all the building blocks selected for the other substituent positions (Figure 9.2). This method has the highest chemical efficiency in that the least number of reagents are needed to make a certain desired number of compounds in the library. Logistically, this is also a better way to conduct combinatorial synthesis. The second scenario is to selectively pick library members to make. This approach is often called cherry-picking method (Figure 9.2), where no combinatorial constraints are placed on the selection of building blocks for making the desired molecules. Thus, depending on how the chemical synthesis will be conducted, in combinatorial fashion or cherry-picking fashion, different design approaches should be adopted. To design libraries in a combinatorial fashion, Monomer Frequency Analysis (MFA)[1,2] or combinatorial optimization methods[3–6] should be used; to design cherry-picking libraries, one should just score and rank the compounds in a virtual library based on certain criteria (*e.g.*, highest predicted activity, or optimal drug-likeness) so that the top ranking compounds are selected for synthesis and testing.

In certain compound acquisition or screening projects, the selection of compounds may only be conducted at the plate level, where a whole plate of compounds is either selected or not selected. This is due to reasons of logistical considerations and concerns of compound stability during freeze–thaw cycles. Thus, compound library design or screening set design should take this constraint into consideration.[7]

To best tackle compound library design problems, one needs to consider the screening platform adopted for a specific project. For example, in a screening project where one specific biological target is used, large numbers of compounds can often be screened in HTS format. In a gene family focused screening (*i.e.*, chemogenomics) project, several members of a particular gene family are often the targets and a Medium-Throughput Screening (MTS) format is often adopted. This often involves testing tens of thousands of compounds against a panel of several tens of biological targets. In this case, any compound that is appropriate for any member of the gene family should be considered. In a chemical genomics project, the whole genome can be the target, where hypothesis-free, high content, phenotypic assays are often adopted.[8–10] In this case, tens of thousands of compounds are tested against the cellular proteome, leading to observations of certain phenotype changes. Since all of these strategies co-exist in pharmaceutical R&D environment, strategies for compound library design should be adopted appropriately according to a particular screening project. However, certain fundamental design factors and their computational models are applicable to all screening platforms.

Factors that need to be considered in designing compound libraries can be broadly classified into two groups: (i) factors related to the biological activity against a target or a family of targets and (ii) factors related to drug developability, drug-likeness and other ADMET (absorption, distribution, metabolism, elimination and toxicity) properties. In the former case, any ligand-based or protein structure-based computational design methods can be employed when appropriate to help guide a library towards the target or family of targets under study. Methods such as similarity to lead molecules, molecular diversity, pharmacophore models derived from a set of active analogs or protein binding site structure, machine learning models and structure-based docking can all be employed. In the latter case, Lipinski's rule of five[11] has become an important factor, as are various drug-likeness, developability properties such as solubility, membrane permeability (Caco-2 or artificial members surrogate), cytochrome P450 inhibition and P-PGP substrates.[12] Also included in the latter case are substructure filters, such as toxicity-causing substructures,[13] common chemical frameworks[14,15] and certain promiscuous substructures.[16]

In this chapter we discuss various computational approaches to compound library design, which include similarity-guided targeted library design, diversity-based compound selection, pharmacophore-guided and statistical model guided methods, as well as protein structure-based methods. We also review how ADMET alerts, models and filters are being used in the design strategy. In each section we first present the fundamental concepts and methods, and then discuss some literature examples.

9.2 Methods for Compound Library Design

9.2.1 Design for Specific Biological Activities

In a targeted screening project, compound library design involves the selection of a subset of compounds from an available pool of chemicals. This subset of selected compounds affords a limited compound library with a high percentage of compounds that are likely to be active against a target or a family of targets. The prediction of biological activity can be based on different types of computational models. As mentioned earlier, similarity to a lead molecule, pharmacophore models, QSAR models or structure-based docking models can all be employed to help drive the library towards desired biological activities. When designing new chemical libraries for synthesis, one needs to select building blocks (or reagents) such that the resulting library can have a high percentage of compounds that are predicted to be active by the computational model.

9.2.1.1   Similarity Guided Design of Targeted Libraries

If we know the chemical structure of a biologically active compound, and want to design a set of compounds that may be active against the same target, we may be interested in selecting or designing compounds that are chemically similar to the known active molecule. This is based on the similarity principle that implies that similar compounds are likely to have similar biological activity. Operationally, this involves several steps: (i) electronically represent the structure of the lead molecule as a connection table; (ii) calculate molecular descriptors for the lead molecular structure; (iii) calculate the same descriptors for every molecule in a database or a virtual compound library; (iv) calculate the similarity values between each member of the database or virtual library and the lead molecule; (v) sort the library members according to their similarity to the lead molecule; (vi) compounds ranked at the top should be considered as having a higher priority; (vii) if a new chemical library is being designed for synthesis in combinatorial fashion, a frequency analysis of the building blocks (reagents) that occur in the top ranking compounds should be conducted. More frequently occurring reagents should be proposed as reagents of choice for the synthesis of the combinatorial library; if a cherry-picking library is being designed, the top-ranking compounds should simply be selected for synthesis.

Several papers have been published on this topic. Brown *et al.* reported a method for designing combinatorial library mixtures using a genetic algorithm.[17] Sheridan *et al.* also reported a method for designing targeted libraries with genetic algorithms.[2] Agrafiotis has published a method on combinatorial library design.[3,18] Zheng and Tropsha also discussed a method (Focus2D) and its application in the design of a focused library.[1] Here, we use Focus2D as an example to illustrate the similarity guided method for the design of a new combinatorial chemical library.

Focus-2D employs several different strategies for the effective sampling of virtual chemical libraries to prioritize compounds. The reported implementation uses a modified Euclidean distance as the measure of chemical similarity to a lead molecule and a Simulated Annealing (SA) algorithm[19] for sampling the combinatorial chemical space. The virtual library is generated by random combinations of available building blocks based on the underlying chemical reaction; and the resulting virtual compounds are represented by Kier–Hall topological descriptors.[20,21] Although the authors used topological descriptors, other descriptors can certainly be implemented as well. Focus-2D then samples the structural space of virtual libraries using Simulated Annealing (SA) and attempts to maximize similarity between virtual molecules and the lead molecule. Finally, frequency distribution analysis of building blocks found in the molecules with the highest similarity to a lead molecule is conducted, and the building blocks found more frequently than random expectation are suggested as the reagents for combinatorial synthesis.

This method was applied to the design of a tripeptoid library. Experimental work on that library had been described by Zuckermann *et al.*,[22] who had shown that a few members of the library had high affinities for α1-adrenergic or μ-opiate receptors. The results of that published work were used as a retrospective test case to evaluate the effectiveness of Focus-2D.

The results indicate that Focus-2D proposed 12 building blocks on a rational basis (when both lead compounds were used as probes), which included all five building blocks found in the three reported active opioid peptoids.[22] Simple evaluation shows that if all combinations of building blocks (24 of them) described in the original publication were explored in a true sense of combinatorial chemical synthesis as many as $24^3 - 13\,824$ compounds would have to be made and tested. However, if the experiments were limited to using only 12 building blocks, as suggested by Focus-2D, only $12^3 = 1728$ compounds would have to be synthesized. These results show that the same active compounds would be a part of this smaller library. Thus, if the suggestions made by Focus-2D were accepted prior to the synthesis and testing, the total number of compounds that need to be screened would be reduced by almost an order of magnitude. This simple example demonstrates the potential effectiveness of similarity guided design of combinatorial libraries.

Similarity methods and their applications in virtual screening and compound collection design (*i.e.*, cherry-picking) have been widely reviewed and interested readers are referred to these publications [23,24]

9.2.1.2 Diversity-based Design of General Screening Libraries

In a general screening project, compound library design or virtual screening involves the selection of a subset of compounds that are optimally diverse and representative of available classes of compounds, leading to a non-redundant chemical library or a set of non-redundant compounds for biological testing. Reported methods can be generally classified into several categories: (i) cluster sampling methods, which first identify a set of compound clusters, followed then by the selection of several compounds from each cluster;[25] (ii) cell-based sampling, which places all the compounds into a low-dimensional descriptor space divided into many cells, and then chooses a few compounds from each cell;[26] (iii) direct sampling methods, which try to obtain a subset of optimally diverse compounds from an available pool by directly analyzing the diversity of the selected molecules.[3,27]

Many reports have been published addressing various aspects of diversity analysis in the context of chemical library design and database mining. Methods have also been published to map molecules from a high-dimensional chemical space to 2D/3D space so that direct visualization of chemical similarity and diversity becomes feasible.[28,29] Low-dimensional diversity space also has the intrinsic advantage over high-dimensional space in that the low-dimensional space can be used to compare different compound collections to illustrate compound collection overlaps.

Mason *et al.* have described an interesting method for library design[26] using BCUT chemistry-space descriptors and multiple four-point pharmacophore fingerprints. The results demonstrated the feasibility of a simulated annealing process for combinatorial reagent selection that concurrently optimizes product diversity in BCUT chemistry-space and in terms of unique four-point pharmacophores. Flower described a widely cited method for chemical diversity analysis called DISSIM.[30] It addresses the problem of selecting diverse subsets from larger collections of chemical compounds, combining a maximum dissimilarity search algorithm and a general measure of chemical similarity based on the combination of different molecular descriptors. Zheng *et al.* reported a method called SAGE that optimizes the selection of compounds based on a designed diversity function[27] that adequately measures the diversity of a subset of selected molecules.

Since each molecule is represented by a vector of molecular descriptors, geometrically it is mapped to a point in a multidimensional space. The distance between two points, such as Euclidean distance, then measures the dissimilarity between the two molecules. Thus, the diversity function should be based on pair-wise distances between molecules in the subset. Another requirement for the diversity function is that, after the diversity value has been maximized by choosing different subsets of molecules, the final subset that corresponds to the maximum function value is most diverse. An objective function was proposed in SAGE. In a related publication, an S-optimal function was used as the diversity function.[4] Then, SA optimization techniques were used to optimize the diversity function by choosing different combinations of compounds or building blocks.

To evaluate the effectiveness of SAGE in terms of diversity selection and chemical space coverage, several simulated datasets were used. For instance, in a geometrical space (2D or 100D), a certain number of cluster centers were generated, which were more than a preset distance away from each other. Some 99 cluster centers were then generated in a 2D space, and 95 cluster centers were generated in a 100D space. Then, a random number (between 1 and 100) of points for each cluster were generated around each cluster center within a cutoff distance, so that no members from two different clusters could overlap. This led to two simulated data sets: one with 951 points distributed in 99 clusters in 2D space, and second with 950 points distributed in 95 clusters in 100D space.

Simulation was conducted to determine how many clusters of points (molecules) SAGE could cover in comparison with random selection. When the number of sampled points was much smaller than that of clusters that exist in the data set, there was virtually no difference between the rational and random selection. This implies that when the number of sampled compounds was very small compared to that of the natural clusters, rational sampling could not provide any advantages over random sampling. As the number of sampled points increases, SAGE begins to cover more clusters than the corresponding random sampling. For instance, when 50 points were sampled, SAGE covered 50 different clusters (100% of what could maximally be obtained) for both data sets, while random sampling covered less than 40 clusters (<80% of maximal coverage). When 95 points were sampled, SAGE covered 90 clusters (roughly 95% of maximal coverage) while random sampling covered only 58–63 clusters (61–66% of maximal coverage).

To test the ability of diversity sampling to improve the hit rates *vs.* random sampling, an experiment was also conducted on simulated datasets containing varying percentages of active compounds. If the percentage of active compounds increases, the number of active compounds obtained by random sampling increases proportionally. This suggests that when the percentage of active compounds in the library is very high, SAGE (or any other cluster sampling) performs no better than random sampling in terms of the individual compound hit rate. There is a common view that the worse performance of cluster sampling strategies is due to non-ideal descriptors. On the contrary, this simulation indicates that it is the nature of this kind of strategies regardless of what descriptors are used, since ideal simulated data sets were used. Nevertheless, when a cluster hit rate (*i.e.*, one hit is counted even though multiple points from the same cluster are found) was considered as the criterion, the SAGE method performed much better than random sampling in all tested cases. This indicated that information content obtained by SAGE (diversity sampling) was always better than random sampling.

Another aspect of diversity-guided design is the visualization and comparison of compound collections. Compound sets comparison can be conducted by employing efficient dimensionality reduction methods. One interesting method, which is called Stochastic Proximity Embedding (SPE), was described by Agrafiotis *et al.*[29] Let us illustrate its application to the comparison of compound collections. In Figure 9.3(a–c) we use the SPE method to map a set of compounds in a proprietary collection onto a 2D space. Two commercially available compound collections are also described by the same set of descriptors and mapped onto the same 2D space. The points in black are internal compound collection. Red points represent compounds from vendor A and green points are for compounds from vendor B. One can see that the internal compound collection has the highest diversity. The green collection covers a little more space than the red collection. This is a simple technique for visual evaluation of compound collections, which is very useful for making practical decisions on compound collection acquisition.

9.2.1.3 Pharmacophore-guided Design of Focused Compound Libraries

The pharmacophore concept has been widely accepted and used in medicinal chemistry as well as in computational molecular modeling. Two closely related definitions of pharmacophore have been employed by both medicinal chemists and molecular modelers, and the root of this concept can be traced back to more than 100 years ago.[31]

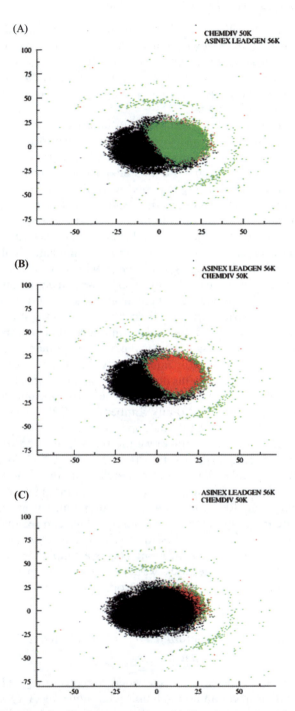

Figure 9.3 Compound collection comparison: black points are proprietary collection; red and green points are from two chemical vendors.

The very first definition of a pharmacophore was offered by Paul Ehrlich in the early 1900s: "a molecular framework that carries the essential features responsible for a drug's biological activity".[31] It is still used today mostly by medicinal chemists. For example, a chemist may refer to a series of compounds derived from the benzodiazepine scaffold as derivatives of the benzodiazepine pharmacophore. The software tool LeadScope (www.leadscope.com) capitalizes on this concept and attempts to automate the perception of molecular frameworks by comparing sets of structures to a predefined set of chemical substructures in the system, and thus categorizes any set of chemical structures into many pharmacophore series. The automated categorization allows the medicinal chemist to analyze large sets of screening data and hunt for interesting chemical series for the follow-up design of compound libraries.

Ehrlich's definition does not consider the fact that different series of molecules may share important chemical features in three-dimensional (3D) space and therefore present similar biological activities, even though they may belong to different molecular frameworks. In the 1970s, Peter Gund defined a pharmacophore as "a set of structural features in a molecule that is recognized at a receptor site and is responsible for that molecule's biological activity".[31] This definition is subtly different from Ehrlich's definition in that it implies the 3D nature of the pharmacophore, and it is more consistent with the present-day knowledge of the ligand–receptor interaction revealed by X-ray structures of ligand–receptor complexes. By this definition, a pharmacophore can be a set of disconnected features in 3D space that are required and recognized by the receptor and could be held together by different molecular frameworks. It was this definition that laid the foundation for many of the state-of-the-art pharmacophore perception algorithms that automate the identification of chemical features shared by active molecules.

Since Gund's definition of a pharmacophore and their first publication on a computer program for pharmacophore research,[32] several pharmacophore perception algorithms have been developed in the past 20 years. The best examples include the commercial package Catalyst distributed by Accelrys (www.accelrys.com) and GASP developed by Willett's group.[33] These tools have made computerized pharmacophore modeling a standard practice in modern rational drug design.

Catalyst has two related techniques for pharmacophore analysis: HipHop and HypoGen. The former identifies feature-based alignments for a collection of molecules without considering activity values, and the latter generates 3D pharmacophore hypotheses that can explain the variations of the activity with the chemical structures. In Catalyst, a pharmacophore consists of a 3D arrangement of chemical functions surrounded by tolerance spheres. Each sphere defines a space that is occupied by a particular chemical feature. The commonly seen features include hydrophobic groups, hydrogen bond acceptors and donors, aromatic fragments, charged groups, and so on. The GASP software was developed by Jones and Willett.[33] Using the Genetic Algorithm (GA), GASP automatically allows for conformational flexibility and maps features among the training set molecules to determine the correspondence between the features in different molecules.

Various examples of pharmacophore-guided virtual screening for compound library design are summarized in a recent book edited by Guner,[31] as well as in several other papers,[34,35] where pre-constructed pharmacophore models were used to filter or score members of virtual libraries or compound collections. The use of Catalyst pharmacophore models for screening large combinatorial libraries was reported by Hecker *et al.*[36] In this report, the two best pharmacophore models were derived from literature compounds and structure–activity data for cyclin dependent kinase 2. Then the models were used in retrospective virtual screening of ten databases containing over 1 000 000 compounds represented by their 3D coordinates. The results were then analyzed in terms of the screening efficiency for compound or library prioritization, and library design. For compound prioritization, one of the models selected active compounds at a rate nearly eleven times that of random compound selection. In library design experiments, most of the key building blocks were over-represented in the hits from at least one of the pharmacophore models. In library prioritization experiments, the two known active libraries both produced a significant number of hits with both pharmacophore models, while none of the eight inactive libraries produced a significant number of hits for both models. This work clearly demonstrates the potential of Catalyst pharmacophore models in the design of compounds or compound libraries.

The past decade has also seen the development of non-classical pharmacophore methods such as the pharmacophore key technique made popular by Mason *et al.* Mason's group applied this technique extensively to diversity assessment, similarity searching, and combinatorial library design.[26,37–39] Other groups have applied the Recursive Partitioning (RP) technique to discover critical features that distinguish the active molecules from the inactive ones. For example, Chen and Young applied RP to analyze the MAO dataset to discover atom pair features as the critical pharmacophores for the MAO inhibitors.[40] In a more recent interesting work, Schneider *et al.* reported a ligand-based combinatorial design method using self-organizing maps.[41] It is based on topological pharmacophore similarity metric and Self-organizing Maps (SOM). They applied this method to optimize combinatorial products functioning as P(1) purinergic receptor antagonists. A SOM map was developed using a set of known molecules to establish a structure–activity relationship. A combinatorial library design was performed by projecting virtually assembled new molecules onto the SOM map. A small focused library of 17 combinatorial products was synthesized and tested. On average, the designed structures yielded a three-fold smaller binding constant and 3.5-fold higher selectivity than the initial library. This result demonstrated that it was possible to design a small, activity-enriched focused library with an improved property profile using the SOM approach. Renner *et al.* also reported a fuzzy pharmacophore modeling method based on molecular alignment of active molecules.[42] It integrates information from 3D molecular alignments into correlation vectors. The pharmacophore model is represented by a number of spheres of Gaussian-distributed feature densities.

Different degrees of "fuzziness" can be introduced to influence the model's resolution. Transformation of this pharmacophore representation into a correlation vector results in a vector of feature probabilities. These feature probabilities are utilized for rapid virtual screening of compound databases or virtual libraries. The approach was validated by retrospective screening for cyclooxygenase 2 (COX-2) and thrombin ligands. Best performance was obtained with pharmacophore models reflecting an intermediate degree of fuzziness, yielding an enrichment factor of up to 39 for the first 1% of the ranked database.

9.2.1.4   QSAR-based Targeted Library Design

Many different approaches to QSAR have been developed since Hansch's seminal work. These include both 2D (two-dimensional) and 3D (three-dimensional) QSAR methods. Most of the 2D QSAR methods employ graph theoretic indices to characterize molecular structures, which have been extensively studied by Radic, Kier and Hall.[20–21,43–49] Similarly, ADAPT system employs topological indices as well as other structural parameters (*e.g.*, steric and quantum mechanical parameters) for QSAR analysis.[50–53] The examples of 3D QSAR include Molecular Shape Analysis (MSA),[54–57] distance geometry and Voronoi techniques.[58–60] These methods have been applied to study structure–activity relationships of many datasets by Hopfinger and by Crippen. Perhaps the most popular example of the 3D QSAR is the Comparative Molecular Field Analysis (CoMFA) developed by Cramer *et al.*, which has elegantly combined the power of molecular graphics and Partial Least-squares (PLS) technique and has found wide applications in medicinal chemistry.[61] More recent development in both 2D and 3D QSAR studies has focused on the development of optimal QSAR models through variable selection. These methods employ either generalized simulated annealing or genetic algorithms as the stochastic optimization tool. It has since been demonstrated that these algorithms combined with various statistical and machine learning tools have effectively improved the QSAR models compared to those without variable selection.

For illustrative purposes, we describe here the kNN QSAR method, which is conceptually simple and quite effective in various applications. Formally, the kNN QSAR technique implements the active analog principle that is used widely by the medicinal chemist. In the original kNN method, an unknown object (molecule) is classified according to the majority of the class memberships of its *k* nearest neighbors in the training set. The nearness is measured by an appropriate distance metric (a molecular similarity measure as applied to the classification of molecular structures). Many variations of the kNN method have been proposed in the past, and new and fast algorithms have continued to appear in recent years. The automated variable selection kNN QSAR technique optimizes the selection of descriptors to obtain the best models.[62] More recently, this technique has been successfully applied to the development of rigorously validated QSAR models and virtual screening of large databases for anticonvulsant agents.[63,64] Model validation was based on several critical statistical criteria, including the randomization of the target property, independent assessment of the predictive power using external test sets, and the establishment of the models' applicability domain. All successful models were employed in database mining concurrently. When these models were applied to search databases containing *ca.* 250 000 compounds, 22 compounds were selected as consensus hits. Nine compounds were synthesized and tested (of these nine, four were exact database hits and five were derived from the hits by minor chemical modifications). Seven of these nine compounds were later confirmed to be active, indicating an exceptionally high hit rate.

Thus, highly validated QSAR models can be used to search and prioritize either large compound collections or new virtual combinatorial chemical libraries. Every molecule in the compound set (real or virtual) will be predicted for its biological activity using the QSAR models. Compound collections can be designed by selecting only the molecules that are predicted to be active. In the case of designing new chemical libraries for synthesis, building blocks can be further prioritized *via* frequency analysis of top ranking molecules.[65]

9.2.1.5 Protein Structure Based Methods for Compound Library Design

Structure-based docking programs have been developed since Kuntz' seminal work on structure-based docking algorithm.[66] These programs are now available as part of popular software packages from companies like Tripos (St. Louis, USA), Accelrys (San Diego, USA), Chemical Computing Group (Montreal, Canada), OpenEye (NM, USA) and Schroedinger (NY, USA) to name a few. It has been well known and demonstrated that many of the generic docking functions do not work well for all protein targets.[67] Many researchers adopted "consensus" scoring schemes to help the accuracy of predictions for both docking poses and binding affinity.[68,69] External validation and consensus scoring are critically important to the success of a docking project, especially when docking tools are applied in a high-throughput fashion. Molecular docking and its application to virtual screening have been well reviewed.[70]

Most recently, empirical structure-based pharmacophore tools and geometric shape comparison technologies have proved to be effective when combined with HTStechnologies. These tools are very appealing to the chemist because of their intuitive nature. Most notable ones are LigandScout and ROCS.[71,72]

LigandScout is a structure-based pharmacophore program that generates pharmacophore centers based on the structure of a given ligand–receptor complex. The pharmacophore model(s) can then be used to virtually screen compound libraries or design new combinatorial libraries.[73] As an example, LigandScout was applied to identify pharmacophore patterns from 3D crystal structures of inhibitors bound to human factor Xa. The resulting chemical-feature based pharmacophores were used for virtual screening of molecular databases such as the WDI database. LigandScout can be used for virtual screening with selective pharmacophore models and *de novo* design molecular scaffolds that were able to adequately satisfy the pharmacophore criteria.

Steindl *et al.* used LigandScout for parallel pharmacophore-based virtual screening[74] to assess bioactivity profiles for organic molecules. In a proof of principle study, they built a set of 50 structure-based pharmacophore models for various viral targets and 100 antiviral compounds. The latter were screened against all pharmacophore models to determine if their known biological targets could be correctly predicted *via* an enrichment of corresponding pharmacophore matches. The results demonstrate that the desired enrichment was achieved for approximately 90% of all input molecules.

ROCS (Rapid Overlay of Chemical Structures) is a shape-based, ligand-centric method that ranks molecules on the basis of their similarity to a known active molecule in three-dimensional shape space, using atom-centered Gaussian functions to allow rapid maximization of molecular overlap. In a recent study that compares the performance of molecular docking and ROCS in virtual screening experiments, the authors use the receptor-bound ligand conformation as the query for ROCS. Fundamentally, they use the bound ligand to represent the protein binding site and search compound collection for molecules that are similar in shape to the query ligand. Direct comparisons between virtual screening results from a significant number of docking programs show that the shape-based ranking method (ROCS) performs at least as well as and often better than docking. Seven different docking programs were compared to ROCS across 21 different protein systems. ROCS provided superior performance even when a bioactive conformation of the ligand was not known. Given the speed, ease of use and predictability, this approach should be considered where compound library design or screening set design is needed.

Similarly, Schuller *et al.* reported a pseudo-ligand approach to compound library design.[75] Based on a receptor-derived pharmacophore model (a pseudo-ligand), which represents an idealized constellation of potential ligand sites that interact with residues of the binding pocket, compound libraries are examined for the potential pharmacophore point matches between the pseudo ligand and the candidate molecule. The method was successfully applied to retrieving factor Xa inhibitors from a Ugi three-component compound library, and yielded high enrichment of actives in a retrospective search for cyclooxygenase-2 (COX-2) inhibitors.

9.2.2 Design for Developability or Drug-likeness

With the much publicized decline in the drug approval rate by the FDA, according to Biomedtracker (www.BioMedTracker.com), there has been a significant effort in the pharmaceutical industry to reduce attrition rates by advancing higher quality candidates to the clinic. As part of the paradigm shift of considering pharmacokinetic and toxicology issues earlier in discovery, the number of computational tools intended to address ADMET related properties has grown markedly. In many ways, library design and compound acquisition processes are ideal venues to utilize these tools. In each of these processes, there exists an opportunity to acquire or make many different compounds.

As discussed above, many methods[18,76–79] have been developed to optimize many properties simultaneously, including ADMET filters.

9.2.2.1   Rule & Alert Based Approaches

Despite the explosion of literature on ADMET modeling, most reported library design efforts[80–83] have relied on simple rule-based methods like the well-known Rule of Five from Lipinski[11] or the flexibility and polar surface area guidelines championed by Veber *et al.*[84] These approaches are somewhat more physico-chemical property based, as opposed to being tailored to specific ADMET activities such as solubility. It is difficult to know if this outcome is dictated more by computational cost, the perceived quality of the ADMET models, or simply reflects a barrier to publishing detailed reports of such designs. Lipinski has suggested that the Rule of Five has been so successful because one rarely finds more than a handful of independent factors that relate to oral absorption.[85]

The Rule of Five specifies that compounds that violate two or more of the following guidelines can be expected to have poor absorption: $MW \leq 500$, clog $P \leq 5$, ten or fewer hydrogen bond acceptors, and five or fewer hydrogen bond donors. These rules were developed by determining the 90th percentile values for each of the properties among orally delivered drugs. Veber *et al.* performed an analysis of rat oral bioavailability data and determined that bioavailability appeared to be greater in compounds with ten or fewer rotatable bonds, regardless of the MW, and a polar surface area below 140 Å^2. This focus on molecular flexibility is not without controversy,[86] however. The more recent focus on lead-like libraries have lead some to use even stricter bounds than those pioneered by Lipinski. A recent review summarizes many of the more common rule-based filtering criteria.[12]

Samiulla *et al.*[81] report the design of natural product scaffolds that were then used in a small library of vasicinone derivatives. A key aspect of the design was the incorporation of parameters intended to result in improved ADME properties by including the filters proposed by Lipinski[11] and Veber *et al.*[84] BCUTs were used to optimize the diversity of the scaffolds while Unity fingerprints were used in the diversity optimization of the vasicinone library. The measured solubility, permeability in MDCK cells and CYP inhibition were measured for the scaffolds. The resulting libraries demonstrated favorable distributions of each of the measured ADME properties.

A genetic algorithm was utilized to optimize a library design by combining diversity, structure-based docking, and a modified Rule of Five. The method uses scores from DOCK4.0 as a measure of binding affinity coupled with a diversity metric[87] and drug-likeness score[88] that are combined using linear equation with user-defined weights. The drug-likeness methodology employed is a derivative of the Rule of Five. In the initial report,[82] example libraries of Cox-2 inhibitors and PRAR-γ inhibitors were designed. For PPAR-γ, the resulting compounds proved more potent than published ligands for this target. A second report using the same methodology[83] discusses a library for human cyclophilin. A inhibitors, a potential target for immunological disorders. Fragments from previously reported inhibitors were used to build a targeted library of 16 compounds. These compounds showed a K_D value as low as 76 nM and an IC_{50} as low as 250 nM. Unfortunately, no experimental verification of the ADME properties was reported.

It is also worth highlighting a recent report of the medicinal chemistry space of oncology targeted compounds.[89] In this report, the authors utilize compounds in the ZINC, WOMBAT, and NCI databases to describe "oncology space" and how it compares with the drug-like space resulting from Rule-of-Five-like filters. The authors demonstrate that applying such cheminformatic filters without careful consideration of the target could dramatically limit the realm of exploration available for library design. This, indeed, is an important consideration when applying filters with the intention of improving ADMET properties.

9.2.2.2 QSAR-based ADMET Models

While most reports in the literature employ a modified version of Lipinski's Rule-of-Five criteria, a few have used models that are more sophisticated as part of the design. Many QSAR models have been reported in the literature for many different ADMET related properties, from solubility[90–95] to permeability[96–98] to hERG inhibition[99–102] to metabolic stability.[103–105] Several recent reviews[106–109] and perspectives[110,111] have also been contributed that nicely summarize the state of the art of ADMET modeling.

A focused library of 320 ketopiperazides was designed by a team at Zentaris GmbH to inhibit tubulin.[112] Several parameters were considered for the design: the similarity of the "best" docked pose to a pose for a known lead, library diversity, and the predicted solubility, CNS activity and log *P* from the program QikProp.[113] The design appears to have been done manually. Several compounds in the resulting library were shown to have superior activity compared to the initial lead, while also having favorable calculated ADMET profiles.

Werner *et al.* utilized QikProp in the design of a pyrrolecarboxamide combinatorial library intended for general screening.[114] The factors considered in the design of the library included regioselective reactivity, avoidance of diastereoselective mixtures, MW, clog *P*, functional group diversity, cost, and several calculated parameters from QikProp. Of the compounds in the final library, 95% had a calculated permeability within the range of marketed drugs. The authors noted that the solubility was somewhat below the range of marketed drugs. Again, based on the manuscript, it appears that the library design was done manually as opposed to employing a computational optimization method.

The predictive models included in the program Volsurf[115] were utilized as part of a library designed to identify non-benzodiazepine ligands for the GABA-A receptor.[116] A virtual library of 500 compounds was enumerated using Accord for Excel. The virtual library was filtered to remove compounds that did not meet the Rule-of-Five. Predicted Caco-2 permeability and blood–brain barrier penetration (BBB) values were generated using Volsurf. Compounds meeting defined thresholds for these properties were then filtered again to contain a particular pharmacophore using Catalyst. The resulting 20 compounds failed to show significant inhibitory activity.

While the number of reports in the literature applying ADMET activity models is fairly limited, we believe that the actual usage of these models is larger in practice. There are several hurdles to publishing library designs from an industrial setting, including the ability to publish activity and ADMET data for a large number of compounds in the project due to competitive considerations. Typically, by the time such publication is possible the individual library designs have faded from memory and the SAR of the lead series seem to take priority. Nonetheless, with the shift to smaller, targeted, libraries we believe ADMET models will become more common. These libraries typically are generated from smaller virtual libraries, making the more expensive computations of most ADMET models well within acceptable burdens.

9.2.2.3   Undesirable Functionality Filters

Many authors have pointed out the importance of filtering out undesirable functional groups during library design, screening collection design, or compound acquisition.[16] Pearce *et al.* discuss in detail the derivation of 180 substructure filters based on compound promiscuity and reactivity. They have developed a Promiscuity Ratio Index (PRI) that identifies functional groups that display activity more frequently than expected based on historical norms. PRI values below 1 indicate a substructure that is present in active compounds less frequently than expected, while those greater than 1 indicate a higher hit rate. The authors also calculate a 95% confidence interval of the ratio to give context to the potential promiscuity for the functional group. Figure 9.4 shows example substructures and the PRI.

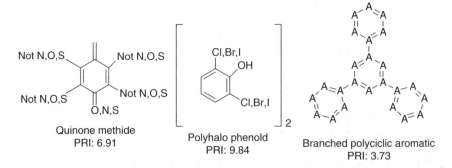

Figure 9.4   Example substructures with high hit rates.[16] PRI=Promiscuity Ratio Index (see text for explanation). "A" represents any atom that can form part of an aromatic structure.

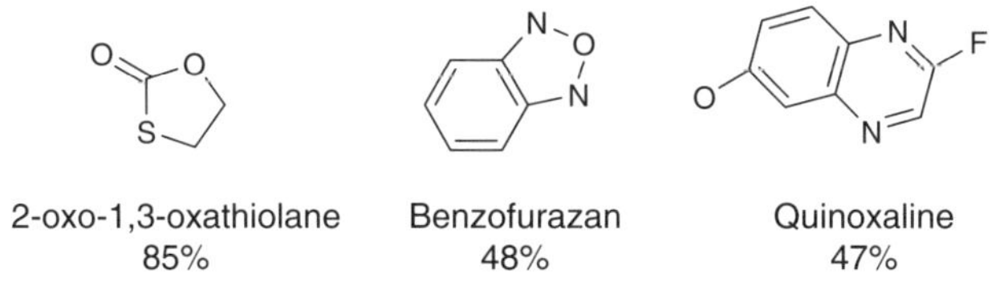

| 2-oxo-1,3-oxathiolane | Benzofurazan | Quinoxaline |
| 85% | 48% | 47% |

Figure 9.5 Example substructures with high rates of reactivity against protein thiol groups.[120] The percentage reflects the percentage of cases with experimentally confirmed reactivity as determined in ref. 120.

Once identified, these functional groups can be optionally excluded from future compound acquisitions or library designs unless a competing interest encourages their use. A report[117] on the compound acquisition process at Novartis discusses how compounds are prioritized or eliminated for acquisition based on similarity to the historical collection, as well as physicochemical properties and undesirable functionality. Many of the same themes are presented in other reports on compound selection.[118]

Two recent reports present substructures that show a high likelihood of reacting with thiol groups (Figure 9.5).[119,120] The original work included an analysis performed manually on data from ALARM NMR[119] to identify several interesting substructures. The follow-up paper described a more automated system using Pipeline Pilot that is capable of continually updating the collection of reactive structures as additional data becomes available. In any event, the substructures contained in the publications represent an additional source of reactive substructures that should be considered for use in library design and compound acquisition.

9.2.3 Design for Multiple Objectives and Targets Simultaneously

As described in previous sections, similarity to leads, molecular diversity, pharmacophore models, QSAR models and structure-based docking or pharmacophore models can all be used to guide compound library design for biological activities. Drug-likeness parameters (Lipinski's Rule-of-Five), liabilities against P450 enzymes, aqueous solubility, cell permeability and other ADMET models should be used to select the best quality molecules for consideration, in addition to the activity concerns. To optimally design combinatorial libraries that have balanced activity and ADMET properties, researchers have developed library design tools that employ Multi-objective Optimization (MOO) methods.[3,4,77]

The combinatorial space is huge when the number of reagents increases. For example, a reaction that involves amine and carboxylic acids could generate 2500 compounds if we have 50 available amines and 50 carboxylic acids. The total number of solutions for an 8×12 library, however, would be on the order of 10^{19}. Thus, Simulated Annealing (SA) is used to sample the space.

In the Piccolo method,[4] the objective function combines all possible penalty functions in a weighted fashion:

$$E(S) = \sum w_i * E_i(S)$$

where, E_i can be Tanimoto "distances" between all the compounds in a given library and the lead, averaged over the size of the library; it can be reagent diversity measured as an S-optimal function; product novelty with respect to an existing compound collection, which can be measured by counting how many of the new library compounds fall into the neighborhood of compounds in the existing collection; developability penalty scores such as the percentage of compounds that violate the Lipinski's rule of five. The term w_i is an adjustable weight that controls the contribution of an objective factor to the penalty score. Libraries designed using the multi-objective optimization procedure tend to have more balanced properties than those designed using only one type of computational models.

In a similar method, Agrafiotis described an algorithm rooted in the principles of multi-objective optimization. They employ an objective function that encodes all of the desired selection criteria, and then use a simulated annealing or evolutionary approach to identify the optimal (or a nearly optimal) subset from among the vast number of possibilities. Many design criteria can be accommodated, including diversity, similarity to known actives, predicted activity and/or selectivity determined by quantitative structure–activity relationship (QSAR) models or receptor binding models, enforcement of certain property distributions, reagent cost and availability, and many others. The method offers the user full control over the relative significance of the various objectives in the final design, and permits the simultaneous selection of compounds from multiple libraries in combinatorial-array or cherry-picking format.

Gillet *et al.* described the program MoSELECT for multi-objective library design that is based on a multi-objective genetic algorithm (MOGA).[77] MoSELECT searches the product-space of a virtual combinatorial library to generate a family of equivalent solutions where each solution represents a combinatorial subset of the virtual library optimized over multiple objectives. The family of solutions allows the relationships between the objectives to be explored and thus enables the library designer to make an informed choice on an appropriate compromise solution.

With the development of chemogenomics strategies to drug discovery, library designers often need to consider modeling multiple targets in the same gene family, such as NHR (nuclear hormone receptors), GPCR (G-protein coupled receptors), kinases, and proteases. Multi-objective optimization methods should be adopted when multiple computational models for the target activities have been developed. Schnur *et al.* have described knowledge-based approaches to target class combinatorial library design.[121] Lowrie *et al.* have described different strategies for designing GPCR and kinase targeted libraries.[122] The tools that have proven to be the most useful are those that can extract trends from the computational data from docking and clustering or data mining of large amounts of structure–activity data. They reported tools to extract the relevant patterns from all available information for a family of targets and tools to efficiently apply models for all targets in the family.

Ideally, when ligand-based or structure-based models are developed for all members of a gene family, the validated models can be used as part of the multi-objective function in a comprehensive chemical library design environment. Like other factors discussed earlier, the predicted activities can be used in the weighted objective function. Compound library design or virtual screening can be achieved by optimizing the total objective function.

9.3 Concluding Remarks

We have reviewed various aspects of compound library design, from background, rationale and significance to its definition to various computational approaches to library design. Table 9.1 lists the various design factors we have discussed in this chapter *versus* three common screening platforms, namely, monolithic screening where one biological target is screened against, platform screening where a panel of targets from a gene family are being screened against, and chemical biology screening where cellular phenotypic screening is conducted. Proper design factors should be considered according to the project goals.

We also discussed factors and models that need to be employed for the design against specific biological activities, and then introduced the importance of ADMET models. We have also reviewed ways to conduct comprehensive, balanced, multi-objective compound library design. These multi-objective methods should be applicable to gene family oriented library design as well, so

Table 9.1 Design factors *vs.* screening platforms.

Design factors	Monolithic screening	Platform screening	Chemical biology
Similarity	Single lead or leads	Leads across members of family	Leads with desired phenotype
Diversity	Maximum diversity	Family directed diversity	Bioactive diversity
Pharmacophore models	Single target pharmacophore	Union pharmacophore or intersection pharmacophore	
QSAR models	Single assay QSAR	Joint QSAR across family	Phenotypic assay QSAR
Docking models	Single target docking	Multi target docking	Structural genome panel docking
Drug-likeness/ developability models	ADMET	ADMET	ADMET

long as individual models can be developed and used in the multi-objective function.

With the recent development of the NIH chemical genomics initiative (Molecular Library Initiative, MLI), development of comprehensive library design methods are crucial for its long-term success. Since *all* targets, not just the "drugable genome", should be considered for their roles in biological pathways and networks, we need to design and continue to expand the NIH library collection in a biologically relevant fashion. This means developing and validating models for the PubChem assay database, and feedback to iterative compound collection design. Using known biologically active compounds of some sort as the seeds to expand the compound collection around biologically active compounds is a good approach to chemical genomics library design. Constructing multiple binding site shapes, pharmacophores and assembling them into a reference panel to profile biologically relevant diversity would be a complementary approach to the above methods.

Although most applications were of the "cherry-picking" type design, the combinatorial design of new chemical libraries should also be feasible. In this case, the scores obtained with the various models can be used to sort the virtual library, followed by building block frequency analysis (*cf.* Focus2D) to determine which reagents should be used in chemical synthesis. Alternatively, combinatorial optimization approaches, such as those in described in ref. 4, can be applied where the model-predicted scores are used as the objective function for optimization.

With continuing development of chemical library design or compound collection design methods, better understanding and modeling of ADMET properties, as well as better validated computational models of individual biological targets in the human genome, we expect to see further improvement in the efficiency of drug discovery and chemical probe development in the foreseeable future.

References

1. W. Zheng, S. J. Cho and A. Tropsha, *J. Chem. Information Comput. Sci.*, 1998, **38**, 251–258.
2. R. P. Sheridan, S. G. SanFeliciano and S. K. Kearsley, *Journal of Molecular Graphics & Modelling*, 2000, **18**, 320–334, 525.
3. D. K. Agrafiotis, *Journal of computer-aided molecular design*, 2002, **16**, 335–356.
4. W. Zheng, S. T. Hung, J. T. Saunders and G. L. Seibel, *Pacific Symposium on Biocomputing*, 2000, 588–599.
5. V. J. Gillet, P. Willett, P. J. Fleming and D. V. Green, *Journal of molecular graphics & modelling*, 2002, **20**, 491–498.
6. V. J. Gillet, W. Khatib, P. Willett, P. J. Fleming and D. V. Green, *J. Chem. Inf. Comput. Sci.*, 2002, **42**, 375–385.
7. D. K. Agrafiotis and D. N. Rassokhin, *Journal of chemical information and computer sciences*, 2001, **41**, 798–805.

8. B. R. Stockwell, *Nat. Rev. Genet.*, 2000, **1**, 116–125.
9. B. R. Stockwell, *Trends in biotechnology*, 2000, **18**, 449–455.
10. B. R. Stockwell, S. J. Haggarty and S. L. Schreiber, *Chemistry & biology*, 1999, **6**, 71–83.
11. C. A. Lipinski, F. Lombardo, B. W. Dominy and P. J. Feeney, *Advanced drug delivery reviews*, 1997, **23**, 3–25.
12. S. R. Johnson and W. Zheng, *AAPS Journal*, 2006, **8**, E27–E40.
13. J. Wang, L. Lai and Y. Tang, *J. Chem. Inf. Comput. Sci.*, 1999, **39**, 1173–1189.
14. G. W. Bemis and M. A. Murcko, *Journal of medicinal chemistry*, 1999, **42**, 5095–5099.
15. G. W. Bemis and M. A. Murcko, *Journal of medicinal chemistry*, 1996, **39**, 2887–2893.
16. B. C. Pearce, M. J. Sofia, A. C. Good, D. M. Drexler and D. A. Stock, *J. Chem. Inf. Model.*, 2006, **46**, 1060–1068.
17. R. D. Brown and Y. C. Martin, *Journal of medicinal chemistry*, 1997, **40**, 2304–2313.
18. D. K. Agrafiotis, *Molecular Diversity*, 2002, **5**, 209–230.
19. S. Kirkpatrick, C. D. Gelatt Jr and M. P. Vecchi, *Science*, 1983, **220**, 671–680.
20. L. H. Hall and L. B. Kier, *Journal of pharmaceutical sciences*, 1978, **67**, 1743–1747.
21. L. H. Hall, L. B. Kier and W. J. Murray, *Journal of pharmaceutical sciences*, 1975, **64**, 1974–1977.
22. R. N. Zuckermann, E. J. Martin, D. C. Spellmeyer, G. B. Stauber, K. R. Shoemaker, J. M. Kerr, G. M. Figliozzi, D. A. Goff, M. A. Siani and R. J. Simon, *et al.*, *Journal of medicinal chemistry*, 1994, **37**, 2678–2685.
23. H. Eckert and J. Bajorath, *Drug discovery today*, 2007, **12**, 225–233.
24. R. P. Sheridan and S. K. Kearsley, *Drug discovery today*, 2002, **7**, 903–911.
25. R. D. Brown and Y. C. Martin, *SAR and QSAR in environmental research*, 1998, **8**, 23–39.
26. J. S. Mason and B. R. Beno, *Journal of molecular graphics & modelling*, 2000, **18**, 438–451, 538.
27. W. Zheng, S. J. Cho, C. L. Waller and A. Tropsha, *J. Chem. Inf. Comput. Sci.*, 1999, **39**, 738–746.
28. D. K. Agrafiotis and H. Xu, *J. Chem. Inf. Comput. Sci.*, 2003, **43**, 475–484.
29. D. K. Agrafiotis, *Journal of computational chemistry*, 2003, **24**, 1215–1221.
30. D. R. Flower, *Journal of molecular graphics & modelling*, 1998, **16**, 239–253, 264.
31. O. Guner, *Pharmacophore Perception, Development, and use in Drug Design*. ed. O.F. Guner, International University Line, La Jolla, 2000, Chapter 4, pp. 254–268.
32. P. Gund, J. D. Andose, J. B. Rhodes and G. M. Smith, *Science*, 1980, **208**, 1425–1431.

33. G. Jones, P. Willett and R. C. Glen, *Journal of computer-aided molecular design*, 1995, **9**, 532–549.

34. O. Guner, O. Clement and Y. Kurogi, *Current medicinal chemistry*, 2004, **11**, 2991–3005.

35. Y. Kurogi and O. F. Guner, *Current medicinal chemistry*, 2001, **8**, 1035–1055.

36. E. A. Hecker, C. Duraiswami, T. A. Andrea and D. J. Diller, *Journal of chemical information and computer sciences*, 2002, **42**, 1204–1211.

37. B. R. Beno and J. S. Mason, *Drug Discov Today*, 2001, **6**, 251–258.

38. J. S. Mason, I. Morize, P. R. Menard, D. L. Cheney, C. Hulme and R. F. Labaudiniere, *Journal of medicinal chemistry*, 1999, **42**, 3251–3264.

39. J. S. Mason and M. A. Hermsmeier, *Current opinion in chemical biology*, 1999, **3**, 342–349.

40. X. Chen, A. Rusinko 3rd, A. Tropsha and S. S. Young, *Journal of chemical information and computer sciences*, 1999, **39**, 887–896.

41. G. Schneider and M. Nettekoven, *Journal of combinatorial chemistry*, 2003, **5**, 233–237.

42. S. Renner and G. Schneider, *J. Med. Chem.*, 2004, **47** (19), 4653–4664.

43. L. H. Hall and L. B. Kier, *Journal of pharmaceutical sciences*, 1977, **66**, 642–644.

44. L. B. Kier and L. H. Hall, *Journal of pharmaceutical sciences*, 1976, **65**, 1806–1809.

45. L. B. Kier, L. H. Hall, W. J. Murray and M. Randic, *Journal of pharmaceutical sciences*, 1975, **64**, 1971–1974.

46. L. B. Kier, W. J. Murray and L. H. Hall, *Journal of medicinal chemistry*, 1975, **18**, 1272–1274.

47. L. B. Kier, W. J. Murray, M. Randic and L. H. Hall, *Journal of pharmaceutical sciences*, 1976, **65**, 1226–1230.

48. W. J. Murray, L. H. Hall and L. B. Kier, *Journal of pharmaceutical sciences*, 1975, **64**, 1978–1981.

49. W. J. Murray, L. B. Kier and L. H. Hall, *Journal of medicinal chemistry*, 1976, **19**, 573–578.

50. R. Guha, J. R. Serra and P. C. Jurs, *Journal of molecular graphics & modelling*, 2004, **23**, 1–14.

51. D. R. Henry, P. C. Jurs and W. A. Denny, *Journal of medicinal chemistry*, 1982, **25**, 899–908.

52. P. C. Jurs, M. N. Hasan, D. R. Henry, T. R. Stouch and E. K. Whalen-Pedersen, *Fundam Appl Toxicol*, 1983, **3**, 343–349.

53. T. R. Stouch and P. C. Jurs, *Environmental health perspectives*, 1985, **61**, 329–343.

54. B. J. Burke, W. J. Dunn 3rd and A. J. Hopfinger, *Journal of medicinal chemistry*, 1994, **37**, 3775–3788.

55. M. G. Koehler, K. Rowberg-Schaefer and A. J. Hopfinger, *Archives of biochemistry and biophysics*, 1988, **266**, 152–161.

56. R. L. Lopez de Compadre, R. A. Pearlstein, A. J. Hopfinger and J. K. Seydel, *Journal of medicinal chemistry*, 1987, **30**, 900–906.

57. J. S. Tokarski and A. J. Hopfinger, *Journal of medicinal chemistry*, 1994, **37**, 3639–3654.
58. L. Boulu and G. M. Crippen, *Progress in clinical and biological research*, 1989, **289**, 267–277.
59. L. G. Boulu, G. M. Crippen, H. A. Barton, H. Kwon and M. A. Marletta, *Journal of medicinal chemistry*, 1990, **33**, 771–775.
60. G. M. Crippen, *NIDA research monograph*, 1991, **112**, 7–20.
61. R. D. Cramer 3rd, D. E. Patterson and J. D. Bunce, *Progress in clinical and biological research*, 1989, **291**, 161–165.
62. W. Zheng and A. Tropsha, *Journal of chemical information and computer sciences*, 2000, **40**, 185–194.
63. M. Shen, C. Beguin, A. Golbraikh, J. P. Stables, H. Kohn and A. Tropsha, *Journal of medicinal chemistry*, 2004, **47**, 2356–2364.
64. M. Shen, A. LeTiran, Y. Xiao, A. Golbraikh, H. Kohn and A. Tropsha, *Journal of medicinal chemistry*, 2002, **45**, 2811–2823.
65. S. J. Cho, W. Zheng and A. Tropsha, *Journal of chemical information and computer sciences*, 1998, **38**, 259–268.
66. B. K. Shoichet and I. D. Kuntz, *Journal of molecular biology*, 1991, **221**, 327–346.
67. J. A. Erickson, M. Jalaie, D. H. Robertson, R. A. Lewis and M. Vieth, *Journal of medicinal chemistry*, 2004, **47**, 45–55.
68. S. Betzi, K. Suhre, B. Chetrit, F. Guerlesquin and X. Morelli, *Journal of chemical information and modeling*, 2006, **46**, 1704–1712.
69. A. Oda, K. Tsuchida, T. Takakura, N. Yamaotsu and S. Hirono, *Journal of chemical information and modeling*, 2006, **46**, 380–391.
70. G. Schneider and H. J. Bohm, *Drug discovery today*, 2002, **7**, 64–70.
71. P. C. Hawkins, A. G. Skillman and A. Nicholls, *Journal of medicinal chemistry*, 2007, **50**, 74–82.
72. G. Wolber and T. Langer, *Journal of chemical information and modeling*, 2005, **45**, 160–169.
73. E. M. Krovat, K. H. Fruhwirth and T. Langer, *Journal of chemical information and modeling*, 2005, **45**, 146–159.
74. T. M. Steindl, D. Schuster, C. Laggner and T. Langer, *Journal of chemical information and modeling*, 2006, **46**, 2146–2157.
75. A. Schuller, U. Fechner, S. Renner, L. Franke, L. Weber and G. Schneider, *Combinatorial chemistry & high throughput screening*, 2006, **9**, 359–364.
76. T. Wright, V. J. Gillet, D. V. S. Green and S. D. Pickett, *Journal of chemical information and computer sciences*, 2003, **43**, 381–390.
77. V. J. Gillet, W. Khatib, P. Willett, P. J. Fleming and D. V. S. Green, *Journal of chemical information and computer sciences*, 2002, **42**, 375–385.
78. V. J. Gillet, *Methods in Molecular Biology*, 2004, **275**, 335–354.
79. V. J. Gillet, P. Willett, P. J. Fleming and D. V. S. Green, *Journal of molecular graphics & modelling*, 2002, **20**, 491–498.

80. C. Liao, B. Liu, L. Shi, J. Zhou and X.-P. Lu, *European Journal of Medicinal Chemistry*, 2005, **40**, 632–640.

81. D. S. Samiulla, V. V. Vaidyanathan, P. C. Arun, G. Balan, M. Blaze, S. Bondre, G. Chandrasekhar, A. Gadakh, R. Kumar, G. Kharvi, H. O. Kim, S. Kumar, J. A. Malikayil, M. Moger, M. K. Mone, P. Nagarjuna, C. Ogbu, D. Pendhalkar, A. V. S. R. Rao, G. V. Rao, V. K. Sarma, S. Shaik, G. V. R. Sharma, S. Singh, C. Sreedhar, R. Sonawane, U. Timmanna and L. W. Hardy, *Molecular Diversity*, 2005, **9**, 131–139.

82. G. Chen, S. Zheng, X. Luo, J. Shen, W. Zhu, H. Liu, C. Gui, J. Zhang, M. Zheng, C. M. Puah, K. Chen and H. Jiang, *Journal of combinatorial chemistry*, 2005, **7**, 398–406.

83. J. Li, J. Zhang, J. Chen, X. Luo, W. Zhu, J. Shen, H. Liu, X. Shen and H. Jiang, *Journal of combinatorial chemistry*, 2006, **8**, 326–337.

84. D. F. Veber, S. R. Johnson, H. Y. Cheng, B. R. Smith, K. W. Ward and K. D. Kopple, *Journal of medicinal chemistry*, 2002, **45**, 2615–2623.

85. C. Lipinski, in *Analysis and Purification Methods in Combinatorial Chemistry*, ed. Y. Bing, John Wiley and Sons, New York, 2004, pp. 407–434.

86. J. J. Lu, K. Crimin, J. T. Goodwin, P. Crivori, C. Orrenius, L. Xing, P. J. Tandler, T. J. Vidmar, B. M. Amore, A. G. E. Wilson, P. F. W. Stouten and P. S. Burton, *Journal of medicinal chemistry*, 2004, **47**, 6104–6107.

87. D. R. Flower, *Journal of molecular graphics & modelling*, 1998, **16**, 239–253.

88. S. Zheng, X. Luo, G. Chen, W. Zhu, J. Shen, K. Chen and H. Jiang, *Journal of chemical information and modeling*, 2005, **45**, 856–862.

89. D. G. Lloyd, G. Golfis, A. J. S. Knox, D. Fayne, M. J. Meegan and T. I. Oprea, *Drug discovery today*, 2006, **11**, 149–159.

90. C. Catana, H. Gao, C. Orrenius and P. F. W. Stouten, *Journal of chemical information and computer sciences*, 2005, **45**, 170–176.

91. S. R. Johnson, X.-Q. Chen, D. Murphy and O. Gudmundsson, *Molecular Pharmaceutics*, 2007, **4**, 513–523.

92. H. Zhang, H. Y. Ando, L. Chen and P. H. Lee, *Molecular Pharmaceutics*, 2007, **4**, 489–497.

93. A. Schwaighofer, T. Schroeter, S. Mika, J. Laub, A. Ter Laak, D. Suelzle, U. Ganzer, N. Heinrich and K.-R. Mueller, *Journal of chemical information and modeling*, 2007, **47**, 407–424.

94. Y.-D. Hu and Y.-L. Wang, *Asian Journal of Chemistry*, 2007, **19**, 407–416.

95. N. T. Hansen, I. Kouskoumvekaki, F. S. Jorgensen, S. Brunak and S. O. Jonsdottir, *Journal of chemical information and modeling*, 2006, **46**, 2601–2609.

96. H. H. Refsgaard, B. F. Jensen, P. B. Brockhoff, S. B. Padkjaer, M. Guldbrandt and M. S. Christensen, *Journal of medicinal chemistry*, 2005, **48**, 805–811.

97. K. Obata, K. Sugano, R. Saitoh, A. Higashida, Y. Nabuchi, M. Machida and Y. Aso, *International Journal of Pharmaceutics*, 2005, **293**, 183–192.
98. Y. C. Martin, *Journal of medicinal chemistry*, 2005, **48**, 3164–3170.
99. O. Roche, G. Trube, J. Zuegge, P. Pflimlin, A. Alanine and G. Schneider, *ChemBioChem*, 2002, **3**, 455–459.
100. A. Cavalli, E. Poluzzi, F. De Ponti and M. Recanatini, *Journal of medicinal chemistry*, 2002, **45**, 3844–3853.
101. R. A. Pearlstein, R. J. Vaz, J. Kang, X.-L. Chen, M. Preobrazhenskaya, A. E. Shchekotikhin, A. M. Korolev, L. N. Lysenkova, O. V. Miroshnikova, J. Hendrix and D. Rampe, *Bioorganic & Medicinal Chemistry Letters*, 2003, **13**, 1829–1835.
102. K. Yoshida and T. Niwa, *Journal of chemical information and modeling*, 2006, **46**, 1371–1378.
103. P. H. Lee, L. Cucurull-Sanchez, J. Lu and Y. J. Du, *Journal of computer-aided molecular design*, 2007, **21**, 665–673.
104. M. M. Ahlstroem, M. Ridderstroem, I. Zamora and K. Luthman, *Journal of medicinal chemistry*, 2007, **50**, 4444–4452.
105. V. K. Gombar, J. J. Alberts, K. C. Cassidy, B. E. Mattioni and M. A. Mohutsky, *Current Computer-Aided Drug Design*, 2006, **2**, 177–188.
106. D. E. Clark, *Annual Reports in Computational Chemistry*, 2005, **1**, 133–151.
107. R. J. Zimmerman, *Comprehensive Medicinal Chemistry II*, 2006, **2**, 559–572.
108. U. Norinder and C. A. S. Bergstroem, *Chemical Biology*, 2007, **3**, 1003–1042.
109. U. Norinder and C. A. S. Bergstroem, *ChemMedChem*, 2006, **1**, 920–937.
110. D. E. Clark, *Expert Opinion on Drug Discovery*, 2007, **2**, 1423–1429.
111. J. C. Dearden, *Expert Opinion on Drug Metabolism & Toxicology*, 2007, **3**, 635–639.
112. M. Gerlach, E. Claus, S. Baasner, G. Müller, E. Polymeropoulos, P. Schmidt, E. Günther and J. Engel, *Archiv der Pharmazie*, 2004, **337**, 695–703.
113. QikProp, Schrödinger, LLC, New York.
114. S. Werner, P. S. Iyer, M. D. Fodor, C. M. Coleman, L. A. Twining, B. Mitasev and K. M. Brummond, *Journal of combinatorial chemistry*, 2006, **8**, 368–380.
115. Volsurf, 4.1.4 edn., Molecular Discovery, Ltd., 2004, http://www.moldiscovery.com (accessed May 2008).
116. J. L. Falco, M. Lloveras, I. Buira, J. Teixido, J. I. Borrell, E. Mendez, J. Terencio, A. Palomer and A. Guglietta, *European Journal of Medicinal Chemistry*, 2005, **40**, 1179–1187.
117. A. Schuffenhauer, M. Popov, U. Schopfer, P. Acklin, J. Stanek and E. Jacoby, *Combinatorial Chemistry and High Throughput Screening*, 2004, **7**, 771–781.
118. M. M. Olah, C. G. Bologa and T. I. Oprea, *Current Drug Discovery Technologies*, 2004, **1**, 211–220.

119. J. R. Huth, R. Mendoza, E. T. Olejniczak, R. W. Johnson, D. A. Cothron, Y. Liu, C. G. Lerner, J. Chen and P. J. Hajduk, *J. Am. Chem. Soc.*, 2005, **127**, 217–224.
120. J. Metz, J. Huth and P. Hajduk, *Journal of computer-aided molecular design*, 2007, **21**, 139–144.
121. D. Schnur, B. R. Beno, A. Good and A. Tebben, *Methods in molecular biology*, 2004, **275**, 355–378.
122. J. F. Lowrie, R. K. Delisle, D. W. Hobbs and D. J. Diller, *Combinatorial Chemistry and High Throughput Screening*, 2004, **7**, 495–510.

CHAPTER 10
Integrated Chemo- and Bioinformatics Approaches to Virtual Screening

ALEXANDER TROPSHA

Laboratory for Molecular Modeling and Carolina Center for Exploratory Cheminformatics Research, CB # 7360 School of Pharmacy, University of North Carolina at Chapel Hill, Chapel Hill, NC 27599, USA

10.1 Introduction

Virtual screening is typically considered an area of computer-aided drug discovery where three-dimensional protein structures are used to discover small molecules that fit into the active site (docking) and have high predicted binding affinity (scoring). Traditional docking protocols and scoring functions rely on explicitly defined three-dimensional (3D) coordinates and standard definitions of atom types of both receptors and ligands. Albeit reasonably accurate in many cases, conventional structure-based virtual screening approaches are relatively computationally inefficient, which typically precludes their application towards screening of very large compound collections. Yet millions of compounds in chemical databases and billions of compounds in synthetically feasible virtual chemical libraries are available for virtual screening, calling for the development of approaches that are both fast and accurate in their ability to identify a small number of viable computational hits.

This chapter discusses the use of structure-based cheminformatics approaches as a powerful alternative virtual screening methodology that is also complementary to traditional approaches. The concepts discussed in this chapter extend cheminformatics approaches typically reserved for ligand-based computational

Chemoinformatics Approaches to Virtual Screening
Edited by Alexandre Varnek and Alex Tropsha
© Royal Society of Chemistry, 2008
Published by the Royal Society of Chemistry, www.rsc.org

drug discovery studies towards novel applications for structure-based virtual screening of very large available collections of chemical compounds. Previous chapters of this book have discussed several major cheminformatics concepts that are employed in ligand-based virtual screening. Methodologies presented in this chapter rely on these important cheminformatics concepts such as representation of molecules using multiple descriptors of chemical structures, advanced chemical similarity calculations in multidimensional descriptor spaces, and machine learning and data-mining approaches.

We consider the emerging use of cheminformatics principles in devising novel protocols for structure-based virtual screening, including scoring of bound ligands and database mining. To set the stage, we begin by describing briefly the major principles of the conventional structure-based approaches. To make a bridge between traditional cheminformatics approaches and the structure-based concepts introduced in the later parts of the chapter we then discuss the use of predictive QSAR (Quantitative Structure–Activity Relationship) models for screening chemical libraries. We stress that cheminformatics methodologies discussed in this chapter are very distinct from conventional structure-based methods. Nevertheless, they should not be viewed by any means as substitutes for those approaches. On the contrary, as we discuss in the final section, cheminformatics-based and conventional 3D structure-based virtual screening methods form a natural symbiotic methodological continuum. We argue that cheminformatics approaches that are discussed in this chapter should serve to enrich and broaden the available repertoire of computational molecular modeling methodologies that are used for prioritizing subsets of available chemical databases and libraries for experimental biological assays.

10.2 Availability of Large Compound Collections for Virtual Screening

The early stages of modern drug discovery often involve screening small molecules for their effects on a selected protein target or a model of a biological pathway. In the past 15 years, innovative technologies that enable rapid synthesis and high-throughput screening of large libraries of compounds have been adopted in almost all major pharmaceutical and biotech companies. As a result, there has been a huge increase in the number of compounds available on a routine basis to quickly screen for novel drug candidates against new targets/pathways. In contrast, such technologies have rarely become available to the academic research community, thus limiting its ability to conduct large-scale chemical genomics research. The NIH Molecular Libraries Roadmap Initiative[1] has changed this situation by forming multiple Chemical Library and Screening Centers. These efforts promise to enhance the academic community's ability to perform large-scale chemistry and biological screening against hundreds of biological targets or pathways, emphasizing tool and probe development in areas frequently disregarded by pharmaceutical industry such as rare diseases.

Overall, numerous compounds are available in many databases, as discussed in a recent review.[2] Since the availability of such datasets is critical to make virtual screening technologies practical, we discuss some of them below.

10.2.1 NIH Molecular Libraries Roadmap Initiative and the PubChem Database

As an essential component of NIH's Molecular Libraries Roadmap Initiative, PubChem is the largest chemical database in the public domain. As of October 2007 it contains 19 600 000 substance records for the Substance database and 10 900 000 unique compound records for the Compound database, with links to bioassay description, literature, references, and assay data for each entry. Its BioAssay Database provides searchable descriptions of nearly 600 bioassays, including descriptions of the conditions and readouts specific to a screening protocol.

10.2.2 Other Chemical Databases in the Public Domain

ZINC[3] is a free database of commercially-available compounds. The 2007 ZINC release (ZINC7) is the current default version, with *ca.* 7 000 000 compounds. Among them, over 4.6 million compounds are in ready-to-dock, 3D formats. ChemNavigator[4] contains the iResearch Library, an up-to-date compilation of commercially accessible screening compounds from international chemistry suppliers. This database currently tracks over 39.8 million chemical samples, with over one million sample record updates per month. ChemSpider[5] is delivered *via* a web site and allows one to search a chemical database containing millions of chemical structures and various associated property information. It includes approximately 20 million unique chemical compounds as of October of 2007. The SureChem database[6] allows one to search from more than 5.4 million unique chemical structures extracted from the full text of US, European and World Intellectual Property Organization (WO) patent documents, including prophetic and excluded compounds not found in manually curated databases. New patents and applications are available for structure search within days of publication. The database covers over 164 million chemical occurrences in over 7.3 million applications and granted patents. Several smaller databases, both commercial and publicly available, are described in a recent survey.[2] The availability of these databases with the content frequently linked to the commercial sources of compounds underscores the importance of virtual screening technologies as enabling tools capable of discovering novel drug candidates.

10.3 Structure-based Virtual Screening

Structure-based virtual screening is fundamental to the field of computer-aided drug design.[7] It entails docking and scoring of libraries of small molecules to find compounds that fit into the binding site and bind tightly to the receptor. Since the first seminal paper was published in 1982 by the Kuntz group,[8] this approach has been used successfully in numerous studies, resulting in many cases in the design of approved drugs.[9] Numerous algorithms and programs have been introduced (for reviews see refs. 10–12). Some examples of widely used docking programs include Dock,[13] FlexE,[14] and Gold.[15] We briefly discuss some of the underlying technologies and challenges faced by the current approaches.

10.3.1 Major Methodologies

Virtual screening usually includes three types of studies:[16] (i) sampling of the ligand's positional, conformational and configurational space to predict the ligand's pose within the binding site of the receptor; (ii) scoring of the ligand pose such that the score reflects binding affinity of the ligand; and (iii) hit identification, *i.e.*, screening compound collections (by means of docking and scoring) with the goal of identifying the top scoring candidates that are expected to bind to the receptor. Note that the latter approach does not necessarily require that the binding affinity is accurately reproduced; the method efficiency is typically evaluated by its ability to recover known experimental hits from chemical libraries. Obviously, an accurate scoring function is critical to the outcome of virtual screening. Generally, scoring functions can be classified into three types: Force-field-based scoring functions rely on explicitly computed electrostatic and van der Waals interaction energies between the ligand and the protein. Empirical scoring functions are defined as the sum of individual uncorrelated energy terms and regression analysis is used to reproduce experimental data such as binding energies. Knowledge-based scoring functions are designed based on various statistical parameters that could reflect the interaction between ligands and receptors such as the statistics of pairwise atomic contacts.[17] Among the three, knowledge-based scoring functions are most effective computationally, allowing fast and efficient scoring of large sets of ligand receptor complexes resulting from docking.

Structure-based cheminformatics virtual screening approaches are probably closest in spirit to scoring functions. The scoring functions that are discussed below are judged either by their ability to correlate with experimental binding affinity or to recover known hits from compound collections.

10.3.2 Challenges and Limitations of Current Approaches

The main challenges in the field of virtual screening include computationally efficiency of current protocols for mining large chemical databases to identify putative ligands and the design of accurate and effective scoring functions. Significant progress has been achieved over many years of research in developing structure-based virtual screening approaches. However, several recent publications comparing many available scoring and docking approaches suggest that their accuracy still needs to be improved considerably to afford their automated and successful application to solve practical problems in drug design.[10,16,18,19]

Scoring functions implemented as part of docking software typically perform poorly in identifying the correct pose accurately, which is why modern approaches have converged to separating docking and scoring. Typically, multiple docking poses are produced initially and then scored independently using different functions, including in some cases consensus scoring.[20–24] Many robust and accurate algorithms are available to fit the molecule into the binding site and produce binding poses that position the ligand presumably very close to its native orientation. However, significant challenges remain in developing scoring functions that can accurately identify the native binding pose among many decoys generated using docking. Indeed, the development of accurate docking and scoring functions continues to be a major limiting factor in ensuring greater success of structure-based virtual screening.[17]

A growing number of evaluations of docking programs and scoring functions have been published in recent years.[16,18,25,26] The most recent studies,[16] conducted by scientists at GlaxoSmithKline Pharmaceuticals (GSK), characterized the state of the art for a wide range of docking algorithms and scoring functions applied to systems of relevance for drug discovery. An evaluation of ten docking programs and 37 scoring functions was conducted against proteins from several protein families. Most scoring functions were not successful in their ability to identify the correct crystallographic conformation from the set of docked poses. Furthermore, none of the approaches were found to be universally successful in the accurate prediction of ligand binding affinity, although reasonable correlations between actual and predicted affinity were achieved for some protein families. This recent study showed unequivocally that significant improvements of scoring functions are needed before docking algorithms can have a consistent major impact on virtual screening and lead optimization.

Another critical factor in traditional docking calculations is their computational efficiency. Acceptance or rejection of a ligand must be done efficiently to screen the large number of candidates. The cheminformatics approaches we consider below do offer the advantage of computational efficiency and reasonable accuracy and could supplement or be employed prior to the theoretically more robust structure-based approaches.

10.4 Implementation of Cheminformatics Concepts in Structure-based Virtual Screening

This entire book is devoted to cheminformatics and virtual screening – many of the chapters have discussed major cheminformatics concepts in the context of virtual screening. The main goal of this chapter is to highlight the applicability of common cheminformatics concepts to structure-based virtual screening. However, before introducing the special methodologies we discuss several concepts that serve as a theoretical basis for our structure-based cheminformatics techniques. One such concept is chemical similarity, which is discussed in the Chapter 4 in great detail and therefore will not be discussed here. Another important concept is the QSAR modeling approach that is intrinsically based on the chemical similarity principle. Thus, QSAR models can be regarded as advanced queries for chemical similarity calculations. We discuss aspects of QSAR modeling that we consider critical to allowing the use of models for virtual screening, and also introduce important concepts that can be implemented, in novel ways, as part of structure-based strategies.

10.4.1 Predictive QSAR Models as Virtual Screening Tools

QSAR modeling has been traditionally applied as an evaluative approach, *i.e.*, with the focus on developing retrospective and explanatory models of existing data. Model extrapolation has been considered, if only in a hypothetical sense, in terms of potential modifications of known biologically active chemicals that could improve compounds' activity. Below, we provide arguments and examples suggesting that current methodologies may afford robust and validated models capable of accurate prediction of compound properties for molecules not included in the training sets. We shall discuss a data-analytical modeling workflow developed in our laboratory that incorporates modules for combinatorial QSAR model development (*i.e.*, using all possible binary combinations of available descriptor sets and statistical data modeling techniques), rigorous model validation, and virtual screening of available chemical databases to identify novel biologically active compounds. Our approach places particular emphasis on model validation as well as the need to define model applicability domains in the chemistry space. We present examples of studies where the application of rigorously validated QSAR models to virtual screening identified computational hits that were confirmed by subsequent experimental investigations. The emerging focus of QSAR modeling on target property forecasting brings it forward as a predictive, as opposed to evaluative, modeling approach.

Any QSAR method can be generally defined as an application of mathematical and statistical methods to the problem of finding empirical relationships (QSAR models) of the form $P_i = \hat{k}(D_1, D_2, \dots D_n)$, where P_i are biological activities (or other properties of interest) of molecules, $D_1, D_2, \dots, D_n$ are calculated (or, sometimes, experimentally measured) structural properties (molecular descriptors) of compounds, and $\hat{k}$ is some empirically established mathematical transformation that should be applied to descriptors to calculate the property values for all molecules. The goal of QSAR modeling is to establish a trend in the descriptor values, which parallels the trend in biological activity. In essence, all QSAR approaches imply, directly or indirectly, a simple similarity principle, which for a long time has provided a foundation for the experimental medicinal chemistry: compounds with similar structures are expected to have similar biological activities.

10.4.1.1 Critical Importance of Model Validation

In our important paper titled "Beware of q²!",[27] we have demonstrated the insufficiency of the training set statistics for developing externally predictive QSAR models and formulated the main principles of model validation. Despite the earlier observations of several authors[28–30] warning that a high cross-validated correlation coefficient R^2 (q^2) is a necessary, but not sufficient, condition for the model to have high predictive power, many studies continue to consider q^2 as the only parameter characterizing the predictive power of QSAR models. In ref. 27 we showed that the predictive power of QSAR models can be claimed only if the model was successfully applied for prediction of the external test set compounds, which were not used in the model development. We have demonstrated that most models with high q^2 values have poor predictive power when applied for prediction of compounds in the external test set. Paying attention only to the training set statistics equates to "*narcissistic*" modeling in the sense that such models appear "beautiful" only in the eyes of their developers but provide little if any utility to potential users ("viewers") of these models. In a subsequent publication[31] the importance of rigorous validation was again emphasized as a crucial, integral component of model development. Several examples of published QSPR models with high fitted accuracy for the training sets, which failed rigorous validation tests, have been considered. We presented a set of simple guidelines for developing validated and predictive QSPR models and discussed several validation strategies such as the randomization of the response variable (Y-randomization) external validation using rational division of a dataset into training and test sets. We highlighted the need to establish the domain of model applicability in the chemical space to flag molecules for which predictions may be unreliable, and discussed some algorithms that can be used for this purpose. We advocated the broad use of these guidelines in the development of predictive QSPR models.[31–33]

At the 37th Joint Meeting of Chemicals Committee and Working Party on Chemicals, Pesticides & Biotechnology, held in Paris on 17–19 November 2004, the OECD (Organization for Economic Co-operation and Development) member countries adopted the following five principles that valid (Q)SAR models should follow to allow their use in regulatory assessment of chemical safety: (i) a defined endpoint; (ii) an unambiguous algorithm; (iii) a defined domain of applicability; (iv) appropriate measures of goodness-of-fit, robustness and predictivity; (v) a mechanistic interpretation, if possible. Since then, most European authors publishing in the QSAR area include a statement that their models fully comply with OECD principles (*e.g.*, see refs. 34–37).

Validation of QSAR models is one of the most critical problems of QSAR. Recently, we have extended our requirements for the validation of multiple QSAR models selected by acceptable statistics criteria of prediction of the test set.[38] Additional studies in this critical component of QSAR modeling should establish reliable and commonly accepted "good practices" for model development.

10.4.1.2 Applicability Domains and QSAR Model Acceptability Criteria

One of the most important problems in QSAR analysis is establishing the domain of applicability for each model. In the absence of the applicability domain restriction, each model can formally predict the activity of any compound, even with a completely different structure from those included in the training set. Thus, the absence of the model applicability domain as a mandatory component of any QSAR model would lead to the unjustified extrapolation of the model in the chemistry space and, as a result, a high likelihood of inaccurate predictions. In our research we have always paid particular attention to this issue.[14,31,39-45] The various definitions of applicability domains have been reviewed in Chapter 8 and we refer interested readers to this excellent chapter for additional information on this important subject.

In our earlier publications[27,31] we recommended a set of statistical criteria that must be satisfied by a predictive model. For continuous QSAR, the criteria we follow in developing activity–property predictors are (i) correlation coefficient R between the predicted and observed activities; (ii) coefficients of determination[46] (predicted *versus* observed activities R_0^2, and observed *versus* predicted activities $R_0'^2$ for regressions through the origin); and (iii) slopes k and k' of regression lines through the origin. We consider a QSAR model *predictive* if the following conditions are satisfied: (i) $q^2 > 0.5$; (ii) $R^2 > 0.6$; (iii) $[(R^2 - R_0^2)/R^2] < 0.1$ and $0.85 \leq k \leq 1.15$ or $[(R^2 - R_0'^2)/R^2] < 0.1$ and $0.85 \leq k' \leq 1.15$; (iv) $|R_0^2 - R_0'^2| < 0.3$, where q^2 is the cross-validated correlation coefficient calculated for the training set, but all other criteria are calculated for the test set.

10.4.1.3 Predictive QSAR Modeling Workflow

Our experience in QSAR model development and validation has led us to establish a complex strategy that is summarized in Figure 10.1. It describes the predictive QSAR modeling workflow, which focuses on delivering validated models and, ultimately, computational hits confirmed by the experimental validation. We start by randomly selecting a fraction of compounds (typically, 10–15%) as an external validation set. The remaining compounds are then divided rationally (*e.g.*, using the Sphere Exclusion protocol implemented in our laboratory[33]) into multiple training and test sets that are used for model development and validation, respectively, using criteria discussed in more detail below. We employ multiple QSAR techniques based on the combinatorial exploration of all possible pairs of descriptor sets coupled with various statistical data mining techniques (termed combi-QSAR) and select models characterized by high accuracy in predicting both training and test sets data. Validated models are finally tested using the evaluation set. The critical

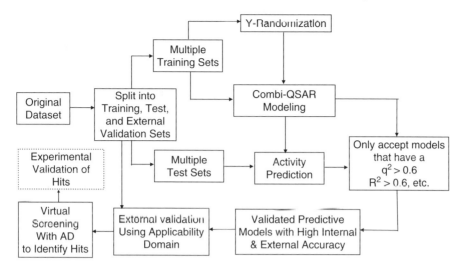

Figure 10.1 Flowchart of predictive QSAR modeling framework based on the validated QSAR models.

step of the external validation is the use of applicability domains. If external validation demonstrates the significant predictive power of the models we use all such models for virtual screening of available chemical databases (*e.g.*, ZINC[3]) to identify putative active compounds and work with collaborators who could validate such hits experimentally. The entire approach is described in detail in several recent papers and reviews (*e.g.*, refs. 31 and 47–49).

In our recent studies we were fortunate to recruit experimental collaborators who have validated computational hits identified through our modeling of anticonvulsants,[43] D1 antagonists[50] and antitumor compounds;[51] some of these studies are described in detail below, preceded by discussion of some methodological aspects of the workflow development. Thus, models resulting from this workflow could be used to prioritize the selection of chemicals for the experimental validation. However, since we can not generally guarantee that every prediction resulting from our modeling effort will be validated experimentally we can not include the experimental validation step as a mandatory part of the workflow on Figure 10.1, which is why we used the dotted box for this component. We note that our approach shifts the emphasis on ensuring good (best) statistics for the model that fits known experimental data towards a focus on generating testable hypothesis about purported bioactive compounds. Thus, the output of the modeling has exactly same format as the input, *i.e.*, chemical structures and (predicted) activities, making model interpretation and utilization completely seamless for medicinal chemists.

10.4.1.4   Examples of Applications

To illustrate the power of validated QSAR models as virtual screening tools we discuss the examples of studies that resulted in experimentally confirmed hits. Such studies could only be performed if there is sufficient data available for a series of tested compounds such that robust validated models could be developing using the workflow described in Figure 10.1.

The first example is anticonvulsant compounds. In the first phase of modeling, we applied kNN[52] and Simulated Annealing–Partial Least-Squares (SA-PLS)[53] QSAR approaches to a dataset of 48 chemically diverse Functionalized Amino Acids (FAA) with anticonvulsant activity that had been synthesized previously, and, thereby, developed successful QSAR models of FAA anticonvulsants.[14] Both methods utilized multiple descriptors such as molecular connectivity indices or atom pair descriptors, which are derived from two-dimensional molecular topology. QSAR models with high internal accuracy were generated, with leave-one-out cross-validated R^2 (q^2) values ranging between 0.6 and 0.8. The q^2 values for the actual dataset were significantly higher than those obtained for the same dataset with randomly shuffled activity values, indicating that the models were statistically significant. The original dataset was further divided into several training and test sets, and highly predictive models providing q^2 values for the training sets greater than 0.5 and R^2 values for the test sets greater than 0.6.

In the second phase of modeling, we applied the validated QSAR models to mining available chemical databases for new lead FAA anticonvulsant agents. Two databases have been thoroughly explored: the National Cancer Institute[54] and the Maybridge[55] databases, including 237771 and 55273 chemical structures, respectively. Database mining was performed independently using ten individual QSAR models that have been extensively validated using several criteria of robustness and accuracy. Each individual model selected some hits as a result of independent database mining, and the consensus hits (*i.e.*, those selected by all models) were further explored experimentally for their anticonvulsant activity. As a result of computational screening of the NCI database, 22 compounds were selected as potential anticonvulsant agents and submitted to our experimental collaborators. Of these 22 compounds, our collaborators chose two for synthesis and evaluation; their choice was based on the ease of synthesis and the fact that these two compounds had structural features that would not be expected to be found in active compounds based on prior experience. Several additional compounds, which were close analogs of these two were either taken from the literature or designed in our collaborator's laboratory. In total, seven compounds were re-synthesized and sent to the NIH for the Maximum Electroshock test (a standard test for the anticonvulsant activity, which was used for the training set compounds as well). The biological results indicated that, upon initial and secondary screening, five out of seven compounds tested showed anticonvulsant activity with an ED_{50} less than 100 mg kg^{-1}, which is considered promising by the NIH standard. Interestingly, all seven compounds were also found to be very active in the same tests performed on rats (a complete set of experimental data on rats for the training set were not available, and therefore no QSAR models for rats were built).

Mining of the Maybridge database yielded two additional promising compounds that were synthesized and sent to NIH for the MES anticonvulsant test. One of the compounds showed moderate anticonvulsant activity, with an ED_{50} of 30–100 mg kg^{-1} (in mice), while the other was found to be a *very* potent anticonvulsant agent, with an ED_{50} of 18 mg kg^{-1} in mice (i.p.). In summary, both compounds were found to be very active in both mice and rats. Figure 10.2 summarizes the results of using validated QSAR models for virtual screening as applied to the anticonvulsant dataset. It presents a practical example of the drug discovery workflow that can be generalized for any dataset where sufficient data to develop reliable QSAR models are available. Importantly, *none* of the compounds identified in external databases as potent anticonvulsants and validated experimentally belong to the same class of FAA molecules as the training set. This observation was very stimulating because it underscored the power of our methodology to identify potent anticonvulsants of novel chemical classes as compared to the training set compounds, which is one of the most important goals of virtual screening.

Anticancer Agents. A combined approach of validated QSAR modeling and virtual screening was successfully applied to the discovery of novel tylophorine derivatives as anticancer agents.[51] QSAR models were initially developed for 52 chemically diverse Phenanthrine-based Tylophorine Derivatives (PBTs) with known experimental EC_{50} using chemical topological descriptors (calculated with the MolConnZ program) and variable selection k nearest neighbor (kNN) method. Several validation protocols have been applied to achieve robust QSAR models. The original dataset was divided into multiple training and test sets, and the models were considered acceptable only if the leave-one-out cross-validated R^2 (q^2) values were greater than 0.5 for the

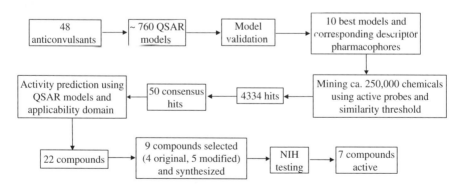

Figure 10.2 Application of the QSAR modeling workflow to the discovery of novel anticonvulsants. The workflow emphasizes the importance of model validation and applicability domain in ensuring high hit rates as a result of database mining with predictive QSAR models.

training sets and the correlation coefficient R^2 values were greater than 0.6 for the test sets. Furthermore, the q^2 values for the actual dataset were shown to be significantly higher than those obtained for the same dataset with randomized target properties (Y-randomization test), indicating that models were statistically significant. The ten best models were then employed to mine a commercially available ChemDiv Database (*ca.* 500 k compounds), resulting in 34 consensus hits with moderate to high predicted activities. Ten structurally diverse hits were experimentally tested and eight were confirmed active, with the highest experimental EC_{50} of 1.8 μM, implying an exceptionally high hit rate (80%). The same ten models were further applied to predict EC50 for four new PBTs, and the correlation coefficient (R^2) between the experimental and predicted EC_{50} for these compounds plus eight active consensus hits was shown to be as high as 0.57.

In summary, our studies have established that QSAR models can be used successfully as virtual screening tools to discover compounds with the desired biological activity in chemical databases or virtual libraries.[43,48,50,51,56] The discovery of novel bioactive chemical entities is the primary goal of computational drug discovery, and the development of validated and predictive QSAR models is critical to achieve this goal. The implementation of some of the chief principles underlying QSAR modeling as part of structure-based drug discovery strategy presents another exciting avenue at the interface between cheminformatics and bioinformatics. Examples of such hybrid approaches are discussed below.

10.4.2 Structure-based Chemical Descriptors of Protein–Ligand Interface: The EnTESS Method

The prediction of the protein–ligand binding affinity is a critical component of computational drug discovery. The rapid growth of the Protein Data Bank[57] and derivative databases such as PDBBind,[21] MOAD[58] or BindingDB[59] provide opportunities to enhance current protocols for molecular docking and scoring, which are at the core of structure-based drug design and hit identification. Accurate estimation of binding affinities, or at least correct relative ranking of different ligands, has proven difficult due to multiple energetic and entropic factors that must be accounted for.[60] As discussed above, the limited accuracy of current scoring functions is one of the problems hampering the broad application of docking and virtual screening in lead optimization.

Structure-based drug design approaches rely on the availability of structural information about protein–ligand complexes. In contrast, ligand-based approaches rely only on the experimental structure–activity relationships for ligands only. As discussed above, QSAR methods are typically used to find correlations between ligands' binding affinities and their chemical descriptors. As an innovative use of QSAR approaches, several so-called receptor-dependent quantitative structure–activity relationship (RD-QSAR) methods have been

developed that rely on the receptor structure information to calculate independent variables.[61,62] Holloway and co-workers[63] have derived a highly significant
3D-QSAR model for HIV-1 protease and its peptidomimetic inhibitors and
used it to predict binding affinities for newly designed ligands. Several other
authors[64-66] have developed new methodologies by considering all of the
enthalpic and entropic contributions as well as solvation effects of the receptor–
ligand interactions and treated them as independent variables in the RD-QSAR
development.

Recently, we have begun to develop a hybrid methodology to predict the
binding affinities for a highly diverse dataset of protein–ligand complexes using
concepts from both structure- and ligand-based approaches. This methodology
is based on four-body statistical scoring function derived by combined application of the Delaunay tessellation of protein–ligand complexes and the definition of chemical atom types using the fundamental chemical concept of
atomic electronegativity. As described in our publications,[67,68] Delaunay tessellation naturally partitions a tertiary structure of a protein or a protein–
ligand complex into an aggregate of space-filling, irregular tetrahedra, or
simplices; the vertices of the simplices are quadruplets of nearest neighbor
residues or atoms (Figure 10.3). Thus, Delaunay tessellation reduces a complex
3D structure to a collection of explicit, elementary atomic quadruplet structural
motifs. Four vertices (atoms) of a simplex form a particular quadruplet composition and the chemical properties of the atom types can characterize the type
of the tetrahedron. As we describe below the properties of quadruplet atomic
compositions of the interfacial tetrahedral formed by both ligand and receptor
atoms could be used as a novel type of chemical descriptors for the protein–
ligand interface.

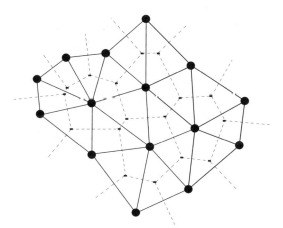

Figure 10.3 Illustration of Voronoi/Delaunay tessellation in 2D space (Voronoi
polyhedra are represented by dashed lines, and Delaunay simplices by
solid lines). For the collection of points with 3D coordinates, such as
atoms of the protein–ligand complex, Delaunay simplices are tetrahedra
whose vertices correspond to the atoms.

10.4.2.1   Derivation of the EnTESS Descriptors

Using Delaunay tessellation, the protein–ligand interface can be defined by tetrahedra formed by both protein and ligand atoms. We used a distance cutoff value of 8 Å to exclude Delaunay simplices with long edges (exceeding the physically meaningful interaction distance) between vertices. As shown in Figure 10.4, we could distinguish three classes of interfacial tetrahedra, *i.e.*, RRRL, RRLL and RLLL, where each R and L corresponds to a receptor and ligand atom, respectively. For each class we further defined 554 types of quadruplet compositions based on our definition of chemical atom types (*vide infra*) without taking into account their order in the quadruplet. For example, all quadruplets with atom types C_L, C_R, S_L and X_L, were assigned to the same [X_L, S_L, C_L, C_R] composition type.

In principle, quadruplet compositions of different atom types could be used as simple metrics to distinguish different protein–ligand complexes. However, we sought some fundamental atomic property that could be attributed to any chemical atom type of either receptor or ligand and could be useful in describing interatomic interactions at the ligand–receptor interface. We decided to use the Pauling electronegativity[69] as a parameter to characterize atom types. Electronegativity is considered to be the main factor determining the atom's polarity and its ability to form a hydrogen bond. For example, oxygen has high electronegativity and a high ability to form hydrogen bonds and it is a polar atom type in most cases. Thus, electronegativity could be used to describe the interactions between protein and ligand atoms. For instance, Hall *et al.* have introduced electrotopological state (E-state) indices, which are indirectly related to electronegativity, and successfully used them in QSAR studies of many datasets.[70] Zefirov and co-workers have used the electronegativity equalization scheme as a source of electronic descriptors to study some types of chemical reactivity and obtained good models for thermodynamic and kinetic data such as proton affinity and Taft's inductive sigma* constants.[71]

To distinguish ligand *vs.* protein atoms, we have classified the protein and ligand C, N, O and S as different atom types. Hydrogen atoms were not considered since, usually, they are not defined explicitly in X-ray structures. Thus, we have defined four atom types for receptor proteins and six atom types for the ligands. In total, there were 554 possible types of interfacial atomic quadruplet compositions, and each of them gave rise to an independent variable

| RLLL | RRLL | RRRL |

Figure 10.4   Topological Tetrahedral Types: RRRL: formed by three receptor atoms and one ligand atom; RRLL: formed by two receptor atoms and two ligand atoms; RLLL: Formed by one receptor atom and three ligand atoms.

(a sum of EN values for composing atom types) for our QSBR studies. Table 10.1 summarizes atom-type definitions.

We have applied this procedure to a training set of 264 protein–ligand complexes selected from the PDBBind dataset[21] and counted the number of occurrences of each of the 554 atom quadruplet types. If the number of times a particular type occurred was higher than 50, we considered this quadruplet type significant. Otherwise, this type was discarded, thereby reducing the number of independent variables for the subsequent analysis; 132 types of quadruplets were found to occur with sufficiently high frequency. For each type of the tetrahedral composition, the EN values of the four composing atoms were added up, and the resulting sums for all of the tetrahedra belonging to this composition type were then added up again. The result of these calculations represented the value of the descriptor (*i.e.*, one of possible 132 descriptors) for the particular protein–ligand complex (Figure 10.5).

Since we employed the concept of electronegativity combined with Delaunay *tess*ellation of protein–ligand complexes we termed these unique characteristics of the protein–ligand interface the *EnTess descriptors*.[38] We have applied the variable selection *k*-nearest neighbor (*k*NN) QSAR approach[52] to establish correlations between binding affinities and the EnTess descriptors as described below.

Table 10.1 Atom-type definitions used in deriving EnTess.

Ligand atom types	*EN*
O	3.4
N	3.0
C	2.5
S	2.4
X	2.0–2.4, 4.0 (P and halogens)
M	0.6–1.6 (metal and all other rare atom types)
Receptor atom types	
O	3.4
N	3.0
C	2.5
S	2.4

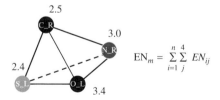

$$EN_m = \sum_{i=1}^{n} \sum_{j}^{4} EN_{ij}$$

Figure 10.5 Calculation of the EnTess descriptors. The same atom type from receptor and ligand is treated differently. In the formulas, m is the mth quadruplet composition type, n represents the number of occurrences of this composition type in a given protein–ligand complex, and j is the vertex index within the quadruplet.

10.4.2.2 Validation of the EnTESS Descriptors for Binding Affinity Prediction

We have employed the EnTess descriptors as independent variables in multivariate correlation analysis of the experimental dataset of 264 diverse protein–ligand complexes with known binding constants selected from the PDBBind dataset.[21] Following the protocols for developing validated and predictive QSAR models (Section 10.4.1.3 and Figure 10.1) we divided the modeling datasets into the training, test, and independent validation sets. We have obtained statistically significant Quantitative Structure-Binding Affinity Relationships (QSBR) models capable of predicting the binding affinities of ligands in the independent validation set with an R^2 of 0.85 (see ref. 38 for additional details).

The results were compared with those obtained earlier using both knowledge-based and empirical scoring functions (Table 10.2). Since there are no standard training and test sets used by different groups, precise comparison of results is impossible. Nevertheless, the results show that, generally, EnTess gives better or at least comparable prediction accuracy of binding affinity as compared to other scoring functions. For example, training sets in EnTess and SMoG96[72] studies have a comparable number of complexes (122–200), while most of the EnTess test sets are larger than that used in SMoG96 studies (40–118 *vs.* 46 complexes). However, EnTess gave significantly higher predictive accuracy for test sets than SMoG (Table 10.2). The R^2 values for the test sets obtained with EnTess are 0.28–0.35 higher than that for SMoG. Even for a much bigger test set including 118 complexes EnTess gave $R^2 = 0.63$. With respect to other published knowledge-based scoring functions, test sets of comparable size and much smaller training sets (two to four times less) have been used in EnTess studies; nevertheless, the R^2 values from EnTess were approximately 0.1–0.4 higher. Importantly, EnTess afforded high and stable prediction accuracy for an external structurally diverse dataset (on average, $R^2_{cons} = 0.81$). Alternative

Table 10.2 Comparison of predictive power of EnTess models with that obtained with alternative scoring functions.

Methods	Training set size	Test set size	R^2 for test sets	Consensus R^2 for external set (10 best models)
BLEEP95,96	351	90	0.53	–
PMF[20]	697	77	0.61	–
SMoG9672,97	120	46	0.42	–
SMoG2001[98]	725	111	0.436	–
SCORE[74]	170	11	0.65	–
XSCORE[99]	200	30	0.36	–
LUDI[100]	82	12	0.45	–
VALIDATE[64]	51	14	0.81	–
ChemScore[101]	82	20	0.63	–
EnTess[38]	199–175	41–65	0.70–0.77	0.85

empirical scoring functions, such as VALIDATE,[73] led to worse or comparable results with relatively smaller training sets (except SCORE[74] and XSCORE[75]), but those test sets are about half the size of those used in EnTess studies. This usually causes an increase of R^2 values. In summary, the models were rigorously validated with test sets, using the additional external prediction set of 24 compounds to simulate the real application of the model, and by performing a Y-randomization test. The results demonstrate the high prediction power of these models and the applicability of the novel geometrical chemical descriptors to receptor–ligand complex binding affinity prediction.

10.4.3 Structure-based Cheminformatics Approach to Virtual Screening: The CoLiBRI Method

Ligand-based approaches rely on series of ligands with known binding affinities to build correlations between ligand chemical structure and target properties of interest, such as binding constants or specific biological activities (see ref. 76 for a review). The ligand structures are typically represented by multiple chemical descriptors,[77] and statistical data modeling techniques are used to establish quantitative correlations between descriptors and binding affinities. Chemical descriptors and various chemical similarity measures (*e.g.*, Euclidean distances between compounds in multidimensional descriptor space) are at the core of chemometric approaches to the analysis of molecular databases.[78] Such approaches afford rapid chemical similarity calculations and are widely used in database mining or rational library design to discover molecules similar to available compounds that are likely to have similar biological activity.[79] Chemical similarity searches are much more computationally efficient than structure-based virtual screening. However, they are more likely to identify false positives that are too bulky or simply not stereochemically complementary to the actual binding site because the binding site information is not typically used as part of the query. Furthermore, chemometric approaches typically identify compounds that are highly similar to the training set compounds, making it difficult to identify novel ligands of a different structural class.

We present below a novel computational drug discovery strategy that combines the strengths of *both* structure- and ligand-based approaches while attempting to surpass their individual shortcomings. In developing this strategy, we sought a representation that would allow us to characterize both receptor active sites and their corresponding ligands in the same universal, multidimensional, chemical descriptor space. We reasoned that mapping of both binding pockets and corresponding ligands onto the same multidimensional chemistry space would preserve the complementarity relationships between binding sites and their respective ligands. Thus, we expected that similar binding sites (where similarity is described quantitatively using one of the conventional metrics, such as Manhattan distance in multidimensional descriptor space) would correspond to similar ligands. This would imply that the relative location of a novel binding site in this chemistry space with respect to other binding sites could be used to predict the location of the ligand(s) complementary to this site in the ligand chemistry space. This virtual ligand(s) could then be used as a query in chemical similarity searches to identify putative ligands of the same receptor in available chemical databases. These considerations formed the basis for the development of the cheminformatics structure-based drug discovery strategy that we termed CoLiBRI (identification of Complementary Ligand Based on Receptor Information). As we discuss below, CoLiBRI relies on the knowledge of the receptor active site to afford highly computationally efficient and accurate identification of its respective ligand(s) within a large compound database.

10.4.3.1 Representation of Three-dimensional Active Sites in Multidimensional Chemistry Space

Several important considerations went into finding the most capable descriptors in the context of our studies. There are two major classes of traditional chemical descriptors that are derived from either 2D chemical graphs (*e.g.*, molecular connectivity indices, charge descriptors, and others[80-86]) or from 3D molecular models using relative atomic positions in addition to atom properties. A major benefit of 2D over 3D chemometric methods is that the former neither requires a conformational search nor structural alignment of molecules. Accordingly, 2D methods are more easily automated and adapted to the task of database searching or virtual screening.[87,88] In fact, 2D descriptors have been shown to be superior to 3D descriptors in database mining.[89] However, most 2D chemical descriptors are typically calculated from only complete molecular graphs. Consequently, they can not be used to characterize active sites that are composed of fragments or individual atoms of amino acid residues that are involved in specific contacts with ligands. A notable exception is the TAE descriptors.

The TAE/RECON method, developed by Breneman and co-workers,[90] is based on Bader's quantum theory of Atoms In Molecules (AIM). The TAE method of molecular electron density reconstruction utilizes a library of integrated atomic "basins", as defined by the AIM theory, to rapidly reconstruct representations of molecular electron density distributions and van der Waals electronic surface properties. RECON is capable of rapidly generating $6-31 + G^*$ level electron densities and electronic properties of large molecules, proteins or molecular databases, using TAE reconstruction. A library of atomic charge density fragments has been assembled in a form that allows for the rapid retrieval of the fragments, followed by rapid molecular assembly. Additional details of the method are described elsewhere.[90-92]

10.4.3.2 Mapping between Chemistry Spaces of Active Sites and Ligands

The calculation of TAE/RECON descriptors for the ligands (extracted from their protein complexes) is straightforward. However, similar calculations for the binding sites first require the identification of individual atoms or amino acid fragments involved in specific ligand–receptor interactions. To this end, we have utilized Delaunay tessellation described above (*cf.* Section 10.4.2) to identify protein atoms that make contacts with bound ligands. The RECON/TAE method was then used to generate a set of descriptors for pseudo-molecules constructed from the active site atoms. In doing so, we relied on the unique feature of this method that calculates molecular descriptors from those for molecular fragments, *i.e.*, in principle, the TAE/RECON method does not require that fragments are connected.

Multiple descriptors were generated for both the receptor binding sites and their corresponding ligands so that each chemical entity is represented as a vector in a multidimensional TAE/RECON chemical space. Each dimension of this space corresponds to specific structural features of the ligands and active sites, but not every feature may be important for determining ligand–receptor complementarity. Thus, special procedures were involved to select the most significant descriptors (see ref. 93 for additional details).

In brief, the CoLiBRI model is a series of ligand–receptor complexes mapped into a descriptor space. Ligands for a test receptor's binding pocket are predicted by positioning the test receptor pocket in the selected descriptor subspace and finding the K most similar receptor pockets from the training set. The known ligands of these K most similar receptor pockets are then used to estimate the position of the test receptor's virtual ligand in the descriptor space (see below). All potential ligands are then ranked based on their distance to this predicted virtual ligand point, and the ligand(s) with the smallest distance are considered the most probable hits. Identifying a potential receptor target for a test ligand occurs in the opposite fashion, whereby the K most similar training set ligands are found and the known receptors of those ligands are used to interpolate what receptor target is the most likely candidate.

The CoLiBRI models were developed using standard leave-one-out cross-validation procedure as follows:

(1) Choose a receptor in the training set and select its k nearest neighbors in the TAE/RECON binding site descriptor space. Identify the ligands of the kNN receptors in the ligand space and use their coordinates to predict the coordinates of the chosen receptor's virtual ligand. The coordinates of the virtual ligand are calculated from Equation (10.1) for $k \geq 2$ (different values of k are explored to find the best model as described below):

$$\overrightarrow{X_{\mathrm{pp}i}} = \sum_{k=1}^{K_{\mathrm{Best}}} \left[\frac{\overrightarrow{X_{\mathrm{L}\neg\mathrm{R}k}}}{K_{\mathrm{Best}} - 1} \cdot \left(1 - \frac{\left\|\overrightarrow{X_{\mathrm{R}k}} - \overrightarrow{X_{\mathrm{R}\,\mathrm{Pred}i}}\right\|}{\sum\limits_{k=1}^{K_{\mathrm{Best}}} \left\|\overrightarrow{X_{\mathrm{R}k}} - \overrightarrow{X_{\mathrm{R}\,\mathrm{Pred}i}}\right\|} \right) \right] \quad (10.1)$$

where $X_{\mathrm{RPred}i}$ is the chosen receptor i, $X_{\mathrm{pp}i}$ is the predicted ligand vector for the receptor i, $X_{\mathrm{R}k}$ is the k nearest receptor, and $X_{\mathrm{L}\leftarrow\mathrm{R}k}$ is the ligand of the

k nearest neighbor receptor. For the case where $K_{\text{Best}} = 1$, then $X_{\text{pp}j}$ is simply the position of the nearest receptor's ligand in the ligand space, $X_{\text{L}\leftarrow\text{R}1}$.

(2) Rank known ligands based on their chemical similarity to the virtual ligand. The similarities are evaluated as Euclidean distances (Equation 10.2) using only the subset of descriptors that correspond to the current N_{var} selection:

$$\text{Dist}_{i,j} = \sqrt{\sum_{d=1}^{N_{\text{var}}} (X_{id} - X_{jd})^2} \qquad (10.2)$$

(3) Repeat steps 1 and 2 until every receptor in the training set has been eliminated once, and the receptor's virtual ligand and the rank order of all compounds are predicted.

(4) Calculate the PMR for the model using Equation (10.3), where NLR is the number of ligand–receptor complexes in the training set, N_{var} is the number of descriptors used to build the correlation, X_{jd} and X_{id} are the d-th selected descriptor for ligands j and i, respectively, and $X_{\text{pp}id}$ is the d-th descriptor of the predicted ligand point.

$$\text{PMR} = \frac{1}{N_{\text{LR}}} \sum_{i=1}^{N_{\text{LR}}} \sum_{j=1}^{N_{\text{LR}}} \begin{cases} 1 & \text{if } \sum_{d=1}^{N_{\text{var}}} (X_{jd} - X_{\text{pp}id})^2 \\ & \leq \sum_{d=1}^{N_{\text{var}}} (X_{id} - X_{\text{pp}id})^2 \cap i \neq j \\ 0 & \text{if } \sum_{d=1}^{N_{\text{var}}} (X_{sjd} - X_{\text{pp}id})^2 \\ & > \sum_{d=1}^{N_{\text{var}}} (X_{id} - X_{\text{pp}id})^2 \cup i = j \end{cases} \qquad (10.3)$$

(5) Repeat steps 1–4 for $k=3, 4, 5$, *etc.* Formally, the upper limit of k is the total number of ligand–receptor pairs in the data set minus one; however, the best value has been found empirically to lie between two and five. The k value that leads to the lowest PMR value is chosen as optimal.

10.4.3.3   *Application of CoLiBRI to Virtual Screening*

A diverse training set of 670 receptor binding pockets was selected from a modeling set of 800 complexes using the Sphere Exclusion Algorithm,[32] as is typically done in our QSAR studies. This set was used by CoLiBRI to build models with the lowest PMR (Equation 10.3). The remaining 130 receptors were used as a test set to evaluate the ability of the optimized model(s) to identify the correct ligand of each test receptor out of the original 800 ligands.

Previous studies from our group in the area of QSAR indicated that the most reliable predictions of the test set data are obtained by using the consensus prediction approach.[43] In this approach, multiple variable selection models are built for the training set and used for the prediction of the test set ligands concurrently. To accomplish a consensus prediction, each model ranked all compounds in our ligand database based on the distance of each ligand to a test receptor's virtual complementary ligand. We then re-ranked the ligands based on those that were most similar to the virtual ligand across multiple models. These studies have shown that the inclusion of variable selection improved the mean rank of the test set from 37 to 24 out of 800. Furthermore, by using 100 models for consensus prediction, the mean rank of the test set was improved from 24 to 18.1 out of 800 (Figure 10.6). This increased the CPU time required to predict the test set by more than two orders of magnitude. Nevertheless, despite the increased CPU time, the calculations were still completed within 15 minutes. Since variable selection and consensus modeling vastly improved test set prediction, these methods were used in all subsequent model developments.

To simulate the use of CoLiBRI for screening large chemical databases, we added the 800 training set ligands to the WDI dataset,[94] which contained *ca.* 54 000 drugs and drug candidates at the time of calculations in 2004. Training set CoLiBRI models were used in a consensus manner to predict the correct ligands for each of the 130 test receptors from of the entire combined database. The results illustrated that, even when searching a large compound database, CoLiBRI is, on average, able to rank known ligands for a test receptor to

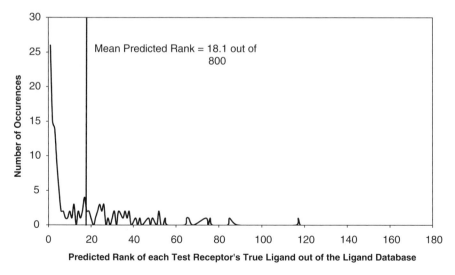

Figure 10.6 Predictive ability of CoLiBRI to identify ligands of 130 test binding pockets out of the original 800 ligands.

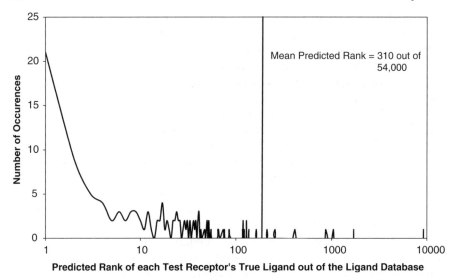

Figure 10.7   Predictive ability of CoLiBRI to identify ligands of 130 test binding
pockets from the WDI (World Drug Index) and the original 800 ligands.

within the top 310 ligands out of *ca.* 54 000, which translates to the top 1% of
all compounds (Figure 10.7).

The entire screening calculation for 130 test receptors took roughly four
hours on a 2.4 GHz Pentium 4 machine. Figure 10.7 illustrates that most of the
ligands were correctly identified within the top 12 ranked compounds; however,
there were two distant outliers that made the average rank much higher. These
two outliers (PDB codes 1BM7 and 1G4J) did not contain a receptor–ligand
complex from the same family as those in the training set, which could possibly
explain the inaccuracy of the predictions. The ligands extracted from 1BM7
and   1G4J,   flufenamic   acid   and   4-(aminosulfonyl)-*N*-[(2,3,4,5,6-penta-
fluorophenyl)methyl]benzamide, respectively, also do not appear to be very
similar to ligands found within the training dataset. This additional dissimilarity may have also played a role in their poor prediction. CoLiBRI appears
to perform best when a receptor of the same family as the test set receptor is
present in the training set. Otherwise, CoLiBRI is best used as a quick, rough
filtering tool that can be used prior to the application of alternative less computationally efficient but perhaps more robust screening methodologies.

10.5   Summary and Conclusions: Integration
of Conventional and Cheminformatics
Structure-based Virtual Screening Approaches

In this chapter, we have considered the use of cheminformatics approaches
towards structure-based virtual screening. (Parenthetically, this terms is

traditionally reserved for approaches that explicitly utilize the knowledge of 3D structure of the protein active site; however, the term "structure" applies equally well to cheminformatics ligand-based approaches as well where the structure of low molecular weight compounds that are known to bind to a specific receptor is used for virtual screening using similarity or QSAR based approaches.) We have reviewed the major tenets and limitations of conventional target structure-based approaches and presented evidence that several typical cheminformatics constructs (*e.g.*, structure representation using multi-dimensional chemical descriptor spaces, QSAR models, similarity searches) could be extended towards novel applications for structure based design. Both types of approaches have their natural advantages and limitations. Thus, most ligand-based methods can not be applied effectively unless the structure and activity of a series of ligands is known. Conversely, structure-based approaches can be used in principle even in the case of orphan receptors where no information about any ligands is available (although, of course, the knowledge of at

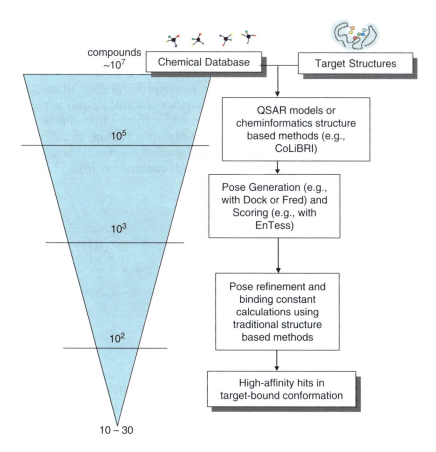

Figure 10.8   Integration of cheminformatics-based and conventional structure-based approaches to virtual screening.

least a few low molecular weight ligands, especially co-crystallized with the target protein, is very helpful). On the other hand, cheminformatics approaches offer the advantage of computational efficiency and in some cases when a significant number of protein ligands are known could perhaps be as effective as structure-based methods in identifying potent ligands from chemical libraries as a result of virtual screening.

As mentioned in the Introduction, the two types of methods should not be viewed as competing. In fact, the most attractive approach is to use both types in concert when possible, *i.e.*, when enough data to enable their use is available. Figure 10.8 illustrates a scenario when both cheminformatics and traditional structure-based methods could be applied in combination to screen large compound collections against a target with known 3D structure (and perhaps known ligands). Thus, one could start using fast cheminformatics approaches (such as CoLiBRI and if possible QSAR models) to filter out a significant fraction of the available compound collection. Then, docking methods (such as FRED or DOCK) could be used to generate poses for the remaining compounds, and scoring approaches (*e.g.*, EnTess or current scoring functions, possibly in consensus fashion) could help to eliminate further those molecules that are unlikely to have an appreciably high binding affinity. Finally, most theoretically robust but relatively inefficient methods such as molecular dynamics or free energy simulations could be applied to a relatively small number of computational hits resulting from the previous steps of the workflow. In the end, we may arrive at a small number of viable ligands in receptor bound conformation with high predicted binding affinity that can be tested experimentally. Strategies that integrate cheminformatics and bioinformatics approaches to virtual screening may become especially useful as the databases of available or synthetically feasible compounds continue to grow.

Acknowledgements

The studies described in this chapter were supported in parts by the National Institutes of Health's Cheminformatics Center planning grant P20-RR20751 and the research grant R01GM066940.

References

1. MLI. http://mli.nih.gov/ (accessed May 2008).
2. T. Oprea and A. Tropsha, Target, chemical and bioactivity databases – integration is key, *Drug Discov. Today*, 2006, **3**, 357–365.
3. J. J. Irwin and B. K. Shoichet, ZINC–a free database of commercially available compounds for virtual screening., *J. Chem. Inf. Model.*, 2005, **45**, 177–182.
4. ChemNavigator. http://www.chemnavigator.com (accessed May 2008).
5. ChemSpider. http://www.chemspider.com (accessed May 2008).

6. SureChem. http://surechem.reeltwo.com (accessed May 2008).

7. N. Brooijmans and I. D. Kuntz, Molecular recognition and docking algorithms., *Annu. Rev. Biophys. Biomol. Struct.*, 2003, **32**, 335–373.

8. I. D. Kuntz, J. M. Blaney, S. J. Oatley, R. Langridge and T. E. Ferrin, A geometric approach to macromolecule-ligand interactions., *J. Mol. Biol.*, 1982, **161**, 269–288.

9. A. Wlodawer and J. Vondrasek, Inhibitors of HIV-1 protease: a major success of structure-assisted drug design., *Annu. Rev. Biophys. Biomol. Struct.*, 1998, **27**, 249–284.

10. C. F. Wong and J. A. McCammon, Protein flexibility and computer aided drug design., *Annual Review of Pharmacol. Toxicol*, 2003, **43**, 31–45.

11. R. D. Taylor, P. J. Jewsbury and J. W. Essex, A review of protein-small molecule docking methods., *J. Comput. Aided Mol. Des*, 2002, **16**, 151–166.

12. I. Muegge, Selection criteria for drug-like compounds., *Medicinal research reviews*, 2003, **23**, 302–321.

13. S. J. Cho, W. Zheng and A. Tropsha, Rational combinatorial library design. 2. Rational design of targeted combinatorial peptide libraries using chemical similarity probe and the inverse QSAR approaches., *J. Chem. Inf. Comput. Sci.*, 1998, **38**, 259–268.

14. M. Shen, A. LeTiran, Y. Xiao, A. Golbraikh, H. Kohn and A. Tropsha, Quantitative structure-activity relationship analysis of functionalized amino acid anticonvulsant agents using k nearest neighbor and simulated annealing PLS methods., *J. Med. Chem.*, 2002, **45**, 2811–2823.

15. G. Jones, P. Willctt, R. C. Glen, A. R. Leach and R. Taylor, Development and validation of a genetic algroithm for flexible docking., *Journal of Molecular Biology*, 1997, **267**, 727–748.

16. G. L. Warren, C. W. Andrews, A. M. Capelli, B. Clarke, J. LaLonde, M. H. Lambert, M. Lindvall, N. Nevins, S. F. Semus, S. Senger, G. Tedesco, I. D. Wall, J. M. Woolven, C. E. Peishoff and M. S. Head, A critical assessment of docking programs and scoring functions., *J. Med. Chem.*, 2006, **49**, 5912–5931.

17. D. B. Kitchen, H. Decornez, J. R. Furr and J. Bajorath, Docking and scoring in virtual screening for drug discovery: methods and applications., *Nat. Rev. Drug Discov.*, 2004, **3**, 935–949.

18. H. Chen, P. D. Lyne, F. Giordanetto, T. Lovell and J. Li, On evaluating molecular-docking methods for pose prediction and enrichment factors., *J. Chem. Inf. Model.*, 2006, **46**, 401–415.

19. S. J. Teague, Implications of protein flexibility for drug discovery., *Nature reviews*, 2003, **2**, 527–541.

20. I. Muegge and Y. C. Martin, A general and fast scoring function for protein–ligand interactions: a simplified potential approach., *J. Med. Chem.*, 1999, **42**, 791–804.

21. R. Wang, X. Fang, Y. Lu and S. Wang, The PDBbind database: collection of binding affinities for protein–ligand complexes with known three-dimensional structures., *J. Med. Chem.*, 2004, **47**, 2977–2980.

22. I. J. Bruno, J. C. Cole, J. P. Lommerse, R. S. Rowland, R. Taylor and M. L. Verdonk, IsoStar: a library of information about nonbonded interactions., *J. Comput. Aided Mol. Des*, 1997, **11**, 525–537.

23. M. Hendlich, A. Bergner, J. Gunther and G. Klebe, Relibase: design and development of a database for comprehensive analysis of protein–ligand interactions., *J. Mol. Biol.*, 2003, **326**, 607–620.

24. S. Vajda and F. Guarnieri, Characterization of protein–ligand interaction sites using experimental and computational methods., *Curr. Opin. Drug Discov. Devel.*, 2006, **9**, 354–362.

25. M. Kontoyianni, G. S. Sokol and L. M. McClellan, Evaluation of library ranking efficacy in virtual screening., *J. Comput. Chem.*, 2005, **26**, 11–22.

26. C. Bissantz, P. Bernard, M. Hibert and D. Rognan, Protein-based virtual screening of chemical databases. II. Are homology models of G-protein coupled receptors suitable targets?, *Proteins, Structure, Function and Genetics*, 2003, **50**, 5–25.

27. A. Golbraikh and A. Tropsha, Beware of q2!, *J. Mol. Graph. Model.*, 2002, **20**, 269–276.

28. E. Novellino, C. Fattorusso and G. Greco, Use of comparative molecular field analysis and cluster analysis in series design., *Pharm. Acta Helv*, 1995, **70**, 149–154.

29. U. Norinder, Single and domain made variable selection in 3D QSAR applications., *J. Chemomet.*, 1996, **10**, 95–105.

30. A. Tropsha and S. J. Cho, Cross-validated R2-guided region selection for CoMFA studies, *In 3D QSAR in Drug Design*. III ed., H. Kubinyi, G. Folkers and Y. C. Martin Eds., Kluwer Academic Publishers, Dordrecht, 1998, pp. 57–69.

31. A. Tropsha, P. Gramatica and V. K. Gombar, The importance of being earnest: validation is the absolute essential for successful application and interpretation of QSPR models., *Quant. Struct. Act. Relat. Comb. Sci.*, 2003, **22**, 69–77.

32. A. Golbraikh and A. Tropsha, Predictive QSAR modeling based on diversity sampling of experimental datasets for the training and test set selection., *J. Comput. Aided Mol. Des*, 2002, **16**, 357–369.

33. A. Golbraikh, M. Shen, Z. Xiao, Y. D. Xiao, K. H. Lee and A. Tropsha, Rational selection of training and test sets for the development of validated QSAR models., *J. Comput. Aided Mol. Des*, 2003, **17**, 241–253.

34. M. Pavan, T. I. Netzeva and A. P. Worth, Validation of a QSAR model for acute toxicity., *SAR QSAR Environ. Res.*, 2006, **17**, 147–171.

35. M. Vracko, V. Bandelj, P. Barbieri, E. Benfenati, Q. Chaudhry, M. Cronin, J. Devillers, A. Gallegos, G. Gini, P. Gramatica, C. Helma, P. Mazzatorta, D. Neagu, T. Netzeva, M. Pavan, G. Patlewicz, M. Randic, I. Tsakovska and A. Worth, Validation of counter propagation neural network models for predictive toxicology according to the OECD principles: a case study., *SAR QSAR Environ. Res.*, 2006, **17**, 265–284.

36. A. G. Saliner, T. I. Netzeva and A. P. Worth, Prediction of estrogenicity: validation of a classification model., *SAR QSAR Environ. Res.*, 2006, **17**, 195–223.

37. D. W. Roberts, A. O. Aptula and G. Patlewicz, Mechanistic applicability domains for non-animal based prediction of toxicological endpoints. QSAR analysis of the Schiff base applicability domain for skin sensitization., *Chem. Res. Toxicol.*, 2006, **19**, 1228–1233.

38. S. Zhang, A. Golbraikh and A. Tropsha, Development of quantitative structure-binding affinity relationship models based on novel geometrical chemical descriptors of the protein–ligand interfaces., *J. Med. Chem.*, 2006, **49**, 2713–2724.

39. A. Golbraikh, D. Bonchev and A. Tropsha, Novel chirality descriptors derived from molecular topology., *J. Chem Inf. Comput. Sci.*, 2001, **41**, 147–158.

40. A. Kovatcheva, G. Buchbauer, A. Golbraikh and P. Wolschann, QSAR modeling of alpha-campholenic derivatives with sandalwood odor., *J. Chem. Inf. Comput. Sci.*, 2003, **43**, 259–266.

41. A. Kovatcheva, A. Golbraikh, S. Oloff, Y. D. Xiao, W. Zheng, P. Wolschann, G. Buchbauer and A. Tropsha, Combinatorial QSAR of ambergris fragrance compounds., *J. Chem. Inf. Comput. Sci.*, 2004, **44**, 582–595.

42. M. Shen, Y. Xiao, A. Golbraikh, V. K. Gombar and A. Tropsha, Development and validation of k-nearest-neighbor QSPR models of metabolic stability of drug candidates., *J. Med. Chem.*, 2003, **46**, 3013–3020.

43. M. Shen, C. Beguin, A. Golbraikh, J. P. Stables, H. Kohn and A. Tropsha, Application of predictive QSAR models to database mining: identification and experimental validation of novel anticonvulsant compounds., *J. Med. Chem.*, 2004, **47**, 2356–2364.

44. S. Zhang, A. Golbraikh, S. Oloff, H. Kohn and A. Tropsha, A novel automated lazy learning QSAR (ALL-QSAR) approach: method development applications, and virtual screening of chemical databases using validated, *ALL-QSAR models., J Chem. Inf. Model.*, 2006, **46**, 1984–1995.

45. A. Golbraikh, M. Shen, Z. Xiao, Y. D. Xiao, K. H. Lee and A. Tropsha, Rational selection of training and test sets for the development of validated QSAR models., *J. Comput. Aided Mol. Des*, 2003, **17**, 241–253.

46. L. Sachs, *Applied Statistics: A Handbook of Techniques*, 2nd ed., Springer-Verlag, Berlin, 1984.

47. A. Tropsha, Predictive QSAR (quantitative structure activity relationships) modeling., In *Comprehensive Medicinal Chemistry II.*, Y. C. Martin Ed., Elsevier, 2006, pp. 113–126.

48. A. Tropsha, *Application of predictive QSAR models to database mining*, In: *Cheminformatics in Drug Discovery.*, T. Oprea Ed., Wiley-VCH, Weinheim, 2005, pp. 437–455.

49. A. Tropsha and A. Golbraikh, Predictive QSAR modeling workflow, model applicability domains, and virtual screening., *Curr. Pharm. Des.*, 2007, **13**, 3494–3504.

50. S. Oloff, R. B. Mailman and A. Tropsha, Application of validated QSAR models of D1 dopaminergic antagonists for database mining., *J Med. Chem.*, 2005, **48**, 7322–7332.
51. S. Zhang, L. Wei, K. Bastow, W. Zheng, A. Brossi, K. H. Lee and A. Tropsha, Antitumor Agents 252. Application of validated QSAR models to database mining: discovery of novel tylophorine derivatives as potential anticancer agents., *J. Comput. Aided Mol. Des*, 2007, **21**, 97–112.
52. W. Zheng and A. Tropsha, Novel variable selection quantitative structure–property relationship approach based on the k-nearest-neighbor principle., *J. Chem. Inf. Comput. Sci.*, 2000, **40**, 185–194.
53. S. J. Cho, W. Zheng and A. Tropsha, Rational combinatorial library design. 2. Rational design of targeted combinatorial peptide libraries using chemical similarity probe and the inverse QSAR approaches., *J. Chem. Inf. Comput. Sci.*, 1998, **38**, 259–268.
54. NCI. http://dtp.nci.nih.gov/docs/3d_database/structural_information/smiles_strings.html (accessed November 2004).
55. Maybridge. http://www.daylight.com/products/databases/Maybridge.html (accessed November 2004).
56. A. Tropsha and W. Zheng, Identification of the descriptor pharmacophores using variable selection QSAR: applications to database mining., *Curr. Pharm. Des*, 2001, **7**, 599–612.
57. H. M. Berman, J. Westbrook, Z. Feng, G. Gilliland, T. N. Bhat, H. Weissig, I. N. Shindyalov and P. E. Bourne, The Protein Data Bank., *Nucleic Acids Res.*, 2000, **28**, 235–242.
58. M. L. Benson, R. D. Smith, N. A. Khazanov, B. Dimcheff, J. Beaver, P. Dresslar, J. Nerothin and H. A. Carlson, Binding MOAD, a high-quality protein–ligand database., *Nucleic Acids Res.*, 2008, **36**, D674–D678.
59. T. Liu, Y. Lin, X. Wen, R. N. Jorissen and M. K. Gilson, BindingDB: a web-accessible database of experimentally determined protein–ligand binding affinities., *Nucleic Acids Res.*, 2007, **35**, D198–D201.
60. Ajay and M. A. Murcko, Computational methods to predict binding free energy in ligand–receptor complexes., *J. Med. Chem.*, 1995, **38**, 4953–4967.
61. W. Deng, C. Breneman and M. J. Embrechts, Predicting protein–ligand binding affinities using novel geometrical descriptors and machine-learning methods., *J Chem. Inf. Comput. Sci.*, 2004, **44**, 699–703.
62. J. S. Tokarski and A. J. Hopfinger, Prediction of ligand–receptor binding thermodynamics by free energy force field (FEFF) 3D-QSAR analysis: application to a set of peptidometic renin inhibitors., *J. Chem. Inf. Comput. Sci.*, 1997, **37**, 792–811.
63. M. K. Holloway, J. M. Wai, T. A. Halgren, P. M. Fitzgerald, J. P. Vacca, B. D. Dorsey, R. B. Levin, W. J. Thompson, L. J. Chen and S. J. deSolms, A priori prediction of activity for HIV-1 protease inhibitors employing energy minimization in the active site., *J. Med. Chem.*, 1995, **38**, 305–317.

64. R. D. Head, M. L. Smythe, T. I. Oprea, C. L. Waller, S. M. Green and G. R. Marshall, VALIDATE: A new method for the receptor-based prediction of binding affinities of novel ligands., *J. Am. Chem. Soc.*, 1996, **118**, 3959–3969.

65. A. R. Ortiz, M. T. Pisabarro, F. Gago and R. C. Wade, Prediction of drug binding affinities by comparative binding energy analysis., *J. Med. Chem.*, 1995, **38**, 2681–2691.

66. C. Perez, M. Pastor, A. R. Ortiz and F. Gago, Comparative binding energy analysis of HIV-1 protease inhibitors: incorporation of solvent effects and validation as a powerful tool in receptor-based drug design., *Journal of Medicinal Chemistry*, 1998, **41**, 836–852.

67. R. K. Singh, A. Tropsha and I. I. Vaisman, Delaunay tessellation of proteins: four body nearest-neighbor propensities of amino acid residues., *J. Comput. Biol.*, 1996, **3**, 213–221.

68. A. Tropsha, C. W. Carter Jr., S. Cammer and I. I. Vaisman, Simplicial neighborhood analysis of protein packing (SNAPP): a computational geometry approach to studying proteins., *Methods Enzymol.*, 2003, **374**, 509–544.

69. L. Pauling, The nature of the chemical bond. IV. The energy of single bonds and the relative electronegativity of atoms., *J. Am. Chem. Soc.*, 1932, **54**, 3570–3582.

70. L. H. Hall, B. Mohney and L. B. Kier, The electrotopological state-an atom index for QSAR., *Quantitative Structure-Activity Relationships*, 1991, **10**, 43–51.

71. A. A. Oliferenko, P. V. Krylenko, V. A. Palyulin and N. S. Zefirov, A new scheme for electronegativity equalization as a source of electronic descriptors: application to chemical reactivity., *SAR QSAR Environ. Res.*, 2002, **13**, 297–305.

72. R. S. DeWitte and E. I. Shakhnovich, SMoG: de novo design method based on simple, fast, and accurate free energy estimates. 1. Methodology and supporting evidence., *J. Am. Chem. Soc.*, 1996, **118**, 11733–11744.

73. R. D. Head, M. L. Smythe, T. I. Oprea, C. L. Waller, S. M. Green and G. R. Marshall, VALIDATE: A new method for the receptor based prediction of binding affinities of novel ligands., *J. Am. Chem. Soc.*, 1996, **118**, 3959–3969.

74. R. X. Wang, L. Liu, L. H. Lai and Y. Q. Tang, SCORE: A new empirical method for estimating the binding affinity of a protein–ligand complex., *Journal of Molecular Modeling*, 1998, **4**, 379–394.

75. R. X. Wang, L. H. Lai and S. M. Wang, Further development and validation of empirical scoring functions for structure-based binding affinity prediction., *J. Comput. Aided Mol. Des.*, 2002, **16**, 11–26.

76. A. Tropsha, recent trends in quantitative structure-activity relationships. In *Burger's Medicinal Chemistry and Drug Discovery*. Sixth Edition ed., D. Abraham Ed.; John Wiley & Sons, Inc: New York, 2003; pp 49–77.

77. D. J. Livingstone, The characterization of chemical structures using molecular properties., *J. Chem. Inf. Comput. Sci*, 2000, **40**, 195–209.

78. P. Willett, Chemoinformatics-similarity and diversity in chemical libraries., *Curr. Opin. Biotechnol.*, 2000, **11**, 85–88.

79. D. B. Turner and P. Willett, Evaluation of the EVA descriptor for QSAR studies: 3. The use of a genetic algorithm to search for models with enhanced predictive properties (EVA_GA)., *J. Comput. Aided Mol. Des*, 2000, **14**, 1–21.

80. L. B. Kier and L. H. Hall, Molecular connectivity VII: specific treatment of heteroatoms., *J. Pharm. Sci.*, 1976, **65**, 1806–1809.

81. L. B. Kier, W. J. Murray, M. Randic and L. H. Hall, Molecular connectivity V: connectivity series concept applied to density., *J. Pharm. Sci.*, 1976, **65**, 1226–1230.

82. L. B. Kier, W. J. Murray and L. H. Hall, Molecular connectivity. 4. Relationships to biological activities., *J. Med. Chem.*, 1975, **18**, 1272–1274.

83. L. B. Kier, L. H. Hall, W. J. Murray and M. Randic, Molecular connectivity. I: Relationship to nonspecific local anesthesia., *J. Pharm. Sci.*, 1975, **64**, 1971–1974.

84. L. B. Kier and L. H. Hall, *Molecular Connectivity in Chemistry and Drug Research.*, Academic Press, New York, 1976.

85. W. J. Murray, L. B. Kier and L. H. Hall, Molecular connectivity. 6. Examination of the parabolic relationship between molecular connectivity and biological activity., *J. Med. Chem.*, 1976, **19**, 573–578.

86. W. J. Murray, L. H. Hall and L. B. Kier, Molecular connectivity. III: Relationship to partition coefficients., *J. Pharm. Sci.*, 1975, **64**, 1978–1981.

87. R. D. Brown and Y. C. Martin, The information content of 2D and 3D structural descriptors relevant to ligand-receptor binding., *J. Chem. Inf. Comput. Sci*, 1997, **37**, 1–9.

88. M. Shen, C. Beguin, A. Golbraikh, J. Stables, H. Kohn and A. Tropsha, Application of predictive QSAR models to database mining: identification and experimental validation of novel anticonvulsant compounds., *J. Med. Chem.*, 2004, **47**, 2356–2364.

89. J. S. Mason, A. C. Good and E. J. Martin, 3-D pharmacophores in drug discovery., *Curr. Pharm. Des*, 2001, **7**, 567–597.

90. C. M. Breneman, T. R. Thompson, M. Rhem and M. Dung, Electron density modeling of large systems using the transferable atom equivalent method., *Comput. Chem.*, 1995, **19**, 161–169.

91. C. B. Mazza, N. Sukumar, C. M. Breneman and S. M. Cramer, Prediction of protein retention in ion-exchange systems using molecular descriptors obtained from crystal structure., *Anal. Chem.*, 2001, **73**, 5457–5461.

92. M. Song, C. M. Breneman, J. Bi, N. Sukumar, K. P. Bennett, S. Cramer and N. Tugcu, Prediction of protein retention times in anion-exchange chromatography systems using support vector regression., *J. Chem. Inf. Comput. Sci.*, 2002, **42**, 1347–1357.

93. S. Oloff, S. Zhang, N. Sukumar, C. Breneman and A. Tropsha, Chemometric analysis of ligand receptor complementarity: identifying Complementary Ligands Based on Receptor Information (CoLiBRI)., *J Chem. Inf. Model.*, 2006, **46**, 844–851.

94. World Drug Index (WDI) is available from Daylight (http://www.daylight.com/products/wdi.html).

95. J. B. O. Mitchell, R. A. Laskowski, A. Alex, M. J. Forster and J. M. Thornton, BLEEP-Potential of mean force describing protein–ligand interactions: II. Calculation of binding energies and comparison with experimental data., *Journal of Computational Chemistry*, 1999, **20**, 1177–1185.

96. J. B. O. Mitchell, R. A. Laskowski, A. Alex and J. M. Thornton, BLEEP-Potential of mean force describing protein–ligand interactions: I. Generating potential., *Journal of Computational Chemistry*, 1999, **20**, 1165–1176.

97. R. S. DeWitte, A. V. Ishchenko and E. I. Shakhnovich, SMoG: De novo design method based on simple, fast, and accurate free energy estimates. 2. Case studies in molecular design., *Journal of the American Chemical Society*, 1997, **119**, 4608–4617.

98. B. A. Grzybowski, A. V. Ishchenko, J. Shimada and E. I. Shakhnovich, From knowledge-based potentials to combinatorial lead design in silico., *Accounts of Chemical Research*, 2002, **35**, 261–269.

99. R. X. Wang, L. H. Lai and S. M. Wang, Further development and validation of empirical scoring functions for structure-based binding affinity prediction., *J. Comput. Aided Mol. Des.*, 2002, **16**, 11–26.

100. H. J. Bohm, Prediction of binding constants of protein ligands: A fast method for the prioritization of hits obtained from de novo design or 3D database search programs., *Journal of Computer-Aided Molecular Design*, 1998, **12**, 309 323.

101. M. D. Eldridge, C. W. Murray, T. R. Auton, G. V. Paolini and R. P. Mee, Empirical scoring functions: I. The development of a fast empirical scoring function to estimate the binding affinity of ligands in receptor complexes., *J. Comput. Aided Mol. Des*, 1997, **11**, 425–445.

Subject Index